*Molecular Genetics
and Developmental Biology*

Society of General Physiologists' Symposia

Molecular Genetics
and Developmental Biology

A symposium held under the auspices of
The Society of General Physiologists
at its annual meeting at
The Marine Biological Laboratory,
Woods Hole, Massachusetts, September 3–
September 6, 1971

Maurice Sussman, *Editor*

Prentice-Hall, Inc.
Englewood Cliffs, New Jersey

4641

ISBN: 0-13-599456-X
Library of Congress Catalogue Card Number: 72-7870
Printed in the United States of America

PRENTICE-HALL INTERNATIONAL, INC., *London*
PRENTICE-HALL OF AUSTRALIA, PTY., LTD., *Sydney*
PRENTICE-HALL OF CANADA, LTD., *Toronto*
PRENTICE-HALL OF INDIA PRIVATE LTD., *New Delhi*
PRENTICE-HALL OF JAPAN, INC., *Tokyo*

Contents

Preface

During the past two decades, we have witnessed an astounding conceptual revolution in Biology sparked by the demonstration that DNA is the primary genetic material and by the elucidation of its structure. In one direction, the structural model led to a stereochemical explanation of DNA duplication and a much more profound understanding of genetic transmission and recombination. In another direction it led to an understanding of how the information coded in DNA is translated into polypeptides with specific amino acid sequences and correspondingly fixed capacities for biochemical and morphological work. These bodies of information and concepts, representing as they did the wedding* of Genetics and Biochemistry, acquired by popular usage the rubric of Molecular Genetics.

In recent years it has become clear that the molecular genetic approach can illuminate many still murky areas of Developmental Biology and, in particular, two of its major mysteries. These are outlined below.

---

*The ceremony failed to satisfy E. Chargaff, who accused Molecular Biologists (i.e., Geneticists) of "practicing Biochemistry without a license."

I. The Molecular Bases of Cellular Differentiation

When a cell is exposed to an appropriate microenvironment, it enters upon a complex, a temporally and quantitatively regulated program of phenotypic alterations that leaves it greatly different in form and function from what it once was and, in the case of multicellular organisms, different from sister cells located elsewhere within the cell assembly. These phenotypic changes can be ascribed to ordered sequences of differential gene expression. Question: What molecular events induce, mediate, and control such sequences?

II. The Inheritance of Differentiative Capacity

By virtue of its position within a developing Metazoan, a given cell may yield a clone whose members are restricted to development only along certain pathways of cellular differentiation. Thus, a hematocytoblast is a stem cell whose progeny are restricted to erythropoietic development. Question: Do such heritable restrictions arise from modifications of the genetic apparatus or from self-maintained feedback circuits within the cytoplasm or both?

This symposium was intended to demonstrate the power of Molecular Genetics as an approach to development and was organized loosely around the framework of the two sets of problems posed above. In my judgment, the intent was amply satisfied by the elegance of the contributions. Five lively sessions were held under the resolute and skillful chairmanship of James Ebert of the Carnegie Institution of Embryology, William Rutter of the University of California, San Francisco, Harlyn Halvorson of Brandeis, Aaron Moscona of the University of Chicago, and Alan Garen of Yale. Written expositions and extensions of all the talks are included in this volume. The easy informality and enjoyable conditions under which the symposium was held were due chiefly to the organizing capacities of David Shepro of Boston University. The National Institutes of Health contributed in a major way by providing funds which partly defrayed the expenses incurred.

With respect to the publication of this volume, it is a pleasure to acknowledge the superb editorial services of Mrs. Karen Meadow and of my secretary, Mrs. Mary Crosby. The cooperation of Prentice-Hall made early publication possible and ensured the timeliness of the contributions. This required that the proofs be read by the Editor rather than be returned to the authors. If in spite of this, errors have crept into the text, they are to be ascribed to supernatural circumstance.

MAURICE SUSSMAN

Waltham, Massachusetts

Molecular Genetics
and Developmental Biology

The Genomes
of Eucaryotic Cells:
Molecular and Developmental
Aspects

Introduction

There are two major problems that presently perplex investigators of Eucaryotic genomes: too many genes and too much repetition.

The first problem has to do with genome sizes: in *E. coli* 4.5×10^6 nucleotide pairs, i.e., about 5000 genes and therefore reasonable; in *Dictyostelium discoideum* 5×10^7 nucleotide pairs, 50,000 genes and worrisome; in man, 8×10^9 nucleotide pairs, 8 million genes and mind boggling since it appears highly unlikely that any cell need employ 8 million different polypeptides during its life cycle. Major portions of this surplus DNA have been thought by one speculator or another to consist of reserve genes, replicate genes, regulatory genes or evolutionary garbage. Which of these or other possibilities is in fact the case bears directly on the nature of the genetic control of developmental programs.

The second problem has to do with the repetitiveness of the DNA. Examination of Procaryotic DNA reveals

that, apart from the rRNA cistrons, small gene sized
fragments have unique nucleotide sequences. In contrast, DNA samples from Eucaryotes have been found to
contain gene sized stretches repeated as often as 10^6
times per genome and the fraction of unique fragments
comprises as little as 20-40% of the total in some
species. Some of the repetitive contingents have been
variously implicated in centromere function or identified as rRNA and 5S RNA cistrons. As will be seen in
the papers that follow, the results of these experiments
have profound implications for the area of developmental
study.

Recent Studies
on Moderately Repetitive
DNA Sequences

Roy J. Britten and Eric H. Davidson

Division of Biology
California Institute of Technology
Pasadena

INTRODUCTION

Recently we presented theoretical models which utilized the repetitive sequences of eukaryotic genomes in specific, key roles in the molecular processes of gene regulation (Britten and Davidson, 1969). The gene regulation model has specific consequences for the areas of development (Davidson and Britten, 1971) and evolution (Britten and Davidson, 1971). Unfortunately, useful data on the function and organization of repetitive DNA sequences do not yet exist in sufficient depth to permit an evaluation of these theoretical proposals. For this reason we are now attempting to measure the

*This work was supported by American Cancer Society grant E334E, by a grant from the U. S. Public Health Service HD-05753, and by the Carnegie Institution of Washington.

5

organization of the repetitive sequences in the genome and their transcription *in vivo*. However, quantitative investigation of repetitive sequence structure and function is still at an early state.

The present report begins with an analysis of repetitive sequence transcription during oogenesis in *Xenopus*. These experiments were performed with Dr. Barbara Hough. Measurements of the degree of interspersion (as opposed to clustering) of repetitive sequences among other sequences in the genomes of several animals are then described.

Transcription of Repetitive DNA in
Xenopus Oocytes

Most of the repetitive DNA in the *Xenopus* genome appears to fall within a narrow range of sequence frequencies, the average repetitiousness of which can be estimated at about 2500 occurrences per haploid genome (Davidson and Hough, 1972). Figure 1 shows the reassociation kinetics of an isolated repetitive DNA fraction of the *Xenopus* genome, which includes most of the repetitive sequences in the DNA (40-45%). Whole *Xenopus* DNA sheared at 50,000 psi was allowed to reassociate in 0.12 M phosphate buffer, at 60°C, to

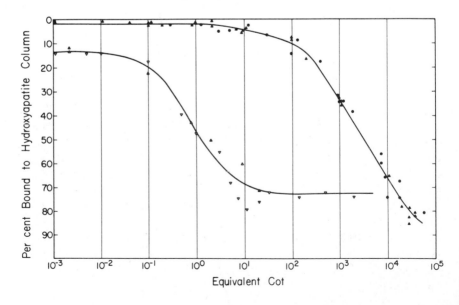

Cot 50, and the reassociated DNA was eluted from a
hydroxyapatite column. We refer to this repetitive
DNA fraction as "Cot 50 DNA." For comparison, reassociation of the isolated nonrepetitive *Xenopus* DNA
fraction is portrayed in Figure 1 as well. Reassociation rates for the Cot 50 repetitive fraction and the
nonrepetitive fraction differ by a factor of several
thousand, illustrating the efficacy of the separation
procedures used in the isolation of these nonoverlap-

Figure 1. (Opposite page.) Reassociation kinetics of
the isolated repetitive DNA of *Xenopus*, plotted together with the curve representing the reassociation
kinetics of the nonrepetitive sequence fractions
studied earlier (Davidson and Hough, 1971). *Xenopus*
DNA was sheared to about 450 nucleotides and was annealed to a Cot of 50. The reassociated fraction,
consisting of repetitive DNA, was harvested from a
hydroxyapatite column (Britten and Kohne, 1968). Reassociation of the nonrepetitive DNA fraction with
itself (▲) and in the presence of several thousand-fold
excesses of whole DNA (●) is displayed. Two Cot 50 DNA
preparations are included. Δ Refers to the reassociation
of an unlabeled Cot 50 repetitive DNA fraction, while
∇ represents the reassociation of a ^{3}H-Cot 50 DNA fraction with whole sheared unlabeled DNA present at 7000X
the concentration of the DNA. The terminal data obtained with the labeled fraction show that reassociate
nonrepetitive DNA or less highly repetitive DNA has
been effectively removed from the preparation since no
reassociation of labeled DNA occurs after about Cot 10,
even though the unlabeled carrier DNA continues to reassociate as usual (not shown). The solid line describing the reassociation of the Cot 50 DNA was fit to
the points by a computer according to a least squares
program. The root mean square of this fit is 0.049.
The reaction rate constant calculated by the computer
for the condition of 100% purity of the major reassociating component is 2.12. In addition to the major
reassociating component a smaller component of 11-14%
of the DNA appears to form duplex structures extremely
rapidly, possibly by self-annealing within each
fragment.

ping frequency components in the DNA. Comparison of
the corrected rate constant for the major fraction of
the repetitive DNA with that of the nonrepetitive DNA
indicates that the complexity of the repetitive fractions—i.e., its sequence diversity—is about 5×10^5
nucleotide pairs. Of course the observed characteristics of the repeated DNA depend strongly on the criterion for duplex stability set by the annealing conditions,
here 60°C in 0.12 M phosphate buffer.

The characteristics of the duplex structures formed
when the isolated repetitive DNA is permitted to anneal
are illustrated in Figures 2 and 3. Duplexes formed by
incubation of denatured Cot 50 DNA to Cot 20 were
trapped on hydroxyapatite and melted from the column by

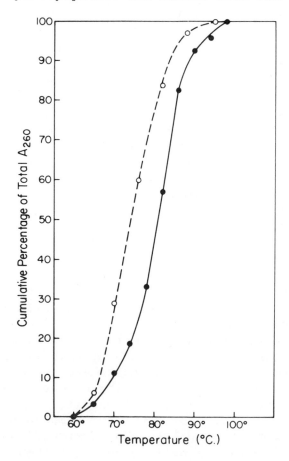

stepwise increases in temperature, as shown in Figure
2. To provide a standard of comparison, a similar
thermal chromatogram of purified *Xenopus* nonrepetitive
DNA is also plotted in Figure 2. The T_m of the latter
is 81.5°C, while that of the Cot 50 DNA duplex is 74°C.
The clear difference in thermal stability between these
DNAs can be attributed to nucleotide mismatch among the
sequences belonging to each repetitive sequence family,
or possibly to the brevity of the duplex regions if the
repetitive sequences are very short—e.g., 60 nucleotides or less (Hayes, Lilly, Ratcliff, Smith and
Williams, 1970).

The presence of unpaired nucleotides in the repetitive DNA duplexes is also suggested by the lower hyperchromicity they display on melting—only 17.5% as
compared to higher values obtained with nonrepetitive
DNA duplexes (25%; Davidson and Hough, 1969; 1972).
Figure 3 presents additional evidence for partial base
pair mismatch in the repetitive DNA duplex fraction.
In Figure 3a equilibrium sedimentation of the Cot 50
DNA duplexes, prepared by hydroxyapatite chromatography
as above, is shown in comparison with native DNA. The
Cot 50 DNA duplexes are about 8 mg/cc heavier than the
native DNA duplexes, while in these gradients denatured
DNA sediments at about 19 mg/cc heavier than does native
DNA (Figure 3b). These data imply that over a third of
the duplex structures trapped by the hydroxyapatite
column consist of single-stranded regions, a result

Figure 2. (Opposite page.) Melting behavior of Cot 50
DNA duplexes. (a) Thermal chromatogram of Cot 50 DNA
duplex prepared by annealing the DNA to Cot 20 and
passing it over a 1 cc hydroxyapatite column at 60°C
in 0.12 M phosphate buffer, O--O. The bound DNA was
melted from the column by raising the temperature in
the indicated steps. At each temperature the DNA released from the column was eluted with 6 ml buffer and
the A_{260} determined. The T_m of the DNA is 74°C. For
comparison a thermal chromatogram of a comparable
amount of nonrepetitive *Xenopus* duplex is also presented, ●—●. The nonrepetitive DNA was melted in
exactly the same manner (Davidson and Hough, 1971) and
its T_m is 81.5°C.

which is consistent with the lowered hyperchromicity displayed by the Cot 50 DNA duplex fraction.

RNA was extracted from mature *Xenopus* oocytes and uniformly labeled with dimethyl sulfate *in vitro* (Smith, McCarthy and Armstrong, 1967; Davidson and Hough, 1969; 1971). This RNA was hybridized with the

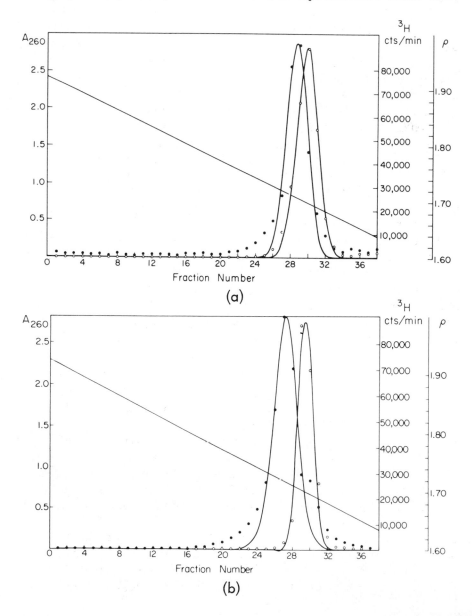

(a)

(b)

Cot 50 DNA to Cot 20 (cf. Figure 1) at an RNA to DNA ratio close to saturation of the Cot 50 DNA. It was found that after ribonuclease treatment of the hybrids only 10% of the hybrid molecules present in the mixture retained sufficient duplex structure to be bound by the hydroxyapatite. The nonbindable hybrids can be demonstrated in Cs_2SO_4 gradients (Davidson and Hough, 1972). This is in sharp contrast to the behavior of nonrepetitive DNA-RNA hybrids, which bind well to hydroxyapatite after the same ribonuclease treatment (Davidson and Hough, 1971; Firtel and Bonner, 1972). The 10% of repetitive DNA-RNA hybrids which do bind are of relatively high thermal stability, melting only 3-4° lower than the Cot 50 DNA, while the hybrids not bound by hydroxyapatite have a significantly lower thermal stability. It is to be stressed, however, that even

---

Figure 3. (Opposite page.) Sedimentation in CsCl gradients of Cot 50 DNA fractions obtained from hydroxyapatite chromatography. Gradients were prepared and calibrated and the optical density (●——●) and radioactivity (O——O) represent Gaussian distributions fit to the peak data by a computer; as indicated in the figures, the breadth of the base of these peaks was ignored in obtaining these fits. The peak specific densities are calculated from the fit curves. (a) Sedimentation of a Cot 50 DNA duplex fraction together with an unsheared ^{3}H-native DNA marker (specific activity 100,000 cts/min/µg). The marker used was HeLa cell DNA, which has the same GC content as *Xenopus* DNA (41%). A sample of the Cot 50 DNA was reassociated to Cot 25 and the duplex fraction harvested from a hydroxyapatite column. The gradient contained 40 µg of this material plus 300,000 cts/min of ^{3}H-HeLa cell DNA. The peak density for the Cot 50 DNA duplex is 0.008 g/cc heavier than the native marker (1.699). (b) Sedimentation of denatured sheared whole *Xenopus* DNA and unsheared native *Xenopus* ^{3}H-DNA. The gradient contained 230 µg denatured DNA which had been immersed in a 100°C bath for 5 min and then quick-frozen in dry ice-acetone, plus about 200,000 cts/min of ^{3}H-DNA (specific activity 650,000 cts/min/µg). The peak density of the denatured DNA, 1.718, is 0.019 g/cc heavier than the native marker.

the ribonuclease-resistant hybrids binding hydroxyapatite are 10° less thermally stable than are nonrepetitive DNA-RNA hybrids obtained in similar experiments.
Figure 4 shows that some of the ribonuclease-resis-

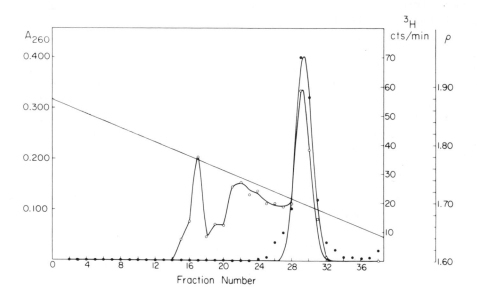

Figure 4.   Bouyant density study of ribonuclease-resistant hybrids which bind to hydroxyapatite.   The annealing mixture contained 600 µg ^{3}H-RNA and 200 µg Cot 50 DNA
in a total volume of 1.78 ml.   Incubation
was to Cot 20 with respect to the DNA.   The
hybrids were ribonuclease-treated (20 µg
RNase A in 0.24 M phosphate buffer, 1 hr),
excluded from Sephadex G-200, and loaded on
a hydroxyapatite column (0.12 M phosphate
buffer, 60°C).   The duplex fraction was adjusted to 0.12 M phosphate buffer and about
38 µg DNA containing 500 cts/min ^{3}H-DNA as
hybrids were present in the gradient.   A
Gaussian curve has been fit to the upper 4/5
of the peak of the A_{260} profile (closed
circle) representing the DNA-DNA duplex, the
density of which is 1.707 g/cc.   The solid
line representing the radioactivity profile
(open circles) is plotted directly.

tant hybrid molecules which bind to hydroxyapatite
consist of DNA fragments appreciably covered with RNA.
In this CsCl gradient the DNA duplex is represented by
optical density, and a portion of the RNA bands with
the DNA. Most of the hybridized RNA bands at higher
densities, ranging in such experiments up to about 1.78
gm/cc, by comparison to the DNA peak. The bouyant density of pure RNA is about 1.9 gm/cc in CsCl (Bruner and
Vinograd, 1965). Thus even after ribonuclease treatment the hybrid structures evidently retain amounts of
RNA ranging from too low to affect their overall density (for example, 5 or 10% RNA) up to what would appear to be complete coverage of the DNA. The structures whose density is close to that of pure DNA probably contain multiple DNA strands, since the reaction
in which these hybrids were formed was a partially
DNA-driven reaction.

To measure the representation of Cot 50 repetitive
DNA in oocyte RNA an RNA-driven reaction was employed.
Here a very large excess of RNA is annealed with trace
quantities of labeled Cot 50 DNA (specific activity
6×10^5 cts/min/mg: the labeled *Xenopus* DNA from which
this was prepared was the generous gift of Dr. Donald
D. Brown). Since the RNA is unlabeled it is unneces-
sary to subject the hybrids to ribonuclease treatment
before they are bound to hydroxyapatite. As labeled
DNA is present in trace quantities only, in the RNA-
driven reaction, the rate of the reassociation reaction
depends on the concentration of hybridizable RNA molecules. Variants of this procedure, which was developed
originally by D. E. Kohne, have been used by a number
of groups studying nonrepetitive DNA transcripts (e.g.,
Brown and Church, 1971; Gelderman, Rake and Britten,
1968; 1971; Davidson and Hough, 1971; Firtel and Bonner,
1972). A series of reactions of this type, carried out
at increasing RNA Cots, is portrayed in Figure 5. This
experiment shows that oocyte RNA contains sequences
related to over 3.5% of the Cot 50 DNA fragments.

To prove that the DNA hybridizing in these samples
is actually the DNA of the major reassociating fraction
of the Cot 50 preparation (cf. Figure 1) the hybridized
DNA was extracted and annealed with a 10^5-fold excess
of unlabeled Cot 50 DNA. These data are presented in

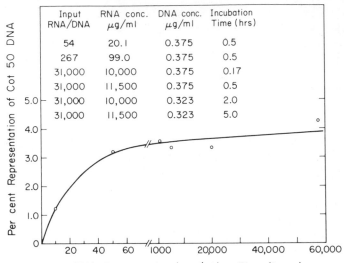

Input RNA/DNA	RNA conc. μg/ml	DNA conc. μg/ml	Incubation Time (hrs)
54	20.1	0.375	0.5
267	99.0	0.375	0.5
31,000	10,000	0.375	0.17
31,000	11,500	0.375	0.5
31,000	10,000	0.323	2.0
31,000	11,500	0.323	5.0

Figure 5. Kinetics of ^{3}H-DNA-RNA hybrid formation in an RNA-driven reaction. The amount of DNA hybridized to RNA has been corrected for DNA-DNA duplex also present, by exposing a sample of each annealing mixture to stringent RNase treatment at low salt to destroy RNA-DNA hybrids, and then measuring the DNA-DNA duplex content. The composition of the individual samples is given in the accompanying table. The ordinate gives the representation of Cot 50 DNA in oocyte RNA. The abscissa is total RNA (μg/ml) times time (in hours), i.e., proportional to the RNA Cot of the reaction.

Table 1. The reassociation rate of the formerly hybridized DNA (i.e., the expressed DNA) is clearly characteristic of the major kinetic component of the Cot 50 DNA, and the repetitiveness of the expressed DNA can therefore be taken to approximate this component, which averages 2500 occurrences of each sequence per genome. Note that a significant amount of RNA complementary to the small extremely rapidly reassociating component appears not to be stored in the oocyte.

These experiments, which are presented in detail

TABLE 1

Reassociation kinetics of hybridized DNA

Cot	Reassociation of Cot 50 DNA carrier (cf. Figure 1) (%)	Calculated reassociation of Cot 50 DNA minus extremely rapidly reassociating fraction (%)	Reassociation of ^{3}H-Cot 50 DNA extracted from RNA–DNA hybrids (%)
0.1	21	7	7.9
0.8	41	27	25
3.0	60	46	40

The annealing mixture contained 200 µg RNA and 0.075 µg ^{3}H–DNA in a total volume of 0.2 ml. The bound hybrid fraction was totally digested with ribonuclease and the ^{3}H–DNA released from duplex form by this treatment was recovered from the hydroxyapatite column, dialyzed and lyophilized. About 0.002 µg of this DNA (approximately 600 cts/min) was mixed with about 200 µg of unlabeled Cot 50 DNA for the samples incubated to Cots 0.1 and 0.8, and about 0.0015 µg (400 cts/min) ^{3}H–DNA was mixed with 150 µg unlabeled Cot 50 DNA for the sample incubated to Cot 3.

elsewhere (Davidson and Hough, 1972), permit some
reasonably good estimates regarding the utilization
during oogenesis of the major repetitive sequence
component in the *Xenopus* genome. We can conclude from
the rate of the RNA-driven reaction in Figure 5 that
the proportion of total oocyte RNA which is hybridizable RNA is about 2%. A minimum value is 0.5%.
Ribosomal RNA, which constitutes at least 95% of the
total oocyte RNA (Davidson, Allfrey and Mirsky, 1964)
is not a significant factor since as much as 3.5% of
the repetitive DNA is represented in the oocyte DNA—
i.e., more than 15 times the proportion of DNA which
is ribosomal. Since the expressed DNA is of the
major frequency class displayed in Figure 1, we
can calculate the complexity, or total sequence diversity represented in the oocyte RNA simply as 3.5% of
5×10^5 nucleotides, or about 2×10^4 nucleotides (as
measured at 60°C, in 0.12 M phosphate buffer). The
RNA transcripts from these sequences are present in the
oocyte 10^5–10^6 times on the average, per hybridizable
DNA sequence. The concentration of these transcripts
is thus surprisingly high within the oocyte, of the
order of 10^{-4} M or about 40 µg/ml.

This numerology differs greatly from that deriving
from our earlier study of nonrepetitive DNA sequence
representation in the same oocyte RNA. The complexity
of the nonrepetitive sequence transcript is much
higher, 20×10^6 nucleotides, while the number of
transcripts per expressed DNA sequence is conversely
more than an order of magnitude lower. The particular
class of repetitive sequences we have studied here fits
no known functional description. The complexity of the
expressed Cot 50 DNA seems too low and the number of
transcripts too high to attribute to them a typical
messenger RNA function. Nor, as noted above, can these
RNAs be accounted for as 5S, 4S or ribosomal RNAs.
Whatever might be their role, it is likely to be exercised during development rather than during oogenesis
itself, since earlier studies have shown that the
repetitive sequence transcripts stored in the mature
oocyte persist well into oogenesis before beginning to
disappear (Crippa, Davidson and Mirsky, 1967).

Interspersion of DNA Sequences in the
Urchin Genome[*]

An exploration of the arrangement of DNA sequences
in the sea urchin *Strongylocentrotus purpuratus* has
been initiated. Figure 6 shows the reassociation
kinetics for total unfractionated urchin DNA as measured with hydroxyapatite in several laboratories. The
right-hand curve shows the results for fragments of
about 450 nucleotides in length (50,000 psi sheared)
incubated and assayed at 60°C in 0.12 M phosphate
buffer.

In addition to these measurements of the reassociation of total urchin DNA, measurements have also been
made of the reassociation and thermal stability of
fractions prepared by binding to hydroxyapatite after
selected incubations leading to partial reassociation.
The fastest component has a Cot for half-reaction of
about 0.1 in the presence of the total DNA. It does
not show clearly on the 50,000 sheared curve of Figure 7 because its quantity is small. There is apparently a group of components which (in the presence of
the total DNA) have half-Cots for reassociation of
about 1, and finally a component or components with a
half-Cot of about 20. For the moment these components
are best identified by their observed reassociation
rates, rather than the frequencies of repetition which
are approximately 10,000, 1,000 and 50 copies. These
estimates may be modified upward when the effect on
the rate of reassociation of the degree of sequence
divergence is accurately known (particularly for the
greater degrees of divergence).

All of these components reassociate to form strand
pairs which melt over a wide range of temperatures.
Thus a wide range of degrees of base pairing appears
to be typical of reassociated urchin repeated DNA.
This observation is similar to that made for the
principal intermediate frequency fractions of the
vertebrates that have been studied (e.g., Figure 2).

_____

[*]The following experimental results were also presented at the Brookhaven Symposium (July 1971) and
published (Britten, 1971) in a somewhat different form.

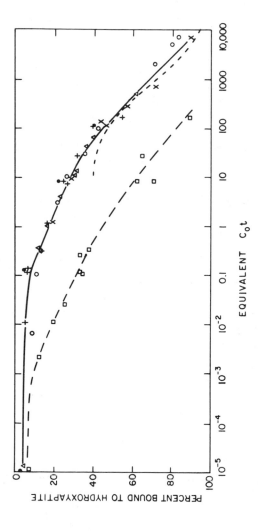

Figure 6. The reassociation of DNA from the sea urchin *Strongylocentrotus purpuratus*. Measurement was made by assaying the fraction of the DNA that was bound to hydroxyapatite at 60° in 0.12 M phosphate buffer after incubation for various times and concentrations in 0.12 M phosphate buffer at 60° except as noted. (——) represents the expected reassociation of the single copy DNA, based on the known genome size of this urchin, assuming that 60% of the DNA is single copy. The right-hand curve shows measurements made with 50,000 sheared DNA (approximately 450 nucleotides average fragment size). The left-hand curve ($\square$--$\square$) shows the binding to hydroxyapatite of an added small quantity of longer labeled DNA fragments (about 4000 nucleotides). Measurements done in several laboratories with 50,000 sheared DNA: (+), DNA from gonads; (O), labeled embryo DNA present in the same reassociation mixture; (△), D. Kohne 0.14 M phosphate buffer rate corrected for salt concentration; (●), A. Aronson; (x), Eric Davidson

18

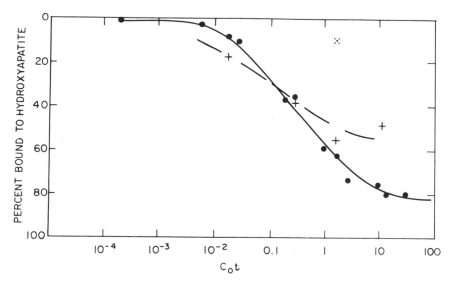

Figure 7. Measurement of the interspersion
of a fraction of the repeated DNA of the
sea urchin *Strongylocentrotus purpuratus*.
The fraction of the urchin DNA corresponding to about 10% of the genome and present
in about 1000 copies was prepared from
sheared ^{3}H-labeled embryo DNA, as shown in
Table 2. The solid circles (●) show the
reassociation of this fraction assayed by
passage over hydroxyapatite (60°C, 0.12 M
phosphate buffer). In addition to the
large quantity of the 450-nucleotide-long
fragments of this selected fraction there
was present a small quantity of long
(about 4000 nucleotides) fragments of
total urchin DNA. The crosses show the
reassociation of the long fragments with
the short ones. A control incubation (x)
of long fragments by themselves showed
very little reassociation at the same
concentration and time of incubation as in
the point below. It appears that representatives of this selected set of repeated
sequences are present on a majority of
4000-nucleotide-long fragments.

The left-hand curve of Figure 6 gives the results of an initial estimate of the interspersion of repeated and nonrepeated sequences. For this measurement a small quantity of much longer labeled fragments (about 4000 nucleotides long) was added to the 50,000 sheared DNA. Controls showed that relatively little reassociation occurred between long fragments as a result of their low concentration. Therefore, the binding of the long fragments to the hydroxyapatite was due, almost entirely, to their reassociation with one or more short fragments. The high degree of binding by Cot 100 of the long fragment DNA shows that much of the nonrepetitive DNA has been bound during the early part of the reaction. This indicates, as previously observed for calf DNA (Britten and Smith, 1970) that a large fraction of the nonrepeated sequences are present on long fragments that also contain repeated sequences. This observation is supported by recent measurements made with moderate length fragments (1000 to 1300 nucleotides average length). Fragments of this size reassociated almost completely by Cot 100, as assayed by hydroxyapatite binding. This shows that almost all fragments of this size have somewhere in their length stretches of repeated sequence.

This rather intimate interspersion of the different frequency classes of sequences is also observed in measurements with an *intermediate frequency* fraction of the repetitive DNA. For this purpose, 50,000 sheared DNA was prepared by successive hydroxyapatite fractionation as indicated in Table 2. A mixture of ^{3}H-labeled embryo DNA (20 hr development) and unlabeled DNA from male gonads was fractionated. Figure 7 shows the reassociation of this intermediate frequency fraction as measured by hydroxyapatite assay at 60°C in 0.12 M phosphate buffer. Also present in the incubation mixture was a small quantity of ^{14}C-labeled total urchin embryo DNA which had been lightly sheared and selected on an alkaline sucrose gradient. This DNA had an average fragment size of about 4000 nucleotides. Clearly a large fraction of the long fragments contain moderately repetitive sequences. This set of measurements leads to the same conclusion as previously published on the basis of experiments with a different

intermediately repetitive fraction of sea urchin DNA
(Britten and Davidson, 1971). The degree of binding

TABLE 2

Fractionation of *Strongylocentrotus purpuratus* DNA

| | | | Fraction bound | |
| | | | Tracer DNA *(embryos)* | Bulk DNA *(gonad)* |
Step	Cot	Material		
1	126.0	Total DNA	45.7	49.4
2	3.6	Bound fraction from step 1	57.0	57.0
3	4.24	Bound fraction from step 2	73.0	72.6
4	0.019	Bound fraction from step 3	22.9	27.1
5	0.025	Unbound fraction from step 4	16.8	14.2

The unbound fraction of the last step is the intermediate frequency fraction. It has a half-reaction
Cot of about 0.3 and reacts to 80% at Cot 30. It is
somewhat heterogeneous. Note that the two DNA preparations behave identically.

of the long DNA and the Cot at which binding occurs
differ significantly from the fraction in Figure 7.
Presumably this is due to differences in the long DNA
fragment sizes and the different steps used in the
preparation of the intermediate fraction.

The class of repetitive DNA utilized in Figure 7
corresponds to little more than 10% of the total DNA
and is present on the average in about 1000 copies.
Nevertheless, 50% or more of the long fragments appear
to have at least one representative of this set of
repeated DNA sequences somewhere in their length.
Consider the number of different fragments in the long
fragment preparation. The genome size of *S. purpuratus*

is about 10^9 nucleotide pairs. Dividing this number by
3000 we find 2.5×10^5 different fragments. About half
of these contain a member of the intermediate frequency
repeated sequence set. Thus there must exist hundreds
of distinct and physically separated short repeated
sequence elements, each one of which is present in
about 1000 copies. Only in this way can we explain
the more than 100,000 repeated sequence elements
present on at least 100,000 different fragments.

These measurements show that the intermediate frequency repeated sequence families are made up of relatively short elements of sequence scattered widely
throughout the DNA. *Short* in this context signifies
that on the average the sequence elements must be very
much shorter (by a factor of more than a hundred) than
the length that can be calculated from the kinetic
complexity of this class of repeated DNA. The kinetic
complexity is simply an estimate of the total sequence
length based on the rate of reassociation of the intermediate fraction. The Cot for half-reaction shown in
Figure 7 is 0.3 or about 15 times faster than would be
observed for *E. coli* DNA under these conditions. Thus
the observed kinetic complexity is about 300,000 nucleotide pairs. The average interspersed sequence
element might be 1/100 of this value, or even less.

Fine Scale Interspersion in Mouse DNA

Figure 8 presents some measurements showing fine-
scale interspersion in mouse DNA. A related observation has been reported previously (Rice, 1971). Intermediate frequency classes of radioactive mouse DNA
(sheared to 500-nucleotide-long fragments) were prepared in the following way. First the satellite DNA
was removed by reassociating to Cot 3×10^{-3} and passing the DNA over hydroxyapatite at 60°C in 0.12 M
phosphate buffer. The DNA that did not bind was
reassociated to Cot 8.9 and again passed over hydroxyapatite under the same conditions. The bound fraction
was recovered, reincubated to Cot 4.9 and again passed
over hydroxyapatite. This final reassociated fraction
(26% of the total DNA) was eluted from the column in
three portions by raising the temperature to 68, 78

and 98°. The 68° and 98° portions used for the experi
epxeriment were representative of the low and high
thermal stability repeated DNA of the mouse.
  The experiment tested these fractions for the pres-

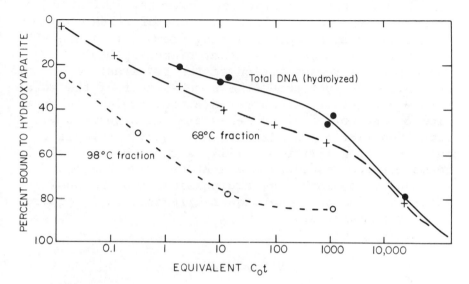

Figure 8. Measurements of the fine-scale
interspersion of repeated and single copy
sequences in the low thermal stability
fraction of reassociated repeated mouse
DNA. In order to achieve the large Cot
required for nonrepetitive DNA reassociation, total unlabeled mouse DNA which had
been similarly sheared and hydrolyzed was
added. (●) binding of total mouse DNA
which had been sheared at 50,000 psi and
then fragmented further by acid depurination; (○) added high thermal stability
labeled fraction; (+) added low thermal
stability labeled fraction prepared as
described in the text.

ence of other classes of sequence after the fragment
size was further reduced by acid depurination and
hydrolysis. The average fragment size was reduced by
incubation at pH 4.2 and 70°C for 30 minutes followed
by 0.1 N NaOH for 5 min (Ullman, 1970). The 98°C

fraction reassociated with the total DNA in just the way a relatively pure preparation of intermediate frequency repeated DNA does. It reaches 85% reassociation and is better than half-reacted by Cot 0.2. The low thermal stability fraction, however, exhibited quite a different behavior. A range of rates of reassociation is observed, extending from that of the 98° repetitive fraction out to that of nonrepeated DNA. The extent of reassociation of the 68° fraction at Cot 25,000 is just about the same as that of the total DNA. This result demonstrates a fine-scale interspersion of the more rapidly reassociating sequences with sequences that reassociate more slowly. A simple interpretation (Britten, 1971) is that the most slowly reassociating sequences are nonrepetitive DNA, but alternative explanations are possible and the positive identification of these sequence regions will require further measurements.

DISCUSSION

The repetitive DNA of both the sea urchin *S. purpuratus* and the amphibian *Xenopus laevis* exhibit a broad range of rates of reassociation up to about 10,000 times that of single-copy DNA. These rates are in the range of those of the major repetitive fractions of mammalian DNA that have been identified as intermediate frequency and which we term moderately repetitive DNA. In addition, the wide range of thermal stability observed for both the urchin and *Xenopus* reassociated repetitive DNA are similar to the range of thermal stabilities previously observed for the principal repetitive DNA of mammals.

Undetected in the urchin and *Xenopus* repetitive DNAs are any analogs to the high frequency, relatively precisely reassociating components of mammalian DNAs that have been termed satellites. In fact, with the possible exception of certain arthropods, such components (identified as being clustered, having low complexity, being relatively homogeneous, present in very high frequency, precisely reassociating, and odd in composition), have only been seen in mammalian DNAs

(Britten and Davidson, 1971). However, these components are not of major concern to us here and are mentioned only to emphasize the contrast between them and the moderately repetitive sets of repeated sequences which apparently form a major fraction of the genome of all species that have been investigated. It is perhaps a coincidence that the components prepared for this work from an amphibian, a mammal, and an echinoderm, each have a half-reaction Cot of about 0.2. While these are major components in each genome, other components with different kinetic complexities could have been prepared in each case.

The evidence presented in this paper deals with two issues: the arrangement and transcription of repetitive DNA. Both approaches are designed to yield evidence on the role or function of the repeated sequences. At present however neither has progressed very far and more questions can now be raised than answered. The measurements on urchin DNA reported here show that most fragments of a few thousand nucleotides length, which contain single-copy sequences, also contain some repetitive DNA. These sequences are moderately repetitive. A similar conclusion has already been drawn for calf DNA (Britten and Smith, 1970). It thus is possible that in general moderately repetitive sequences are interspersed among sequences belonging to other frequency classes. From a quite different type of measurement —*in situ* hybridization—there is evidence that some repeated DNA sequences are interspersed throughout the genome (Hennig, 1970) but of course on a much larger scale than in the cases reported here. The meaning of the existence of sequence interspersion is unknown. It seems to us that the widespread interspersion is directly related in some way to the function of the DNA sequences involved. Thus we take it that the evidence for widespread interspersion implies that at least some of these interspersed repetitive sequences are carrying a significant role in the structure and function of the genome.

The evidence that intermediate frequency repeated sequences are transcribed during oogenesis in *Xenopus* and that the transcription products are preserved in the egg cytoplasm and through early stages of development (Crippa, Davidson and Mirsky, 1967) is suggestive that these transcription products are a part of the regulatory system that controls early development. They might be an unusual mRNA required in the early stages, or they could play some trivial role related to pools of stored metabolites. Experimental information is required on the fate of the egg cytoplasmic RNA. It is crucial to know whether this RNA reaches the nuclei of the developing embryo or the cytoplasmic polyribosomes. Unfortunately at the present time evidence is very limited on the fate of the transcription products from repeated sequences of any cell type. It appears to us that the most incisive evidence will come from studies combining the two approaches described in this paper, in other words, measurements of the arrangement of the transcribed repetitive sequences. The fate and function of their transcripts thus becomes a crucial issue as well.

SUMMARY

Some recent experiments are described which deal with the transcription and the arrangement in the genome of repetitive DNA sequences. Repetitive sequences are shown to be transcribed during oogenesis in *Xenopus*. Oocyte RNA-DNA hybridization experiments were carried out with an isolated repetitive DNA fraction. The related sequences in the major component of this fraction occur an average of 2500 times per genome when measured under standard annealing conditions, and at least 0.5% of the total RNA of the mature oocyte is transcribed from sequences of this frequency class. Initial measurements of the degree of interspersion of some classes of repetitive sequences among other sequences have been made for a number of eukaryotic DNAs, including calf, sea urchin and mouse DNA. Moderately long DNA fragments are reacted with repetitive DNA fractions isolated from highly sheared DNA, and the fraction of molecules bearing the repetitive sequences is measured for different DNA fragment sizes. In each case studied it was found that a major fraction of the DNA fragments containing representatives of a given repetitive sequence family contain other types of sequence, for example, nonrepetitive sequence.

REFERENCES

Britten, R. J. (1971). *Brookhaven Symposium*, in press.

Britten, R. J. and Davidson, E. H. (1969). *Science* *165*, 349.

Britten, R. J. and Davidson, E. H. (1971). *Quart. Rev.* *Biol.* *46*, 111.

Britten, R. J. and Smith, J. (1970). *Carnegie Inst. of Wash.* *Year Book 68*, 378.

Bruner, R. and Vinograd, J. (1965). *Biochim. Biophys.* *Acta 108*, 18.

Crippa, M., Davidson, E. H. and Mirsky, A. E. (1967). *Proc. Nat. Acad. Sci.* *U.S. 57*, 885.

Davidson, E. H., Allfrey, V. G. and Mirsky, A. E. (1964). *Proc. Nat. Acad. Sci.* *U.S. 52*, 501.

Davidson, E. H. and Britten, R. J. (1971). *J. Theoret.* *Biol. 32*, 123.

Davidson, E. H. and Hough, B. R. (1969). *Proc. Nat. Acad. Sci. U.S. 63*, 342.

Davidson, E. H. and Hough, B. R. (1971). *J. Mol. Biol. 56*, 491.

Davidson, E. H. and Hough, B. R. (1972). Submitted to *J. Mol. Biol.*

Firtel, R. and Bonner, J. (1972). Submitted to *J. Mol. Biol.*

Gelderman, A. H. Rake, V. A. and Britten, R. J. (1971). *Proc. Nat. Acad. Sci. U.S. 68*, 172.

Hayes, F. N., Lilly, E. H., Ratcliffe, R. L., Smith, D. A. and Williams, D. I. (1970). *Biopolymers 9*, 1105.

Hennig, W., Hennig, I. and Stein, H. (1970). *Chromosoma 32*, 31.

Rice, N. (1971). *Carnegie Inst. of Wash. Year Book 69*, 479.

Smith, L. D., Armstrong, J. L. and McCarthy, B. J. (1967). *Biochim. Biophys. Acta 142*, 323.

Ullman, J. S. (1970). Ph.D. Thesis, University of Washington, Seattle.

Williams, A. E. and Vinograd, J. (1971). *Biochim. Biophys. Acta 228*, 423.

Closing the Ring

C. A. Thomas, Jr., C. S. Lee,
R. E. Pyeritz, and M. D. Bick

Department of Biological Chemistry
Harvard Medical School
Boston

Last year we reported that a variety of eukaryotic DNA fragments would form rings and other circular structures by "folding" or "slipping" (Thomas *et al.*, 1970). These observations have been extended to mouse satellite DNA (Pyeritz, Lee and Thomas, 1971), and to DNAs with very little identifiable satellite or rapidly reassociating sequences (Lee and Thomas, 1971, in preparation). In this paper, we summarize these experiments and interpret them according to two general models for the mononeme eukaryotic chromatid. One model asserts that a large portion of the DNA is organized into many thousands of different regions composed of tandemly repeating sequences; the other asserts that eukaryotic DNA is essentially like bacterial or phage DNA, that is, nonrepetitious, when considered in blocks of 15 - 20, but containing sequences that are identical at frequent intervals. These models will be named the "tandem repetition"

model, and the "intermittent repetition" model, respectively. While there is room for debate, we believe
that the evidence is in perfect accord with the "tandem
repetition" model, and conflicts with models based on
"intermittent repetitions" unless one pictures these
repetitions to be of unusual character, and densely
spaced. Whether the tandemly repeating sequences are
exact or inexact replicas of one another is a question
beyond the resolution of these experiments. By the
same token, we still do not know whether the terminal
repetitions of phage DNA molecules are *exact* sequence
repetitions, but we suppose they are from genetic
evidence, and because we can imagine how these chromosomes might be replicated. Unfortunately, no such
ancillary concepts exist for eukaryotic DNA.

The "folding" and "slipping" experiments are essentially the same as those employed to identify terminal
repetition and cyclic permutation in unbroken bacteriophage DNA molecules (Thomas, 1967). Folding involves
partial degradation of the terminals of a DNA molecule,
exposing opposite single chains at each end. If the
original duplex molecule had the same sequence at both
ends, the resulting single chains would be complementary and upon annealing would unite to form a region
of double helix, thereby "folding" the formerly linear
molecule into a ring. "Slipping" involves no enzymes,
merely total denaturation and chain separation. Upon
annealing, complementary regions unite. If the original collection of sequences was composed of various
circular permutations of a common sequence, circular
molecules are expected, and found. Fragments of phage
DNA do not form rings by either treatment.

We became interested in the origin of these species
of nonpermuted and permuted phage DNA molecules and
studied the intermediate products of their replication.
Summarizing briefly the work of many laboratories, it
is known that prior to the maturation of mature-sized
DNA molecules, infected cells contain very long DNA
molecules that correspond to many genomes linked end-
to-end. These concatemers have been identified in
cells infected with φ×, λ, T7, P22 and T4, and seem to
be a general feature of phage reproduction, irrespective of whether the final linear chromosomes were permuted or nonpermuted.

We imagined that eukaryotic chromosomes were organized along the same lines as these concatemers that
seemed to be the key structure of phage chromosomes.
The prospect of this possible unity to chromosome
structure, extending from the smallest to the largest
chromosomes presented a fascinating vision. As it
turned out, exactly this kind of model for the chromosome had already been proposed by H. G. Callan from an
entirely different point of view (Callan and Lloyd
1960; Callan, 1967). Within the framework of this
theory it is possible to account for many unexplained
features of eukaryotic chromosomes (Thomas, 1970).

We realized that if there were any merit to these
ideas, a significant portion of the fragments of any
eukaryotic DNA should be cyclizable by folding or
slipping. Our first experiments during the winter of
1968-1969 were with a commercial preparation of salmon
sperm that happened to be in our cold room.

Folded Rings

Presuming that hydrodynamic shear breaks the double
helix at random, one would expect that the fragments
would be cyclizable by folding provided that they were
longer than a repeating unit and that both ends were
broken within a given tandemly repeating region (Figure
1). In general, fragments about 1 - 3 μ in length are
resected or partially degraded by exonuclease III or
λ-exonuclease to expose an average of 500 to 1000
nucleotides at a single end. Upon annealing, many
rings are found (Figure 2). As is apparent from this
photograph, more complex circular structures are also
generated. These turn out to be expected from the
tandem repetition model as shown in Figure 3. Some
examples of the various types of circular structures
can be seen in Figure 4. About one-half of the circular
structures are simple rings, about one-third are lariats
and the remainder are double and polyrings. All of
these various structures have been encountered with all
of the eukaryotic DNAs we have studied. Insofar as we
have analyzed the double rings, we find that the contour length around the large loop is a simple multiple,

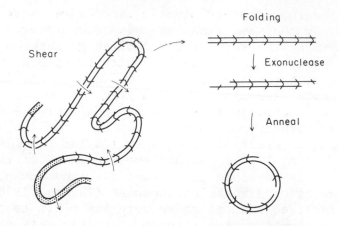

Figure 1. The random fragmentation of DNA
containing regions of tandemly repeating
sequences and the production of folded
rings.

generally 2:1, of the length around the small loop.
As explained in Figure 3, this is expected from tandemly repeating sequences.

The frequency of circular structures for the various
DNA fragments we have studied to date are shown in
Table 1. To this list we can add the qualitative information that DNA fragments from the dinoflagellate
Gonyaulax will form rings. Recent experiments by Jean
Rochaix in Professor R. P. Levine's laboratory have
demonstrated that nuclear DNA from *Chlamydomonas rein-
hardi* will form rings and, surprisingly, so will
chloroplast DNA. Many of these DNA samples have been
exhaustively purified, and in some cases the exonuclease resected fragments have been re-extracted with
phenol before annealing to form rings. It is very
unlikely that nonspecific proteins are causing ring
formation, but we have considered seriously this possibility (Thomas *et al.*, 1970).

Resection. Ring frequency depends on the extent of
resection and the nature of the DNA. If the DNA is
sheared at 0° in 2.5 M NaCl few (< 2%) rings are seen
following annnealing. If the DNA is sheared improperly
(at room temperature in 0.10 M NaCl) the shear breakage

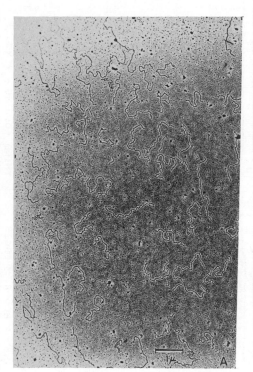

Figure 2. Folded rings of *Necturus* DNA
fragments. A sample of sheared DNA fragments 2 - 3 μ long was purified by CsCl
banding. A fraction from the band center
was treated with exo III until about 10%
degraded as estimated by hyperchromicity.
Samples were removed, mixed to 2XSSC and
annealed at 60°. A. Sample removed at
zero time. B. Sample removed at 120 min.
Electron micrographs by aqueous protein
film technique.

apparently does not break both chains at the same position. Under these conditions about 20% of the mouse
satellite DNA will form rings. This is undoubtedly
related to the short repeat length of this DNA. In
contrast, DNA from the main band will not cyclize after
improper shearing unless treated with exonuclease
(Pyeritz *et al.*, 1971). With most of the DNAs we have

FORMATION OF FOLDED CIRCULAR STRUCTURES
FROM TANDEMLY-REPEATING SEQUENCES

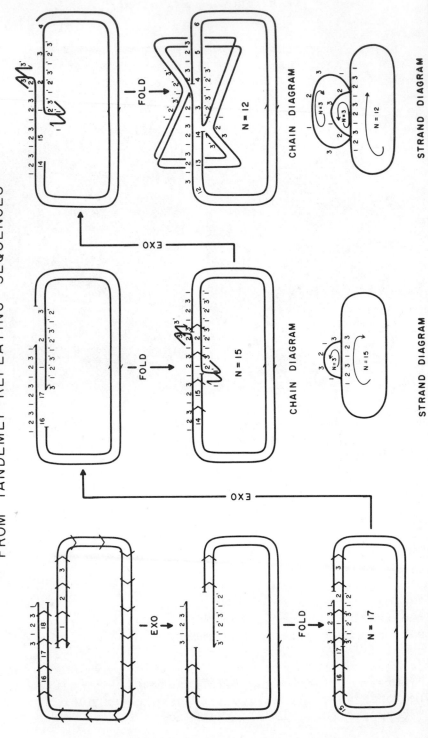

FIRST FOLDING — 1 NODE, 1 EDGE

SECOND FOLDING — 2 NODES, 3 EDGES

THIRD FOLDING — 2 NODES, 4 EDGES

STRAND DIAGRAM

CHAIN DIAGRAM

34

studied, the frequency of rings increased to a maximum
of about 20% circular structures when about 5 to 15%
of the nucleotides have been rendered acid soluble.
For a fragment 2 μ long, this would correspond to exposing terminal single chains, an average of 300 to
900 nucleotides in length. One may estimate the extent of degradation by hyperchromicity, by acid solu-
bility in the case of a labeled DNA, by admixing
labeled T7 DNA to unlabeled fragments and measuring
acid soluble radioactivity presuming all ends are
equally sensitive, or lastly by measuring the decrease
in the length of duplex fragments in the EM. All of
these procedures are wanting in some respect. Most
disturbing, we know that all fragment ends are not
sensitive to the exonuclease. In an inventory of fragments which were partly degraded by exo III, Dr. M.
Fuke has found that only 40 - 60% of the fragments
were resected in such a way that circular structures
could be formed even if all exposed regions were con-

Figure 3.   (Opposite page.)   The formation of folded
circular structures from tandemly repeating sequences.
The diagrams illustrate how randomly broken, tandemly
repetitious DNA fragments would be expected to form
circular structures if their terminals had been partly
degraded by exonuclease III.   The double helix is depicted by two parallel lines, and repetitious sequences
by the broad arrows ──⟩.   The numbers──1, 2, 3──repre-
sent the nucleotide sequence, and the primed numbers
──1', 2', 3'──the complementary sequence.   The numbers
written between the chains── ⟩ 17 ⟩ 1 ⟩ ──represent the
different copies of the identical sequences.   The arrow
──→ represents a 5' end and the bar ──│ a 3' end of
a polynucleotide chain.   The strand diagrams show the
double helix as a single line.   The contour length of
each closed loop shown should be a simple multiple, N,
of a common value, S, the length of the identical copy
(see caption to Figure 8).   It is these figures, and
their degraded products, that are to be expected in the
electron microscope (see Figure 4).   For clarity, the
digestion and folding have been presented in a stepwise manner, however, the same structures could be
produced in a single cycle of digestion and annealing.

sidered complementary. For example, a large proportion of the linear fragments only had single-chained regions at *one* end, suggesting that the other end was being protected from the action of exo III. This heterogeneity has two effects: It causes one to underestimate the extent of resection for the remaining sensitive terminals, and it drastically reduces the abundance of cyclizable fragments, because it is necessary (but not sufficient) that *both* ends be resected to an appropriate extent. Therefore, taking everything together it must be concluded that our quantitation is

FOLDED CIRCULAR STRUCTURES

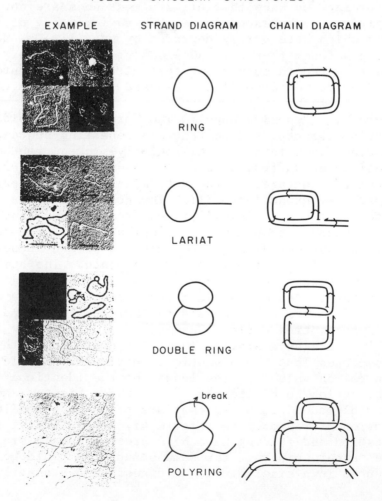

not good—certainly no better than a factor of two.
When these various inefficiencies are worked out, we
think the estimated frequency of cyclizable fragments
will be revised upward by about two-fold. However,
this will still be significantly less than the optimum
cyclization efficiency of intact T7 molecules, which
can produce 80% rings under favorable circumstances.

Annealing. A period of annealing is necessary for
maximum ring formation. For example, for *Necturus* DNA
fragments of lengths shown in Figure 6, the half
time to attain maximum ring frequency is 15 minutes.
After 60 minutes of annealing little increase is seen
(Figure 5). Satellite fragments form rings about twice
as fast, a fact which is probably related to the exposure of multiple short repeating sequences.

The contour length of rings. We have found rings
as small as 0.2 μ and as large as 20 μ in contour
length but they are most abundant at 1 - 3 μ. The
contour lengths of the rings depend, in the first instance, on the length of the fragments from which they
were derived. This has been repeatedly demonstrated
with several different DNA species (Figure 6B). This
means that rings are formed by *intramolecular* associations. These annealing reactions have been conducted
at up to 5 - 10 μg/ml. At this concentration resected
T7 will form abundant concatemers due to *intermolecular*
associations (Ritchie *et al.*, 1967). Clearly all exposed fragment terminals are not equivalent; there must

---

Figure 4. (Opposite page.) Examples of folded circular structures and their possible interpretations in
terms of tandemly repeating sequences. The most abundant structures are rings which make up about half of
all circular structures seen. Lariats are easy to
find, particularly with exonuclease III degraded fragments. Double rings are infrequent, but not surprising. Polyrings are rare and often difficult to identify with confidence. The example shown is the clearest, but we presume it to be broken once. The drawings
follow the same rules as described below Figure 4. The
length of the scale bar in the micrographs is 0.5 μ.

be more than 5 - 10 (perhaps many thousands) of different sequences at the fragment terminals.

TABLE 1

The frequency of folded rings among all fragments
counted in EM

	Per cent rings
Prokaryotes	
E. *coli*	0.5 - 1.0
B. *subtilis*	0.5 - 1.0
T7	1.4 - 2.0
Eukaryotes	
Necturus sperm	20 - 35
liver	25
Trout sperm	17
Salmon sperm	17
Calf thymus	10
HeLa	10
Mouse mainband	20
satellite	70
Drosophila virilis	18
salivaries	17

The values here represent the average of several determinations. Circular and linear molecules are identified and counted on the fluorescent screen of the EM
during a systematic search of the grids. The reprodu-
cibility of measured frequencies with different preparations of fragments is only fair, and variations by
as much as 5% are not uncommon. A series made from a
single fragment preparation shows less variation.   In
most experiments 50 - 100 or more rings were counted.

Length dependence of ring frequency. What information we have is shown in Figure 7. Here we see an
increase in ring frequency on passing from 0.5 to 2.0 μ
with *Drosophila*, *Necturus* and mouse satellite DNA
fragments. Above 2.0 μ there is a gentle decline in

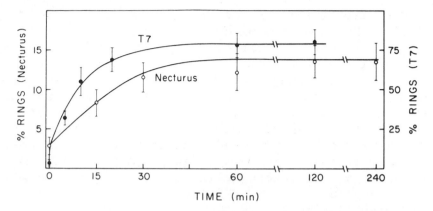

Figure 5. Rate of cyclization of *Necturus*
DNA fragments. Exonuclease III was used
to remove an average of 1200 nucleotides
(~10%) from the 3'-termini of *Necturus* DNA
fragments which were approximately 4 μ in
length. Samples were annealed at 61° in
2XSSC (0.39 M Na+) for the indicated times,
when grids were prepared. The circular
structures observed before annealing at 61°
(0 time) arise during the course of exonuclease treatment at 37°. Such rings are
not seen in samples incubated without exonuclease. T7 DNA was digested with exonuclease III so that 240 - 360 nucleotides
were removed from the 3'-termini. Annealing was carried out in 0.5 M NaCl at 61°.
The data for *Necturus* (O) and T7 (●) are
plotted ± one standard deviation, based on
the number of rings counted for each time
point.

the frequency of larger rings. In order to interpret
this as a length dependence, we must be sure that the
other factors governing ring frequency, such as extent
of resection and annealing conditions, are not limiting.
Presuming that all other things are equal, the decrease
between 2.0 and .05 μ could mean that many of the repetitive lengths are longer than these values. Interpreting this on the basis of an intermittent repetition
model, it could mean that the shorter fragments fall

between repetitions. Since the satellite DNA seems to
display the same kind of decrease, albeit at a much
higher level of cyclization, it could be possible that
the intrinsic stiffness of DNA retards ring closure.
The persistence length of DNA may be longer than previously estimated (Hays, Magar and Zimm, 1969) and may
approach 0.2 μ (Eisenberg, 1969). Since the persistence length is approximately the length of DNA that
will freely make a 90° bend, and four such bends are
required to make a (square) ring, it may be that the
stiffness of the double helix is retarding ring closure
in the region of 0.8 μ or longer.

The decline in cyclization frequencies with longer
fragments could be related to the length of the repetitious regions, or it could reflect the greater difficulty of closing longer rings.

Slipped Rings

In these experiments, one merely denatures the DNA
by heat, alkali, or formamide and anneals at a concentration of 2 to 10 μg/ml for a period of a day or two in

---

Figure 6. (Opposite page.) Length histogram of folded
rings and total fragment population. (A) A mixture of
various size classes of sheared *Necturus* DNA fragments
was treated with exonuclease III. An average of 500
nucleotides were removed from each 3'-terminus and the
fragments were annealed until the cyclization reaction
reached equilibrium (30 - 240 minutes at 61° in 2XSSC).
Specimens prepared for electron microscopy by the aqueous techniques were viewed and random photographs
taken. All fragments in a given photograph were measured to give the histogram populations. (OOO) folded
rings; (——) all fragments. (B) Contour length histo-
grams of *D. virilis* salivary gland DNA fragments with
various extents of exo III digestion. DNA preparations
were sheared with a hypodermic syringe and subjected to
the cyclization treatments. (1) 0% digestion; (2) 5%
digestion; (3) 9% digestion; (4) 13% digestion. Average contour lengths of linear molecules (——) and rings
(OOO) are shown.

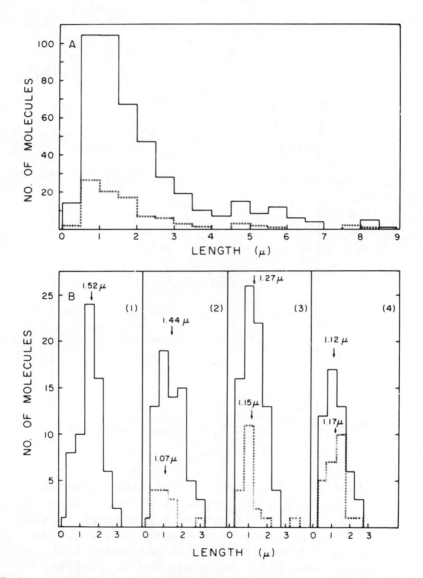

50% formamide or at elevated temperature. The logic is shown in Figure 8, and some examples in Figure 9. Again, all the eukaryotic DNAs we have studied give many slipped rings, including the DNA from *Drosophila* salivary chromosomes. Quantitation in the EM is difficult and, with exceptions, we have not estimated the fraction of reassociated material. Approximately 5 - 10% of the structures that contain some duplex regions are

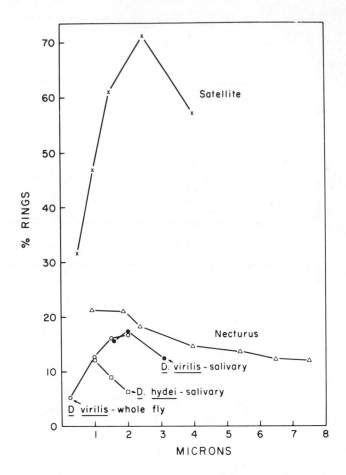

Figure 7.   The frequency of rings depends
on the length of fragments.   Frequencies of
rings are plotted *vs* varying average lengths
of DNA fragments from several sources.   Sucrose sedimentations were performed with
mouse satellite, *Necturus* and *D. virilis*
whole fly DNAs.   Different fractions, corresponding to the calculated length shown,
were selected and subjected to the cyclization treatments.   *D. virilis* and *D. hydei*
salivary gland DNA preparations were sheared
with appropriate gauge needles to give
lengths determined in the EM after resection.
The heterogeneity in lengths is about 25%.

circular. This contrasts sharply with sheared bacterial or phage DNA fragments treated in an equivalent
fashion; they produce no (rare) rings.

Tentative Conclusion

The finding that eukaryotic fragments form rings
and complex circular structures while prokaryotic
fragments do not means that the sequences in the higher
forms are organized in a manner quite unlike those in
bacterial DNA. The results are in qualitative accord
with the idea that much of the chromosome is organized
into many thousands of tandemly repeating sequences.
Therefore, Callan's model must still be considered
seriously. This view conflicts with current interpretation of reassociation rate experiments which
suggest to these workers that a majority of the sequences are so-called single-copy or unique sequences
(see Britten and Kohne, 1968; Laird, 1971).
On the other hand, one might propose that our results are only in superficial accord with the tandem
repetition model. Upon closer inspection it may be
that some other model will also account for the circular structures. In the following sections we raise
a number of objections to the tandem repetition model
and attempt to invalidate it.

Why are Ring Frequencies Nearer to
20% than 100%?

We considered 20% fairly high, but the question
still remains. Even if our observed frequencies are
eventually doubled, still they would be two-fold lower
than what is possible with T7. We think that this is
understandable and expected even if the entire chromosome were organized into tandemly repeating regions.
Briefly, for 100% ring formation, the length of all
DNA fragments must be longer than all repeating sequences ($1 > s$), yet shorter than the length of all
tandemly repeating regions ($1 < g$). This is made
clearer in Figure 10 showing (a) the tandem and (b)
the intermittent repetition model. The tandemly repeating sequence is called s, and the number of nucleo-

FORMATION OF "SLIPPED" CIRCULAR STRUCTURES
FROM TANDEMLY-REPEATING SEQUENCES

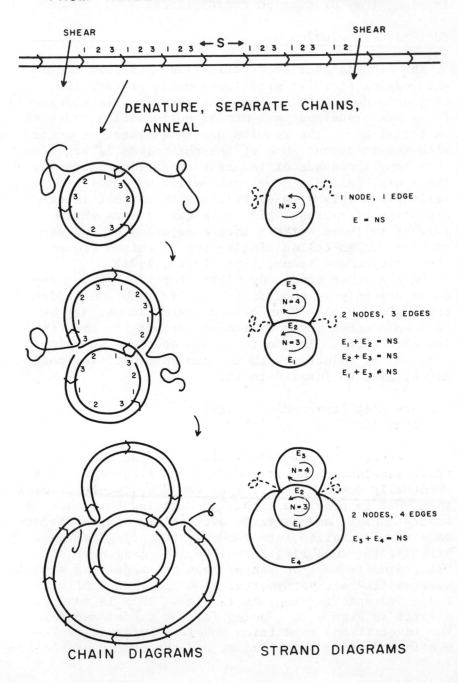

CHAIN DIAGRAMS   STRAND DIAGRAMS

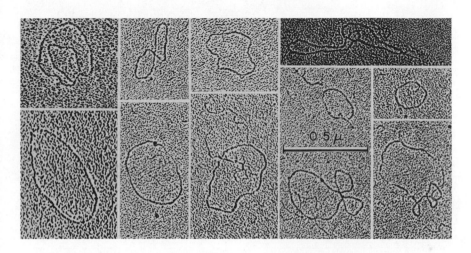

Figure 9. Some examples of slipped circular structures. Salmon or trout sperm DNA
was denatured by heat or alkali, then annealed at 10 - 20 μg/ml for 12 hr or more
at 60°. Alternatively, samples were annealed for several days at room temperature
in standard saline citrate containing 50%
formamide.

tide pairs in a region of a given repetitive sequence
is called g. Clearly both s and g can vary widely

Figure 8. (Opposite page.) The formation of slipped
circular structures from tandemly repeating sequences.
When a double helix consisting of repeating sequences
is denatured and the component single chains are separated, then allowed to anneal, it is probable that
they will reunite out of register or slipped by one or
more unit lengths. The resulting duplex molecule has
complementary terminals that may anneal repeatedly to
form rings, double rings, and polyrings that are
nearly equivalent to those depicted in Figure 3 even
though they were formed by a different route. The segments of length, E_1, E_2, etc. that form closed loops
consisting of an integral number, N, of identical
copies of length S, are indicated by the equations
shown near the related diagram.

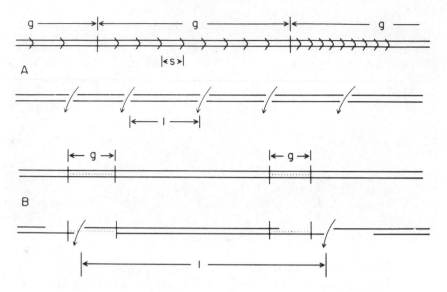

Figure 10. Two models for the mononemic
chromatid. A. Tandem repetition model; B.
Intermittent repetition model. The quantity S, is the number of nucleotide pairs
in a tandemly repeating sequence. The entire group of such tandemly repeating sequences is g nucleotides. Shear or endonuclease randomly breaks the double helix
into fragments of length l, and exonuclease
removes (resects) a certain number, r, of
nucleotides from each terminal. The intermittent repetition model uses the same
symbols; g is the number of nucleotide
pairs in the intermittently repeating sequence, which is pictured as being composed
by a highly repetitious (shorts) sequence.

from region to region. It is conceivable that s can
be as small as 2 (AT polymer) or as long as 8000 (the
length of the histidine operon) or longer! The value
of g could vary from a small value to perhaps 30,000 –
100,000 nucleotide pairs (the amount of DNA in the
average to dense chromomeres in *Drosophila*). Clearly,
a given preparation of fragments is not going to be
longer than all the various values of s yet shorter
than all the various values of g at the same time.

Therefore, the frequency of rings will be significantly less than 100%. When given regions of the chromosome, such as nucleolar DNA (Polito *et al.*, 1971) are cyclized we will begin to separate the s, l, g, and r effects. If this reasoning is correct, it suggests that if it were possible to separate the rings generated from fragments of different length, it would be possible to isolate certain regions of the chromosome.

Perhaps the Rings are Derived from Satellite DNA, or other Special Regions

We have approached this question in two ways. The first was to examine fractions from a CsCl density gradient to see whether certain fractions cyclized very efficiently. The results are shown in Figure 11. Visible satellite DNA did cyclize very efficiently indeed. For example, 70% of mouse satellite DNA fragments would form rings after exo III treatment, but a uniform 20% of the fragments collected from the main band would also cyclize. The same can be seen with *Drosophila* DNA: That region of the density gradient, where satellite fragments are known to exist, appears to cyclize more efficiently. However, apart from these notable exceptions, the remaining regions of the gradient seem to produce a constant fraction of rings over the measurable range. This argues against there being a few special components distinguished by their density.

One might imagine that many satellite DNAs of different densities are hiding under the main band. This indeed could be true and would be compatible with our conclusion that a large portion of eukaryotic DNA fragments is tandemly repetitious. However, the mouse satellite DNA will cyclize after improper shearing, while the DNA from the main band will not (Table 2). As mentioned in the first section, this is undoubtedly related to the short repeat length of mouse satellite DNA. Assuming that the same extent of unequal chain breakage occurs with main band as well as satellite DNA, we conclude that we would have seen (22.3/69.7) x 21.8 = 7.0% rings from the main band fragments, while

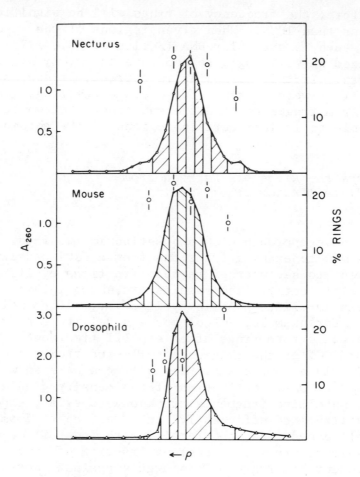

Figure 11. Cyclization frequency of fragments of differing density. DNA from *Necturus* liver, mouse liver (main band only),
and *Drosophila virilis* (whole fly) was
sheared and banded to equilibrium in prepa-
arative CsCl gradients. Fractions spanning
the gradient were dialyzed, treated with
exonuclease III, and annealed as described
in the text. Note that the satellite DNA
was removed from the mouse preparation
shown, but not from the *Drosophila* DNA
preparation.

the observed frequency is 0.8%. These values have been
observed at all density fractions before exo III treat-

TABLE 2

The formation of folded rings from
sheared DNA fragments

DNA	Length	Digestion	Rings	Number of rings seen
	(μ)	(%)	(%)	
Satellite	2.3	0	22.3	73
fragments	2.3	5.9	69.7	166
Satellite fragments (pretreated with *Neurospora* endonuclease)	1.5	0	4.6	82
	1.5	5	46.2	499
Main band	1.5	0	0.8	4
fragments	1.5	2.6	21.8	92

DNA was sheared under conditions which favor the
formation of single-chain termini (0.1 M NaCl, 24°).
Digestion was performed with exonuclease III and annealing was for 2 hr at 65° in 2XSSC.

ment. Therefore, at least 80% of the hypothetical
hidden satellite could not contain repeat lengths as
short as does the visible satellite.

Perhaps Rings are Only Formed from
Rapidly Reassociating DNA, and Unique
Sequences are Noncyclizable

This objection caused us some difficulty because
the only way to separate rapidly reassociating DNA was
to break it into small pieces, let it reassociate for
a prescribed period, then isolate the portion which had
reassociated from the portion which had not. This material would be useless to form folded rings first because it had been broken into fragments only one-fifth
to one-third of a micron, and secondly because the
structure of the reassociated product is totally different from the undenatured fragments. Professor J.

G. Gall suggested that we look at the DNA from the polytene chromosomes of *Drosophila*, first because only the euchromatic portions (the portion of the chromosome known to contain all, or most of, the known genetic map) were polytenized (see Gall, Cohen and Polan, 1971), and secondly because the DNA extracted from salivary glands appeared to be largely free of significant quantities of rapidly reassociating single polynucleotide chains. About 95 - 98% of the DNA appeared to be single-copy or unique sequences by this test (Laird, 1971; Dickson, Boyd and Laird, 1972). It appears that much of the satellite DNA is underrepresented in the polytene chromosomes, some appears not to have been endoreduplicated at all. If this polytene DNA were truly single-copy DNA, it should form no more rings than do *E. coli* DNA fragments. The results were just the opposite as shown in Table 3. The cyclization

TABLE 3

Formation of folded rings following
treatment by exo III

		% Circular structure
D. melanogaster (salivary)		17
D. virilis	(whole fly)	18
	(salivary)	17
D. Hydei	(whole fly)	16
	(salivary)	12

All samples degraded 5 - 10% corresponding to an average resection of 250 to 500 nucleotides at each terminal.

frequency of salivary gland DNA is 15% or 30 times higher than that from *E. coli* fragments. The abundance of circular structures appears to be only 5% lower than what we routinely observed from any other DNA including the DNA from *Necturus* which has a DNA content per haploid nucleus about 500 times greater than *Drosophila*. Therefore, we think that this slowly reassociating DNA

has a good many repetitive sequences, and is not as
unique as the word implies.

Perhaps Shear Breaks Occur at or Near
Special Sequences that are Sufficiently
Identical to Produce Complementary
Chains upon Resection

If specific break points exist in or near special
sequences, then this may explain the origin of the
simple rings (but not the double and polyrings or
slipped rings). In this event, much of the evidence
for tandem repetition collapses. However, we doubt
whether it is true. First, shearing under a variety
of salt concentrations and temperatures produces the
same frequency of rings. Producing the initial breakage using endo I, the double-chain-breaking endonuclease from *E. coli*, leads to about the same frequency of
rings. Lastly, partial degradation by the specific
endonuclease from *Hemophilus* (endo R·H) produces fragments that are fully capable of cyclization at the
same frequency. Also, the material resistant to repeated treatments by this enzyme can be sheared and
cyclized. While the case is not iron-clad, it seems
fairly sure that all these different methods of cleavage would not break at the same imaginary sites, and
that the assumption of random fragmentation is a good
one.

Perhaps there is Some Other Model that
Will Account for the Folded Rings
(Still Assuming Random Shear Breakage)

With this objection we must agree; of course there
may be some other model or combination of models that
accounts for these results. The problem is to devise
an alternative model. The nearest contender seems to
be the intermittent repetition model. This alternative is depicted in Figure 10B. In order to make this
model as strong as possible, we suppose that the intermittent repeats are composed of internally repetitious
homopolymers, the repeating units of which are just a
few nucleotides long. If b is the minimum length for

the formation of a stable hybrid, then any resection
exposing b or more nucleotides of an intermittent repeat will be able to hybridize to its complement at
the opposite terminal to form a ring. For purposes of
calculation, we imagine that these intermittent repeating regions are *irregularly* arranged such that
equation 3 (below) is satisfied. At low densities
of intermittent repeating regions, this means that
$Pg = \alpha\Lambda$, where α is the fraction of nucleotides in
such regions, P their number, and Λ the total number
of nucleotides in the chromosome. At higher densities
it means partial overlaps are precluded: They are considered to be totally overlapping or nonoverlapping.
The problem becomes the following. If a random shear
breakage is followed by a resection of r nucleotides,
how many different shear breakage positions will lead
to the exposure of b or more nucleotides, provided the
intermittent repeat region is a total of g nucleotides
in length? The answer is

$$r + g - 2b + 1 \tag{1}$$

If there are P such regions in a mononemic chromosome
composed of Λ nucleotide pairs, then, neglecting chromosome end effects, the mean number of intermittent
repeat regions having b or more nucleotides exposed
per resected region is:

$$m = \frac{P}{\Lambda} (r + g - 2b + 1) \tag{2}$$

This presumes that P can assume any value. The irregular arrangement of the intermittent repetitions is
such that the following equation relates the number of
such regions to the fraction of the chromosome they
represent:

$$\alpha = 1 - e^{-Pg/\Lambda} \tag{3}$$

This gives $P/\Lambda = - (1/g) \ln (1 - \alpha)$, which is combined
with equation 2.
   If $1 - e^{-m}$ is the chance that one resected terminal
will expose one or more intermittent repetitions, then
the probability that both terminals will have one or

more intermittent repetition regions exposed to the extent of b or more nucleotides is

$$C = \left[1 - e^{-m}\right]^2 \qquad (4)$$

where C is the frequency of rings. The final equation is a combination of (4) and (2) to give

$$C = \left[1 - e^{\{(1/g)\,\ln\,(1 - \alpha)(r + g - 2b + 1)\}}\right]^2 \qquad (5)$$

Now this equation is not as complicated as it appears. First of all it has no meaning for values of $g < b$, otherwise stable rings would never form. Next the number of nucleotides resected must be at least b or greater, otherwise no rings will form. So with these restrictions, equation 5 should give the expected frequency of cyclizable fragments for any other values of r, g, b, and α provided the intermittent repeats are homopolymers. If we assume that the abundance of cyclizable fragments is 25%, then we can calculate the values of g, b, and α that would satisfy this number. This has been done for an experiment using *D. virilis* salivary gland DNA fragments that had a mean length of 1.5 μ. These fragments were degraded to 10% corresponding to a resection r of 450 nucleotides. The results are shown in Figure 12. From this graph we can see that if 15 complementary nucleotides were considered to be sufficient to close the ring, we could explain the observed frequency of rings by supposing that less than 5% of the DNA was organized into intermittent repetitions of length 15 - 30. If conditions were such that 100 complementary nucleotides were required to close the ring, then 20% of the DNA must be in the form of intermittent homopolymer repeats. As g approaches and becomes greater than r, we must devote an increasing percentage of the chromosome to intermittent repetitions, no matter how many complementary nucleotides are required for ring closure. This can be understood intuitively by remembering that the chance that a random shear breakage will occur in an intermittent-repeat region is just α, the fraction of the DNA composed of

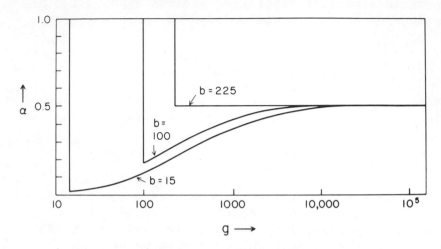

Figure 12. The fraction, α, of the chromosome that must be in the form of intermittent homopolymer repetitions of length, g,
to account for 25% cyclizable fragments,
presuming that b or more contiguous, complementary nucleotides are required for
closure.

such regions. The chance that the two shear breaks
forming the ends of the fragment occur in such regions
is α^2; thus the frequency of cyclizable fragments is
simply $C = \alpha^2$. If $C = 25\%$, then α must be 50%.

We believe that there are at least two kinds of experiments that can illuminate this subject; one deals
with the thermal (or formamide) stability of the rings,
the other with visualization of the length of the
overlap or the reannealed region that closes the ring,
this sets a lower limit on g which in turn demands a
higher α. As the required value of α increases, the
intermittent repetition model merges with a purely
tandem repetition model.

The Stability of Folded Rings

In order to estimate the minimum number of complementary nucleotides closing the rings, we have subjected solutions to an increasing series of tempera-

tures, or to increasing formamide concentrations and
measured the percentage of circular structures in the
EM. In the experiment summarized in Figure 14, folded
T7 rings, separately prepared, were added to a preparation of folded *D. virilis* polytene DNA fragment rings.
The mixture was heated to increasing temperatures.
After a few minutes at each temperature, samples were
taken, plunged into cold cytochrome c, spread for
electron microscopy, and the frequency of linear and
circular T7 DNA measured by counting molecules of each
type. On the same grids the percentage of folded
fragment rings was measured (Figure 13). The T7 rings,
which have a longer (12.5 μ) contour length, are closed
by a terminal repetition of about 260 nucleotide pairs
which are thought (for no reason) to be perfectly complementary (Ritchie *et al.*, 1967). In Figure 14 it can
be seen that they are converted into linear forms at a
mean temperature, T_ℓ, of 81°, which is just slightly
below the mean melting temperature, T_m, of T7 DNA in
this solvent (85°). We expect T_ℓ (T7) to be 3° *lower*
than T_m (T7) because closure is made by 260 nucleotide
pairs rather than an indefinitely large number. Moreover, the terminals denature at a lower concentration
of formamide than do other regions of the molecule
(R. Davis, personal communication) and are probably
somewhat richer in AT than the average. Thus these
observations are in accord with expectation.

The folded polytene DNA rings have a T_ℓ of 81° with
a broad transition. The melting temperature of *Drosophila* DNA in this solvent is 82.5°, calculated from
its buoyant density in CsCl (Schildkraut, Marmur and
Doty, 1962). Thus, on the basis of this experiment,
the linearization temperature of these folded rings
cannot be distinguished from the melting temperature
of the bulk DNA. The spread in T_ℓ is expected from
the heterogeneity in composition of the closure regions
on the basis of what we know about the heterogeneity
that can exist in phage and bacterial DNAs at the level
of 300 nucleotide pairs (Miyazawa and Thomas, 1965). A
further heterogeneity is expected on the basis of the
expected distribution in the number of nucleotides in
the overlap. For example, in the case at hand, r = 450.
This means that one-half the rings are closed by about
250 or less—one-tenth by about 45 or less. This will

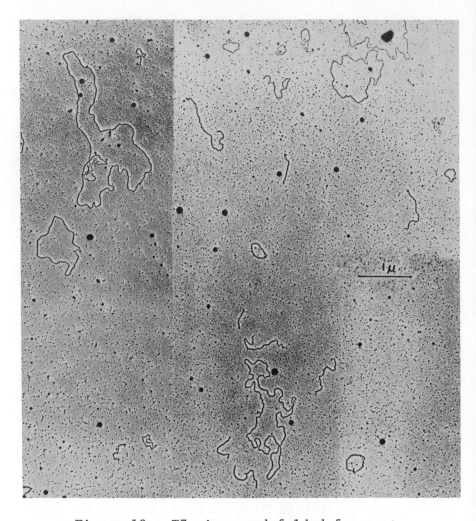

Figure 13. T7 rings and folded fragment rings. Electron micrographs of circular structures of *D. virilis* salivary gland DNA fragments. Large molecules, linear and circular, are T7 DNA. This preparation was made during a thermal stability study of the circular structures. The parlodion grids were stained with a uranyl salt before a rotary shadow.

have the effect of skewing the transition to lower temperatures. The precision of this determination of

T_ℓ is now the question. A transition curve drawn 4.1°
lower is clearly a bad fit to these points, and one
drawn 8.2° lower is unacceptable. The lowering of the
melting temperature is (conservatively) related to the
number of nucleotides in the ring closure for rings
2 - 3 μ long as

$$n \cong \frac{820}{-\Delta T} \qquad (6)$$

(Kallenbach and Crothers, 1965). This means that
values of n = 200 might be compatible with our data,
and n = 100 would fail to fit the observations. Thus
the length of the hypothetical intermittent repetition
must be at least 100 - 200 nucleotides long. From the
graph in Figure 12, we calculate that perhaps 40, and
definitely more than 20% of the polytene DNA must be in
homopolymer intermittent repetitions.

In Figure 14B an equivalent stability experiment
has been performed with T7 rings and *Necturus* rings
by diluting them into formamide at increasing concentrations. After incubation at 20° the sample was
mixed with cytochrome c, spread on water, and the
number of circular and linear fragments counted in the
the EM. The calculated equivalent temperature is
derived from the relation

$$T_{eq} = T_{obs} + 0.72 \ (\% \ Formamide)$$

(McConaughy, Laird and McCarthy, 1969), and plotted
on the abscissa. At this salt concentration T_m (T7)
= 79°, the observed T_ℓ = 77 - 78°. Assuming *Necturus*
DNA has the same composition as *Xenopus* (41%) T_m (Nect)
= 75°, while the observed T_ℓ (Nect) = 72 - 75°. Again,
so far as the precision permits, the rings have a
stability nearly equivalent to the undenatured double
helix.

Direct observations of closure lengths. When the
resected terminals of a fragment anneal, it is to be
expected that the reformed double helix will be frequently bracketed by single chains, either in the
form of gaps or whiskers (Figure 15). Under appropri-

ate conditions of formamide, these single chains will
be extended and the length of the intervening duplex
region, the closure length, can be measured. At this
point we cannot reliably determine in the EM the point
at which a duplex fiber terminates and a single chain
continues, however, it is possible to determine the
length of double helix between two whiskers. Assuming
these lengths are the same as those that cannot yet be
measured, the histogram shown in Figure 16 gives the
closure lengths for 37 folded and slipped rings. They
range in length from 100 to 1000 nucleotide pairs,
with about 80% of them falling between 100 and 600.
This is in complete accord with the thermal or formamide stability experiments mentioned above. If we
are to interpret the rings in terms of intermittent

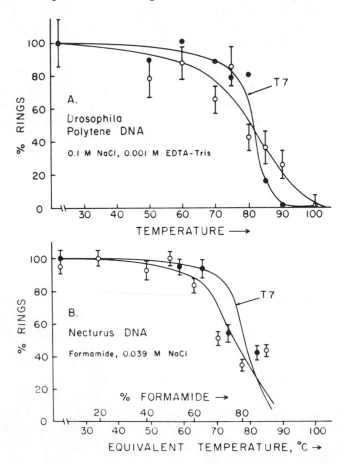

CLOSURE LENGTH

gaps   whiskers

Figure 15. The closure length. Folded
rings will be expected to be closed by the
complementary association of a certain number of nucleotide pairs, called here the
closure length, and depicted by hatching in
the diagram. In general, the closure
length will be bracketed by single chains,
but these may be more visible in the case
of whiskers.

Figure 14. (Opposite page.) The stability of folded
rings. A. *Drosophila virilis* salivary gland DNA. A
preparation of fragments 1.5 μ long, were resected to
an average of 450 nucleotides and annealed. In a parallel experiment intact T7 DNA was resected and folded,
then mixed with the salivary DNA preparation. At the
outset 13% of the *Drosophila* fragments were circular
and 74% of the T7 molecules were folded. The preparation was heated slowly and samples removed, plunged into
cold cytochrome c and spread for election microscopy,
and the number of circular and linear structures of
both types was counted. The number of rings counted
determines the error flags drawn at ± 1 σ. B. *Necturus*
DNA fragments, 4 μ long, were resected to an average of
1200 nucleotides and annealed for 60 min at 61° in
2XSSC (0.39 M Na^+) resulting in 14% rings. The sample
was diluted 1 + 9 in formamide to give the concentration shown. A separate sample of T7 rings was treated
in a similar way, and the number of rings and linear
units counted.

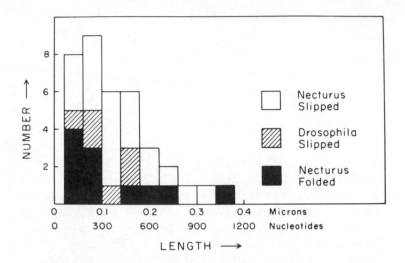

Figure 16. Observed closure lengths. The
length of the double-helical portion of a
ring falling between two whiskers has been
measured in favorable cases from three different DNA preparations.

repetitions, the repetitious regions, g, must be long.
This means they must be abundant (making up 20 - 50%
of the DNA) in order to account for the high frequency
of rings, unless we reject the notion of random fragmentation.

Since the hypothetical intermittent repetitions are
long enough to be visible, we should be able to distinguish the intervening regions of noncomplementary
single chains. This we have sought to do in vain;
most circular structures appear unblemished even at
relatively high concentrations of formamide. Both
folded and slipped rings remain totally duplex while
neighboring fragments or rings have begun to partially
denature. This observation, although a negative one,
is very persuasive to us.

SUMMARY AND PROSPECT

We have attempted to stake out the limits on the
kinds of models that are compatible with the abundance,

stability, and morphology of folded and slipped circular structures. All our observations can easily be interpreted by a chromosome largely composed of regions of tandemly repeating sequences. The intermittent repetition model becomes increasingly difficult to maintain because the rings are abundant, stable, closed by substantial numbers of nucleotides, and appear uninterrupted by intervening regions of noncomplementary single chains. In the absence of any other interpretation, we are left with the tandem repetition model. These sequences must constitute between 20 and 100% of the chromosome. If nonrepeating sequences exist, they could be organized in a single block, perhaps at one end of the chromosome, or they could be dispersed, perhaps surrounding each tandemly repeating region. The proportion and location of these sequences must be identified.

Finally, one overriding question remains: Are structural genes in eukaryotes each represented by a single copy of a nucleotide sequence, or are they each represented by multiple copies of the same sequence? Are genes represented by these tandemly repeating sequences or are they represented in the hypothetical nonrepeating DNA? It is rare in biology that such a clearly defined question comes to the surface.

ACKNOWLEDGMENTS

It is a pleasure to thank Dr. Motohiro Fuke for use of his unpublished experiments, and for preparing exo III, and Dr. Helena Huang for her help with microscopy and measurements. Professor Bruno Zimm and Dr. Lynn Klotz have clarified our quantitative thinking on tandemly repeating DNA, and we thank them. We thank Dr. Barry Dancis for his critical reading of this paper, and his contributions to the calculations on ring frequency. This work was made possible by grants from the National Science Foundation (GB 8611) and the National Institutes of Health (A109186). We also thank the Jane Coffin Childs Memorial Fund, the NSF, and the Damon Runyan Memorial Fund for supporting the authors (CSL, REP, MDB).

REFERENCES

Britten, R. J., Kohne, D. E. 1968. Repeated sequences in DNA. *Science 161*:529-540.

Callan, H. G. and Lloyd, L. 1960. Lampbrush chromosomes of crested newts *Triturus cristatus* (Laurenti). *Phil. Trans. Roy. Soc. (London)*, *ser. B 243*:135-219.

Callan, H. G. 1967. The organization of genetic units in chromosomes. *J. Cell Sci. 2*:1-7.

Dickson, E., Boyd, J. B. and Laird, C. D. 1972. Sequence diversity of polytene chromosome DNA from *Drosophila hydei. J. Mol. Biol.*, in press.

Eisenberg, Henryk 1969. On the inherent flexibility of DNA chains. *Biopolymers 8*:545-551.

Gall, J. G., Cohen, E. H. and Polan, M. L. 1971. Repetitive DNA sequences in *Drosophila. Chromosoma 33*: 319-344.

Hays, J. B., Magar, M. E. and Zimm, B. H. 1969. Persistence length of DNA. *Biopolymers 8*:531-536.

Kallenbach, N. R. and Crothers, D. M. 1966. Theory of thermal transition in cohered DNA from phage lambda. *Proc. Nat. Acad. Sci. U.S. 56*:1018-1025.

Laird, C. D. 1971. Chromatid structure: relationship be-between DNA content and nucleotide sequence diversity. *Chromosoma (Berl.) 32*:378-406.

McConaughy, B. L., Laird, C. D. and McCarthy, B. J. 1969. Nucleic acid reassociation in formamide. *Biochemistry 8*:3289-3295.

Miyazawa, Y. and Thomas, C. A., Jr. 1965. Nucleotide composition of short segments of DNA molecules. *J. Mol. Biol. 11*:223-237.

Polito, L., Graziani, F., Boncinelli, E., Malva, C. and Ritossa, F. 1971. DNA circularization: a molecular approach for the detection and isolation of tandemly duplicated genes. *Nature New Biology 229*:84-86.

Pyeritz, R. E., Lee, C. S. and Thomas, C. A., Jr. 1971. The cyclization of mouse satellite DNA. *Chromosoma (Berl.) 33*:284-296.

Ritchie, D. A., Thomas, C. A., MacHattie, L. A. and Wensink, P. C. 1967. Terminal repetition in nonpermuted T3 and T7 bacteriophage DNA molecules. *J. Mol. Biol. 23*:365-376.

Schildkraut, C. and Lifson, S. 1965. Dependence of the melting temperature of DNA on salt concentration. *Biopolymers* 3:195-210.

Schildkraut, C., Marmur, J. and Doty, P. 1962. Determination of the base composition of deoxyribonucleic acid from its buoyant density in CsCl. *J. Mol. Biol.* 4:430-443.

Thomas, C. A., Jr. 1970. The theory of the master gene. In *Neurosciences: A Second Study Program*. F. O. Schmitt, ed., The Rockefeller University Press, New York, pp. 973-998.

Thomas, C. A., Hamkalo, B. A., Misra, D. N. and Lee, C. S. 1970. Cyclization of eukaryotic DNA fragments. *J. Mol. Biol.* 51:621-632.

Thomas, C. A. 1967. The rule of the ring. *J. Cell. Physiology Supplement 1* to 70:13-33.

Molecular Cytogenetics

M.L. Pardue and J.G. Gall

INTRODUCTION

Recent discoveries of the presence of reiterated sequences in the DNA of eukaryotes (Britten and Kohne, 1968) have given new insights into questions of chromosome structure. The eukaryotic chromosome is now known to contain classes of repeated DNA sequences which would not have been predicted from genetic studies. The presence of reiterated sequences does help to reconcile the large DNA contents of eukaryotic genomes with the much smaller DNA contents expected from extrapolation from bacterial genetics. However we are still faced with the problem of the significance of these reiterated sequences and their relation to the genetic activity of the nucleus.

One new approach which has been successfully applied to the study of reiterated DNA sequences is the technique of nucleic acid hybridization in cytological preparations (Gall and Pardue, 1969; John, Birnstiel, and Jones, 1969; Buongiorno-Nardelli and Amaldi, 1970; Jacob, Todd, Birnstiel, and Bird, 1971). In this technique the DNA of a conventional cytological preparation is denatured and hybridized with radioactive RNA or single-stranded DNA. The sites of hybridization are then detected by autoradiography, thus permitting the localization of specific nucleotide sequences.

Technical Aspects of Cytological Hybridization

Although we are still exploring the technical details of *in situ* hybridization, the experimental evidence to date indicates that the reaction is comparable to more conventional nucleic acid hybridization techniques (Gall and Pardue, 1969). The denaturation of DNA in a cytological preparation, which still contains much of the nonhistone protein of the chromosome, might be expected to differ somewhat from strand separation in purified DNA. However, agents which denature purified DNA—acid, base, heat, formamide, and urea—are all effective on cytological preparations as measured by the amount of hybridization in preparations treated with any one of these. Denaturing agents are chosen to give the most hybridization possible while still maintaining good cytological morphology. Acid denaturation gives the best cytological preservation but, as expected, it causes some depurination of the DNA.

Our few experiments on the kinetics of *in situ* hybridization give results similar to those obtained with hybridization on nitrocellulose filters. Cation concentrations and incubation temperatures used to give maximum specificity of annealing in other systems give good hybridization in cytological preparations. The specificity of the reaction can be demonstrated both by conventional competition experiments (Gall and Pardue, 1969) and by the lack of cross-species hybridization of species-specific RNAs (Arrighi *et al.*, 1970; Pardue, 1970). The extreme localization of the binding of many RNAs provides another criterion of specificity.

At present cytological hybridization is limited to the study of repetitive DNA for two reasons. The first

limitation is imposed by the sensitivity of autoradiographic detection. Enough radioactive nucleic acid
must be hybridized to a given region of the chromosome
to produce detectable grains during a reasonable exposure time. The second limitation comes from the kinetics of the hybridization reaction. It is necessary to
have a sufficient concentration of both the DNA of the
cytological preparation and its complementary sequences
in the hybridizing mixture to obtain hybridization
within the incubation period used.

Britten and Kohne (1968) have defined a measure of
the rate of DNA renaturation, $C_0t_{1/2}$[1], which is expressed
in moles of nucleotides times seconds per liter and
gives a measure of the complexity of the DNA. As a
general approximation the DNA of eukaryotes can be
divided into "fast," "intermediate" and "slow" fractions on the basis of renaturation kinetics. The most
rapidly renaturing fraction consists of sequences which
reassociate at $C_0t_{1/2}$ less than one. The sequences of
the intermediate fraction renature at $C_0t_{1/2}$ between 1
and 100. The sequences of the slow fraction have
$C_0t_{1/2}$'s of greater than 100 and are thought to be
either unique or to have only a very few repetitions.
As shown by Melli and Bishop (1969) the conditions of
concentration and incubation time used in most conventional hybridization studies on RNA sequences limit
them to RNA coded for by DNA of fast and intermediate
renaturation rates. Naturally occurring RNA shows
little, if any, hybridization to the fast fraction.
Recently techniques have been developed which permit
hybridization of RNA to DNA of the slow fraction.
These experiments show that the slow fraction does
indeed code for many natural RNA sequences (Davidson
and Hough, 1969; Gelderman, Rake and Britten, 1971;
Melli et $al.$, 1971).

Although $C_0t_{1/2}$ is a useful measure for comparing the
complexities of DNA fractions, in assessing the feasibility of localizing RNA sequences in cytological prep-

---

[1] $C_0t_{1/2}$ = the product of the concentration of nucleotides of renaturing DNA and the incubation time necessary for half of the sequences to renature.

arations the more relevant parameters are $C_r t_{1/2}$[2] and the amount of DNA complementary to the RNA being studied. $C_r t_{1/2}$ (Birnstiel, Sells and Purdom, 1971) is a measure of the base sequence complexity of the RNA in filter hybridization experiments in which RNA is in vast excess. C_r is the molar concentration of ribonucleotides in solution and $t_{1/2}$ is the time in seconds taken to reach half saturation of the complementary DNA. Purdom and Birnstiel (unpublished) have shown that at low DNA concentrations the amount of complementary DNA does not influence the relative rate of hybrid formation. Although extremely high local concentrations of particular base sequences within a metaphase chromosome may not be exactly comparable to the more general distribution of DNA immobilized on a nitrocellulose filter, $C_r t_{1/2}$ seems a useful approximation in considering *in situ* hybridization experiments. Purdom and Birnstiel have measured the $C_r t_{1/2}$ for *Xenopus* 5S RNA as 0.45 moles·sec/ℓ; for *Xenopus* rRNA[3], 15.04 moles·sec/ℓ; and for rabbit rRNA, 14.5 moles·sec/ℓ. All are easily obtainable in cytological hybridization experiments.

The total amount of DNA complementary to the hybridizing RNA within a nucleus, and in fact within a definable region of the nucleus, is important because the signal from the hybridized RNA must be strong enough to be detected above background noise. In filter hybridization experiments the amount of hybridized RNA can be increased by increasing the DNA on the filter but the amount of DNA within a nucleus in a cytological preparation cannot be changed. In some *in situ* experiments it is possible to use the increased local DNA concentrations in polytene chromosomes to overcome this difficulty.

A limitation in the study of RNA sequences is the relatively low specific radioactivity of the RNA which can be prepared *in vivo*. We have used the genes coding for ribosomal RNA, isolated from the toad *Xenopus*, to

---

[2]$C_r t_{1/2}$ = the product of the concentration of ribonucleotides in an RNA-DNA hybridization and the time of incubation necessary to permit half of the DNA sequences to hybridize.

[3]rRNA = ribosomal RNA.

transcribe cRNA[4] for some experiments (Pardue *et al.*, 1970) but with the present techniques for gene isolation this approach has limited application. However, the recent discovery of RNA-dependent DNA polymerases (Baltimore, 1970; Temin and Mizutani, 1970) suggests the possibility that such enzymes might be used to prepare radioactive DNA from purified RNAs. While an RNA polymerase might also be used to transcribe RNA, a DNA polymerase would permit the radioactive product to be separated from the nonradioactive template which would otherwise act as a competitor in the hybridization reaction.

Localization of Repetitive Sequences

The DNA fractions which have been most extensively studied by cytological hybridization are those containing very highly repetitive sequences. In their simplest form these experiments have utilized either total radioactive DNA or radioactive RNA transcribed *in vitro* from total DNA. Autoradiographs developed after successively longer exposures give an approximate visualization of a C_0t curve since the more reiterated sequences will be detected after shorter exposures than will sequences of lower reiteration. Experiments in three laboratories (Jones and Robertson, 1970; Rae, 1970; Gall, Cohen and Polan, 1971) have shown that RNA complementary to total DNA hybridizes first to chromocentral heterochromatin in *Drosophila* salivary chromosomes, indicating that the most repetitive sequences are concentrated in this region. Autoradiographs which have been exposed for longer periods show binding of the RNA to sites distributed over the chromosome arms, reflecting the distribution of sites which contain either smaller amounts of highly reiterated DNA or DNA sequences with lower degrees of repetition. Radioactive *Xenopus* DNA (Pardue and Gall, 1969) hybridized to many regions widely distributed over the *Xenopus* chromosomes. The *Xenopus* slides did not show the heavy hybridization over centromeric heterochromatin which we might expect from

---

[4] cRNA = complementary RNA produced by *in vitro* transcription of DNA.

more recent studies on other organisms. However, the
DNA used had been fractionated to remove the ribosomal
cistrons. This procedure undoubtedly removed two other
satellite fractions (Birnstiel *et al.*, 1968) which may
well contain the sequences of the centromeric heterochromatin.

Mouse Satellite DNA as an Example of
a Highly Repetitive DNA

In some organisms highly repetitive DNA fractions
appear as minor peaks, somewhat separated from the bulk
of the DNA on caesium salt density gradients. Such DNA
fractions are referred to as satellite DNAs because of
their relation to the main band of DNA in the caesium
gradient. Satellite DNAs from the mouse and the guinea
pig have been studied extensively and each appears to
consist of a family of related sequences which have
diverged slightly from a hypothetical common ancestral
sequence. The repeating unit in mouse satellite DNA
probably consists of 8-13 nucleotide pairs (Southern,
1970). From studies on the reassociation of mouse
satellite DNA Sutton and McCallum (1971) have concluded
that the repeating units contain an average of 3.5%
base substitutions. The guinea pig satellite apparently has a basic unit of only six nucleotide pairs, but
has more base pair changes among the repeats than does
the mouse (Southern, 1970). Since pure fractions of
both satellites can be isolated from high molecular
weight DNA the repeating sequences must be clustered
in large groups. Large blocks of such short sequences
seem unlikely templates for protein message transcription. Although satellite DNA makes up 10% of the mouse
genome it is still unclear whether RNA is copied from
satellite sequences. Flamm, Walker and McCallum (1969)
were unable to detect any hybridization of mouse satellite DNA to RNA from mouse liver, spleen, or kidney.
Harel *et al.* (1968) have reported binding of rapidly
labeled mouse RNA to satellite DNA but the level of
hybridization was low and the possibility remains that
their satellite DNA included a second fraction of rapidly transcribed DNA. Another striking feature of mouse
satellite DNA is its species-specificity. Its sequences

show no detectable homology with the DNA of other ro-
dents, although several of the species examined also
have satellite bands (Flamm, Walker and McCallum, 1969).

Maio and Schildkraut (1969) separated isolated mouse
metaphase chromosomes into several groups by size and
found that each group contained approximately the same
proportion of satellite DNA as did the entire genome,
suggesting that satellite DNA makes up a constant proportion of the individual chromosomes. Satellite DNA
was also shown to be associated with the heterochromatic fraction isolated from mouse liver and brain cells
(Yasmineh and Yunis, 1970). These biochemical localization studies have been extended by *in situ* hybridization experiments (Pardue and Gall, 1969, 1970; Jones,
1970) which show that mouse satellite DNA sequences are
present in the centromeric heterochromatin of all of
the mouse chromosomes except the Y.

It is customary to divide the heterochromatin into
two classes. Regions of constitutive heterochromatin
are those which are regularly more condensed and stain
more heavily than the rest of the chromatin. Certain
chromosomes or parts of chromosomes show heterochromatic staining only in particular tissues and are considered to be facultatively heterochromatic. *In situ*
hybridization experiments have demonstrated a biochemical difference between these two types of heterochromatin. In mouse pachytene spermatocytes the sex chromosomes have undergone facultative heterochromatization
yet only the constitutive heterochromatin of the centromeres binds satellite DNA (Pardue and Gall, 1970).

Satellite Sequences from Species
Other than the Mouse

The localization of mouse satellite DNA within the
centromeric heterochromatin raises two questions: are
centromeres generally associated with highly reiterated
DNA sequences? does constitutive heterochromatin in
other parts of the chromosome contain repetitive DNA?
Experiments on a variety of organisms are beginning to
suggest an affirmative answer to both these questions
and to show complexities within the centromeric heterochromatin which were not detected in the mouse studies.

M. L. Pardue and J. G. Gall

TABLE 1

Satellite DNA fractions localized by cytological hybridization

Organism	% of genome in satellite	Buoyant density of satellite*	$C_0t_{1/2}$ of satellite	Preparative method
Mus musculus	10	1.691	6.6×10^{-4}	CsCl or $Ag^+Cs_2SO_4$
Rhynchosciara hollaenderi	10 (in 3 satellites)	1.680 1.675	< 1	CsCl
Drosophila neohydei	3.4	1.720		CsCl
Drosophila pseudoneohydei	4.9 (in 2 satellites)	1.684 1.691		CsCl
	7.5	1.717		CsCl
Plethodon cinereus	2	1.728		CsCl
Drosophila melanogaster	8 (in 2 satellites)	1.689	< 0.03	CsCl
Drosophila virilis	I 25	1.692		CsCl
Homo sapiens				$Ag^+Cs_2SO_4$
Homo sapiens	II 2			$Ag^+Cs_2SO_4$

*For native DNA in neutral CsCl.

TABLE 1

Satellite DNA fractions localized by cytological hybridization

Chromosomal localization	Reference
Centromeric heterochromatin on all chromosomes except the Y.	Waring, Britten, '66; Pardue, Gall, '69,'70; Jones, '70
Centromeric heterochromatin on each of the four chromosomes and certain densely staining bands in the telomere regions of two chromosomes.	Eckhardt, Gall, 1971
Sites distributed over the salivary chromosomes.	Hennig, Hennig, Stein, 1970
α-Heterochromatin in chromocenters of salivary chromosomes and some bands on chromosome arms.	
Kinetochore regions of salivary chromosomes (possibly α-heterochromatin).	Hennig, Hennig, Stein, 1970
Centromere regions of all chromosomes.	Macgregor, Kezer, 1971
Chromocentral heterochromatin in salivary gland preparations.	Gall, Cohen, Polan, 1971
Centromeric heterochromatin of mitotic chromosomes and primarily α-heterochromatin in salivary gland chromocenters.	Gall, Cohen, Polan, 1971
Principally located centromeric heterochromatin of chromosome 9.	Arrighi, Getz, Saunders, Saunders, Hsu, 1971
Centromeric regions of chromosomes 1, 9, 16 and possibly other chromosomes.	Jones, Corneo, 1971

Table 1 lists the satellite DNAs which have now been studied by *in situ* hybridization. These satellites differ markedly in amount and in base composition as measured by buoyant density in CsCl. However, they all show physical characteristics which indicate that they are composed of homogeneous repetitive DNA. Since the most extensive biochemical analyses have been made on satellite DNAs from mouse, human, *Rhynchosciara*, *D. melanogaster* and *D. virilis* (Waring and Britten, 1966; Flamm, McCallum and Walker, 1967; Corneo, Ginelli and Polli, 1970; Eckhardt and Gall, 1971; Gall, Cohen and Polan, 1971) these can be used as a measure of the characteristics which might be expected of similar DNA fractions from other organisms. The satellite DNAs from these five organisms form sharply defined bands in neutral CsCl density gradients and have sufficient asymmetry in the distribution of guanine and thymine in their two strands to cause the separated strands to band at different densities in alkaline CsCl gradients. These satellite DNAs also show sharp melting transitions and renature quickly after strand separation. Although some of these properties may be essential to the bio- logical functions of satellite DNAs, they might simply be a result of the clustering of many repeats of similar short nucleotide sequences. Most of these properties have not yet been tested for the other satellite DNAs in Table 1.

Another characteristic of mouse satellite DNA is its species-specificity. Mouse satellite sequences show no cross hybridization to the DNA of closely related ro- dents. Of the other satellites listed in Table 1 only those from *Drosophila neohydei* and *Drosophila pseudo- neohydei* have been tested for species-specificity. The specificity was measured by saturation experiments which compared the level of hybridization of RNA tran- scribed from the heavy satellites of the two species to homologous total DNA as well as to total DNA from the other species. In both cases the level of heterologous hybridization was only a fraction of the homologous reaction, yet these species are so closely related that there is precise pairing along the length of homologous chromosomes in the salivary gland nuclei of *neohydei- pseudoneohydei* hybrids.

Since the satellites listed in Table 1 share the
unusual physical characteristics of the mouse satellite
it is most interesting to find that, with one exception,
these satellites are all localized in centromeric heterochromatin. This evidence suggests that the association of centromeres with highly reiterated DNA is a
general phenomenon. It is well known that centromeric
heterochromatin varies in amount even between related
species and satellite DNA variations undoubtedly reflect this cytological fact. The variations in GC content of satellite DNAs from different species suggest
that, though repetitiveness is important, a number of
different sequences can perform the function of DNA in
centromeric heterochromatin.

Centromeric Heterochromatin

The more recent studies on repetitive sequences
suggest that centromeric heterochromatin is more complex than might have been expected from the studies on
the mouse. Some of the results of the hybridization
experiments on *Drosophila* chromosomes were predicted in
the work of Heitz (1934a, b) who studied the large
heterochromatic regions which characterize the mitotic
chromosomes of *Drosophila*. In polytene nuclei of *Drosophila* the centromeric heterochromatin of the mitotic
chromosomes fuses to form a chromocenter. Heitz found
that the chromocenter contained a densely staining
central region, which he named α-heterochromatin and a
less densely staining peripheral region, which he named
β-heterochromatin. *In situ* hybridization experiments
have shown that the distinction which was originally
based on staining reactions is also reflected in the
DNA sequences. The *Drosophila* satellites which have so
far been studied hybridize primarily or exclusively in
the α-heterochromatin. The one possible exception is
the heavy satellite of *D. pseudoneohydei* which has not
yet been localized exactly within the heterochromatin.
β-Heterochromatin also contains highly repetitive DNA,
but the reiterated sequences are different from those
of the α-heterochromatin satellites. Though none of
the sequences of the β-heterochromatin have been isolated these regions are shown to contain repetitive DNA

when hybridized to RNA copied from total *Drosophila* DNA
(Hennig, Hennig and Stein, 1970; Jones and Robertson,
1970; Rae, 1970; Gall, Cohen and Polan, 1971).

Centromeric heterochromatin in the salamander,
Plethodon cinereus, seems to contain repetitive sequences other than those found in the heavy satellite.
Macgregor and Kezer (1971) report that RNA transcribed
from main band DNA hybridized generally over the chromosomes and also hybridized heavily with the same centromeric regions to which RNA,copied from the heavy
satellite,bound. No contaminating satellite sequences
could be detected in the main band cRNA when attempts
were made to hybridize it to fractions of a CsCl gradient containing satellite DNA.

The studies on *Rhynchosciara* also demonstrate the
complexity of centromeric heterochormatin and, in addition, give the first evidence that centromere regions
of nonhomologous chromosomes may differ in the sequences of repetitive DNA which they contain. Eckhardt and
Gall (1971) found that the two satellites from a neutral CsCl gradient of *Rhynchosciara* DNA produced six
separate strands when denatured and centrifuged to
equilibrium in alkaline CsCl. Thus the combined satellites must contain three different families of reiterated sequences, yet these satellites hybridize only
with centromeric heterochromatin and two telomeric
bands. More recently Eckhardt (personal communication)
has been studying the distribution of the isolated components of the satellites. He finds that one of the
families of sequences is present on all four pairs of
Rhynchosciara chromosomes, but a second set of sequences is missing from the B chromosome though present in
the other regions labeled by the combined satellites.
The *Rhynchosciara* satellites studied by Eckhardt and
Gall also hybridized to bands near the telomere on two
pairs of chromosomes, showing that the sequences of
centromeric heterochromatin might be present in other
regions of the chromosome.

The human satellite DNA reported by Arrighi *et al*.
(1971) shows uneven distribution among the centromeres
of the different chromosome pairs and most concentrated
in chromosome 9. Jones and Corneo (in preparation)
have reported a human satellite DNA which hybridizes

with chromosomes 1, 9 and 16. Jones and Corneo find
that this satellite hybridizes most heavily with chromosome 1.

The specific localization of certain satellite DNAs
within centromeric heterochromatin permits the study
of centromere distribution within interphase nuclei by
in situ hybridization. This technique has been used
to follow the arrangement of centromeres during the
stages of spermatogenesis in the mouse (Pardue and Gall,
1970, 1971) and the salamander *P. cinereus* (Macgregor
and Kezer, 1971). These studies show that each cell
type has a characteristic centromere distribution
ranging from that of the early spermatogonia, where
centromeres are generally distributed over the nucleus
with little tendency to associate, to the single central aggregation which is found in the spermatocyte and
persists after the differentiation of the spermatids.

Highly Repetitive DNA Sequences which
do not Form Satellites in CsCl
Gradients

The DNA of many organisms contains no obvious satellite fractions when centrifuged to equilibrium in
CsCl gradients. However certain repetitive DNA fractions show differential binding of heavy metal ions
and may be isolated on Cs_2SO_4 equilibrium gradients by
means of this property (Corneo *et al.*, 1968). In other
cases highly repetitive DNA has been separated from the
rest of the genome by allowing denatured DNA to renature under conditions in which less repetitive DNA does
not reassociate. The renatured sequences are then separated from the single-stranded DNA by hybroxyapatite
column chromatography. Arrighi *et al.* (1971) have
used hydroxyapatite fractionation to purify highly
repetitive sequences from human DNA for cytological
localization. They concluded that rapidly renaturing
DNA sequences hybridized most readily with the centromeric and terminal regions of the chromosome though
there was significant binding to the interstitial segments also. Rae (1970) isolated the rapidly reannealing DNA of *D. melanogaster*, by hydroxyapatite fractionation and found that these sequences hybridized both

with the chromocenter and with regions in the chromosome arms. The DNA in these experiments was not subjected to fractionation other than the hydroxyapatite chromatography and so undoubtedly contained several families of reiterated sequences. The relation between the sequences hybridized in the different chromosomal regions remains to be studied.

Arrighi *et al*. (1970) have also used hydroxyapatite fractionation to study the distribution of repeated sequences in *Microtus agrestis* chromosomes. Localization of these sequences was difficult because the autosomes are quite small, but the hybridization appeared to be centromeric. In addition the constitutive heterochromatin of the Y and the long arm of the X were heavily labeled, in contrast to the unlabeled facultative heterochromatin of the short arm of the X. These experiments also show that the Y chromosome shares some repetitive sequences with other chromosomes since the repetitive DNA used as a template for the hybridizing RNA was derived from a female.

Since the centromeric satellites which have been studied have a wide range of base compositions, it would be expected that in some species the DNA sequences in centromeric heterochromatin have base compositions which band with the bulk of the DNA in CsCl gradients. The studies on rapidly renaturing nonsatellite DNAs discussed above show that highly repetitive DNAs are found near centromeres in organisms which do not have demonstrable DNA satellites as well as in organisms which do have satellites.

*Replication of Highly Repeated
Sequences*

In his early work on the heterochromatin of *Drosophila*, Heitz (1934a, b) suggested that heterochromatin fails to replicate during the formation of polytene chromosomes. This suggestion has been strongly supported by microspectrophotometric measurements on *D. melanogaster* (Rudkin, 1964, 1969) and *D. hydei* (Berendes and Keyl, 1967; Mulder, van Duijn and Gloor, 1968). The studies of Gall, Cohen and Polan (1971) provide additional confirmation and permit identifica-

tion of at least some of the sequences which are under-
replicated in the polytene chromosomes. DNA prepared
from the predominantly diploid tissues of either larval
brains or imaginal discs from *D. melanogaster* was found
to have a satellite comprising approximately 8% of the
total. DNA similarly prepared from *D. virilis* had three
satellites making up some 41% of the total. RNA complementary to the largest satellite of *D. virilis* hybridized to the centromeric heterochromatin of mitotic
chromosomes and to the α-heterochromatin in salivary
gland preparations. However, the number of silver
grains over the polytene nuclei was approximately the
same as the number over the diploid nuclei in the same
preparation, suggesting that the α-heterochromatin had
not taken part in the extra replications of the polytene nucleus. Furthermore when DNA was made from the
polytene larval salivary glands of either *Drosophila*
species or from polytene larval gut tissue of *D. virilis* the satellites were not detected at normal load-
ings in the analytical centrifuge.

The experiments indicate that β-heterochromatin
differs from α-heterochromatin not only in its nucleotide sequences but also in its replication during poly-
tenization. Experiments using cRNA copied from unfractionated DNA showed considerably higher hybridization
to the β-heterochromatin of the chromocenter than to
diploid nuclei on the same slide, indicating that
β-heterochromatin undergoes some, if not all, of the
extra replications of the polytene nucleus.

Nonreplicating sequences in polytene nuclei are
found in a number of species. Satellite DNA fractions
in both *Sarcophaga bullata* (Swift and Rae, in preparation) and *Rhynchosciara hollaenderi* (Eckhardt, personal
communication) are greatly reduced in polytene tissues.

The control of the replication of the centromeric
satellite DNA in polytene tissues apparently does not
extend to all sequences in the regions of α-heterochromatin. The nucleolus organizer of *Drosophila* has been
localized in the α-heterochromatin of the X chromosome
by classical cytogenetic studies (Kaufmann, 1934; Heitz,
1934a; Cooper, 1959) and more recently this region has
been shown to contain the rDNA (Ritossa and Spiegelman, 1965; Ritossa *et al.*, 1966). In cytological ex-

periments ribosomal RNA hybridizes with the nucleolus
at a level which indicates that the rDNA[5] has undergone
numerous replications (Pardue *et al.*, 1970). A similar
situation obtains for *Rhynchosciara hollaenderi*. Here
the ribosomal RNA (Pardue *et al.*, 1970) hybridizes to
regions either very close to or within regions containing the centromeric satellite sequences (Eckhardt and
Gall, 1971). The *Rhynchosciara* satellite is under-rep-
licated in polytene tissues (Eckhardt, personal communication) yet Gerbi (1971) has shown that DNA from larval salivary glands hybridizes to rRNA to the same extent as does DNA from adult carcass.

Cytological hybridization experiments have shown a
correlation between very high repetitive DNA sequences
and DNA with atypical replication patterns in a number
of organisms. Nearly all of the repetitive sequences
which have been studied have been localized in heterochromatic regions. Heterochromatic regions are characterized by late replication (Lima-de-Faria and Jaworska,
1968) though, at least in the case of facultative heterochromatin, the same DNA sequences may replicate earlier
in cells in which the region is not heterochromatic
(Nicklas and Jaqua, 1965; Priest, Heady and Priest,
1967). The localization of mouse satellite DNA in late-
replicating centromeric heterochromatin is supported by
biochemical studies showing that the mouse satellite
sequences are preferentially synthesized in the later
part of the period of DNA replication (Tobia, Schildkraut and Maio, 1970; Bostock and Prescott, 1971; Flamm,
Bernheim and Brubaker, 1971). In some nuclei the replication of the sequences of centromeric heterochromatin may not keep pace with the replication of other DNA
sequences, as shown in the *Drosophila* polytene cells
discussed above. A further indication of differential
replicative control for certain highly repetitive sequences comes from the work of Smith (1970) who found
that, when contact-inhibition of growth in baby mouse
kidney cell cultures was released by infection with

---

[5]rDNA = the DNA sequences coding for rRNA including
both the sequences actually transcribed and the interspersed spacer regions.

polyoma virus, synthesis of satellite DNA preceded rather than followed replication of the rest of the DNA.

Evidence on the Function and Origin of Centromeric DNA

Two major questions arise from these studies on highly repetitive sequences. What is the function, if any, of these sequences? How can we explain the evolution of repetitive sequences which, because of their strict species-limitation, must have arisen relatively recently and yet are distributed throughout the chromosome complements? At this point any answer to either question must be highly speculative. All organisms which have been studied by cytological hybridization have concentrations of repetitive DNA near the centromeres. This suggests that these sequences do have a function. The argument is also supported by the fact that the proportion of mouse satellite DNA is maintained in established cell lines even though the chromosomes have undergone much morphological change. We have found the same pattern of satellite hybridization in all of the mouse cell lines which we have studied.

In our first *in situ* hybridization experiments on mouse satellite we used chromosomes from a tissue culture line which did not have a Y chromosome. In these preparations satellite sequences were detected at the centromere of every chromosome. This suggested that satellite sequences might play a role in binding the spindle microtubule proteins during mitosis. However, preliminary attempts to bind purified microtubule subunits to mouse satellite DNA *in vitro* showed no affinity between the two under conditions of our assay. We later studied mouse satellite localization in mouse testis cells and found that the Y chromosome did not bind satellite DNA (Pardue and Gall, 1970). The failure to detect satellite sequences on the Y chromosome makes it less probable that satellite sequences play a role in organizing the mitotic spindle. On the other hand the apparent lack of satellite sequences on the Y offers some support to theories implicating satellite DNA in chromosome pairing since the Y is the only chromosome in the mouse which never undergoes a normal pairing with its homologue.

Attraction between heterochromatic regions of non-homologous chromosomes is very common. In the polytene nuclei of *Drosophila*, the proximal heterochromatin of all the chromosomes unites to form the chromocenter (Heitz, 1934a, b; Prokofyeva-Belgovskaya, 1935; Schultz, 1939). Other heterochromatic regions at the telomeres and within the euchromatic regions may show ectopic pairing with each other and with the chromocenter (Slizynski, 1945; Kaufmann *et al.*, 1948). Similar associations between nonhomologous centromeres and telomeres are seen in the polytene nuclei of other insects, such as *Rhynchosciara*, although typical chromocenters are not formed in all species. Nonhomologous associations of centromeric and telomeric regions have been described during both mitosis and meiosis in vertebrates (Ohno, Christian and Stenius, 1963). If repetitive sequences do play a causitive role in this association they might function in an orientation of the chromosomes preliminary to homologue pairing. Macgregor and Kezer (1971) have shown that in *Plethodon cinereus* the behavior of the centromeres during spermatogonial meiosis is consistent with a role in chromosome orientation. The present evidence on the distribution of satellite sequences between chromosomes shows few qualitative or quantitative differences which might suggest that satellite sequences could mediate the association of homologous chromosomes. However, it is becoming apparent that centromeric heterochromatin contains several types of sequences and a possible mechanism for homologue specificity cannot yet be ruled out.

The polytene nuclei, in which there is no replication of centromeric heterochromatin, will degenerate at the end of larval life without going through another mitosis. It would be interesting to know whether other types of cells which have become polyploid and are destined never to divide again, such as nurse cells in insect ovaries, are also deficient in the sequences of centromeric heterochromatin. In this context it might be noted that Lindsley and Novitski (1958) concluded from studies on dicentric anaphase bridges in *Drosophila* that the strength of a centromere was determined by the associated heterochromatin. This suggests that these repetitive sequences might play a role in anaphase movements in both mitosis and meiosis.

The evolution and distribution of satellite sequences has recently been discussed by Walker (1971a, b).
He suggests that sequences which had undergone amplification might then spread throughout the chromosome
complement provided that the possession of the repeated
sequences gives chromosomes a selective advantage over
their homologues at meiosis.

Repetitive Noncentromeric DNA Sequences

There are repetitive sequences widely distributed
over the noncentromeric regions of the chromosomes
(Pardue and Gall, 1969; Jones and Robertson, 1970; Rae,
1970; Arrighi et al., 1971; Eckhardt and Gall, 1971;
Gall, Cohen and Polan, 1971). In these experiments the
hybridizing material contained many families of reiterated sequences, so it is difficult to draw conclusions
concerning the relationships of the regions of repetitive DNA to each other and to the centromeric heterochromatin. However, there is suggestive evidence that
the X chromosome in Drosophila may contain a higher
proportion of repeated sequences than the autosomes
(Gall, Cohen and Polan, 1971). To date only the two
bands near the telomeres of chromosomes A and C in
Rhynchosciara have been localized by hybridization with
purified families of repetitive sequences (Eckhardt and
Gall, 1971).

The Ribosomal Cistrons in Amphibians

The genes coding for ribosomal RNA in eukaryotes are
among the DNA sequences with intermediate renaturation
rates, having in most cases a $C_0t_{1/2}$ in the range of
8-40. In most animals the double stranded DNA coding
for rRNA makes up something less than 1% of the genome;
in some plants the ribosomal cistrons may make up nearly 2% of the total. In oocytes of certain insects and
amphibia the ribosomal genes undergo differential replication, or amplification, and so make up a much larger proportion of the total (reviewed by Gall, 1969).
We chose oocytes of the toad Xenopus laevis as a

test system for developing a technique of cytological hybridization (Gall and Pardue, 1969) because the late pachytene oocyte of *Xenopus* contains some 3000 copies of the nucleolus organizer, or 1.8×10^{13} daltons of the total ribosomal DNA (Macgregor, 1968). The amplified rDNA in the pachytene oocyte forms a cytologically distinct cap on one side of the nucleus. In hybridization experiments using tritiated ribosomal RNA prepared from *Xenopus* tissue culture it is possible to detect specific labeling of the rDNA in the pachytene oocyte after an overnight autoradiographic exposure. Such a system is well suited for experiments on technical details of *in situ* hybridization. For quantitative comparisons we use the diplotene nuclei because here the rDNA has spread evenly over the nucleus minimizing the problems of self-absorption and rapid saturation of the film found when DNA is in a more compact configuration. The follicle nuclei contain only 0.03% as much rDNA as the diplotene nuclei and, with the pachytene chromosomes, serve to measure the nonspecific binding to DNA.

In addition to serving as test systems in the study of technical details, the *Xenopus* ovary preparations have given new evidence on the amplification of the ribosomal genes. Measurements of the synthesis of extra DNA in *Xenopus* ovaries by both microspectrophotometry and ^{3}H-thymine incorporation (Macgregor, 1968) have shown that amplification occurs in the leptotene, zygotene, and early pachytene stages of meiosis. However, cytological hybridization has shown that some extra DNA is, in fact, present earlier in the oogonia (Fig. 1). We found that a single oogonium might have as many as ten or more nucleoli each with associated rRNA. Nearby follicle nuclei had only one or two nucleoli. The level of hybridization over follicle cell nucleoli was well below that over each of the oogonial nucleoli. Although amplified rDNA does not become permanently incorporated into the genome, it persists through two or three mitotic divisions in the oogonia of several insects (Giardina, 1901; Hegner and Russell, 1916; Lima-de-Faria, 1962). Our evidence suggests that a similar situation occurs in *Xenopus*.

Early spermatogonia are morphologically similar to

early oogonia and also contain multiple nucleoli.
Neither our experiments nor those of Brown and Blackler
(1971) have detected extra ribosomal genes in the DNA
extracted from the testes of young *Xenopus* toads. However, cytological hybridization experiments show that
each of the spermatogonial nucleoli does have associated amplified rDNA (Fig. 2). Since it is possible to
produce phenotypically normal females from genetic
males by hormone treatment (Chang and Witschi, 1955)
the early stages in the germ cell development of the
two sexes might be expected to be similar; however, the
function of the extra DNA in the male line is obscure
since the sperm contains only the normal haploid amount
of rDNA. The fate of the amplified rDNA in the spermatogonia is currently under study.

The chromosomal ribosomal DNA in the haploid genome
of *Xenopus laevis* consists of some 600 repetitions of
the sequences coding for 18 and 28S RNA separated by
spacer sequences which are apparently untranscribed.
Using ribosomal RNA labeled in tissue culture these
sequences can be detected in the nuclei of follicle
cells after several months of autoradiographic exposure.
When the hybridization is done with cRNA transcribed
from rDNA detection requires only a few days.

The Ribosomal Cistrons in Polytene
Nuclei of Dipterans

We have also used the hybridization of ribosomal
RNA to study the ribosomal cistrons in the polytene
nuclei of the salivary glands of several Dipterans
(Pardue *et al.*, 1970). Here we find that the organization of the rDNA takes a variety of forms. Only in
Chironomus tentans did the ribosomal RNA hybridize with
a typical puffed chromosomal band situated in a euchromatic region of the chromosome (Fig. 5). The hybridized DNA showed little ramification into the large
nucleolus formed around the nucleolus organizer. In
the nucleoli of *C. tentans* RNA transcription is apparently confined to the portion near the chromosome.
Pelling (1959, 1964) found that the earliest uridine
incorporation occurred in this region. Only after
longer pulses was uridine label found over the periph-

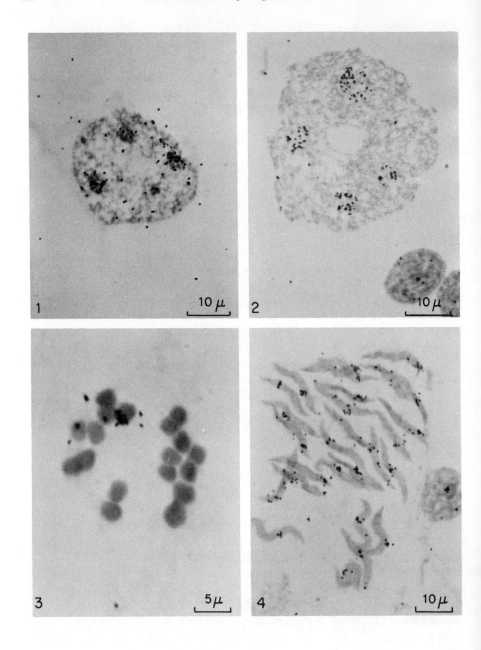

ery of the nucleolus. As predicted by the studies of
Beerman (1960) the ribosomal genes are found in the
nucleolus organizing regions of the second and third
chromosomes in *C. tentans*.

In *Drosophila hydei* the ribosomal genes are localized within the body of the nucleolus. The rDNA tends
to be most concentrated on one side of the nucleolus
and it is from this region that a Giemsa-staining
strand frequently extends towards the chromocenter. In
one case we observed heavy hybridization to the portion
of the strand nearest the nucleolus (Fig. 6).

We have reported that in the nucleolus organizers of
the Sciarids, *Sciara coprophila* and *Rhynchosciara hollaenderi*, the ribosomal DNA spreads widely into the
body of the nucleolus which is formed at the end of the
X chromosome. Some of the rDNA is found in the many

Figures 1-4. (Opposite page.) Autoradiographs of nuclei from the toad *Xenopus laevis* after hybridization
with ^{3}H-cRNA transcribed from *Xenopus* ribosomal DNA.
The DNA of the preparations was denatured with 0.07 N
NaOH for 3 min and hybridized for 8 hr. The cRNA had
a specific activity of 10^9 dpm/μg. All slides stained
with Giemsa. (1) Oogonium from a young *Xenopus* female.
This nucleus contains at least four nucleoli and the
binding of the cRNA indicates that each nucleolus has
some ribosomal genes adjacent to it. Exposure 5 days.
X 1200. (2) Spermatogonium from a young *Xenopus* male.
Like the oogonium this nucleus contains multiple nucleoli, each with associated ribosomal genes. Comparison with the follicle nuclei (lower right) shows that
each of the four spermatogonial nucleoli binds much
more rRNA than a typical diploid nucleus. Exposure 5
days. X 1200. (3) Second meiotic metaphase from the
testis of a mature *Xenopus* male showing the 18 chromosomes of the haploid *Xenopus* complement. Only one
chromosome (arrow) has bound the ribosomal cRNA.
Studies on the anucleolate *Xenopus* mutant (Wallace and
Birnstiel, 1966) have shown that all of the ribosomal
DNA of the haploid *Xenopus* genome is localized on one
chromosome. Exposure 31 days. X 2000. (4) *Xenopus*
spermatozoa showing localized hybridization of the
ribosomal cRNA. Exposure 31 days. X 1000.

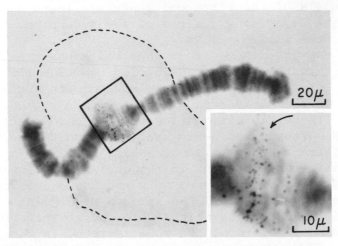

Figure 5. Autoradiograph of chromosome 2
from a salivary gland preparation of *Chironomus tentans* hybridized with ^{3}H-cRNA
copied from *Xenopus* rDNA. The ribosomal
cistrons are seen as a chromosomal puff
(insert) with little evidence of extensive
ramification into the nucleolus. The extent
of the faintly staining nucleolus is indicated by a dotted line. The DNA of these
chromosomes was denatured with 0.07 N NaOH
for 3 min and then hybridized with complementary RNA for 10 hr. The RNA had a calculated specific activity of 10^9 cpm/µg.
Slide stained with Giemsa. Exposure 15 days.

micronucleoli which are scattered among the chromosomes.
Despite this wide morphological distribution of the ribosomal cistrons the relative amount of ribosomal DNA
in the polytene nuclei seems to be about that of the
other somatic cells (Gerbi, 1971).

In older *Rhynchosciara* larvae the nucleolus dis-
perses but rDNA is still found on the heterochromatic
end of the X where the nucleolus once was. In *R. hollaenderi* this end of the X contains two amorphous regions separated by a small beaded puff. Although rRNA
hybridization did not occur evenly over the entire
heterochromatic region we were unable to localize the

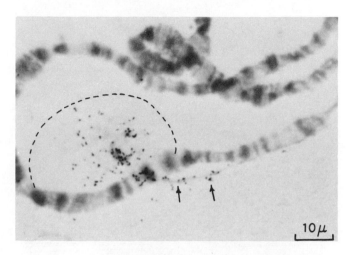

Figure 6. Autoradiograph of a salivary
gland preparation from *Drosophila hydei*
showing hybridization of radioactive *Xenopus*
ribosomal RNA to the DNA within the nucleolus (circled). A strand of DNA which runs
from the nucleolus toward the chromocenter
has bound the rRNA in the region nearest
the nucleolus (arrows). The DNA of the
preparation was denatured with 0.07 N NaOH
for 20 min and then hybridized with rRNA
for 14 hr. The RNA had a specific activity
of 5 x 10^5 cpm/μg. Slide stained with
Giemsa. Exposure 94 days.

rDNA more exactly in these preparations. Recently, in
collaboration with F. Lara of Sao Paulo, we have extended these observations to *R. angelae*. Here the rDNA
regions of the X are confined to a small number of
knoblike projections extending from the end of the
heterochromatic region (Figs. 7-9). In both *Rhynchosciara* species the heterochromatic knob on the end of
the C chromosome appears to contain ribosomal cistrons
though we have seen no nucleolus associated with this
region (Fig. 10). In our studies on *R. hollaenderi* we
noticed variable binding of rRNA to the heterochromatic
end of the B chromosome and suggested that some rDNA
had been pulled with the B from one of the other chro-

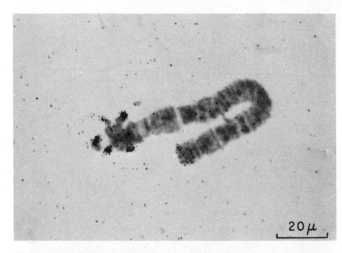

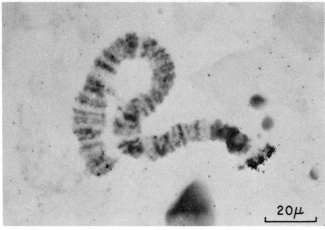

Figures 7 and 8. Autoradiographs of the X chromosome of *Rhynchosciara angelae* hybridized with cRNA transcribed *in vitro* from *Xenopus* ribosomal DNA. The cRNA was bound to the many small knob-like projections on the heterochromatic end of the X. These projections are easily pulled away from the chromosome in squashing as seen in Figure 8. The binding of cRNA to the telomere region of the X shown in Figure 7 is frequently seen in these preparations. Hybridization conditions as in Figure 5. Exposure 44 days.

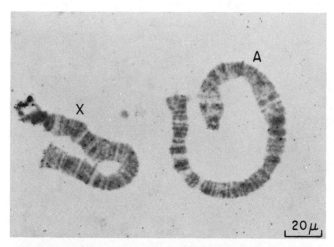

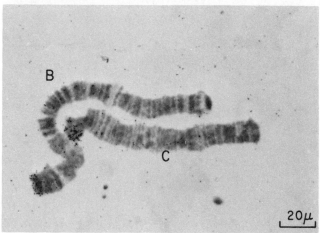

Figures 9 and 10. Autoradiographs of the
four chromosomes of *Rhynchosciara angelae*
hybridized with cRNA transcribed *in vitro*
from *Xenopus* ribosomal DNA. The cRNA has
bound to the knobs on the end of the X, to
the heterochromatin on the end of the C,
and to a well-defined region at the extreme
end of the heterochromatic portion of the B
chromosome. Pictures taken from the same
slide as Figures 13 and 14.

mosomes since in many cases the ends of the B, C and X chromosomes are tightly associated. The B chromosomes in our preparations of *R. angelae* also show rRNA hybridization at the extreme end of this region (Fig. 10) and this too might represent a secondary association similar to the preferential binding which several other chromosome regions show for micronucleoli.

Localization of Genes Coding for 5S RNA

The genes coding for the 5S ribosomal RNA have not been localized by conventional cytogenetic analysis although biochemical studies have shown that they are not lost in deletions of the nucleolus organizing regions of *Xenopus* (Brown and Weber, 1968). From similar experiments Tartof and Perry (1970) have concluded that the sex chromosomes of *Drosophila* do not contain the 5S cistrons. Recently Wimber and Steffenson (1970) have shown that the genes for 5S RNA in *Drosophila melanogaster* could be localized by *in situ* hybridization. They find the 5S cistrons in region 56 E-F of the salivary chromosomes. 5S RNA has a molecular weight of 4×10^4 daltons and the haploid genome of *D. melanogaster* contains only 200 repetitions of these sequences. Thus the 1000-fold replications of the salivary chromosomes give 8×10^9 daltons of DNA at the locus for 5S RNA. In the experiments of Wimber and Steffenson hybridization with *Drosophila* 5S RNA having a specific activity of 4×10^6 cpm/µg gave 15-30 grains over the band after a 2 month exposure.

Our studies on 5S RNA localization (Figs. 11-14) have confirmed those of Wimber and Steffenson (Pardue and Birnstiel, unpublished). Our experiments involved the heterologous hybridization of 5S RNA from *Xenopus* tissue culture to the polytene chromosomes of *Drosophila melanogaster*. *Xenopus* 5S RNA binds specifically to the 56 E-F region of the right arm of the second chromosome of *D. melanogaster* apparently in the proximal bands of 56F as noted by Wimber and Steffenson. Because our preliminary experiments showed heavy nucleolar hybridization we included an excess of unlabeled 18 and 28S rRNA in the incubation mixture. This successfully eliminated all annealing to nucleolar DNA.

Amaldi and Buongiorno-Nardelli (1971) have recently
reported that in interphase nuclei of Chinese hamsters
5S RNA binds to the nucleolus-associated chromatin. It
is possible that, whatever the chromosomal locus of the
5S, the 5S genes would lie close to an active nucleolus
in interphase cells. On the other hand, evidence that
even carefully purified 5S fractions contain fragments
of the 18 and 28S rRNA (Wimber and Steffenson, 1970)
suggests that the nucleolar hybridization might be due
to 18 and 28S rRNA.

Further Applications of In Situ
Hybridization

It may be possible to localize some genes coding for
messenger RNA by cytological hybridization. Kedes and
Birnstiel (1971) have recently studied the characteristics of the DNA coding for a 9S RNA fraction isolated
from the polysomes of cleaving sea urchin embryos.
These 9S RNAs are presumed to be messengers for the
synthesis of histones (Kedes and Gross, 1969). The 9S
RNA binds to DNA renaturing at a C_0t_{12} of 15-40 from
which Kedes and Birnstiel calculate that the DNA sequences have a 400-fold repetition in the sea urchin
DNA. Other experiments showed that the 9S RNA bound to
fractions which were on the heavy side of the DNA peak
in a CsCl gradient. This suggests that the DNA sequences coding for this RNA are clustered, a condition
which would facilitate cytological detection.
With maximum optimization of all variables it may
become possible to detect hybridization of messenger
RNAs coded for by unique DNA sequences in cytological
preparations of polytene chromosomes. If one assumes
a sequence of 10^5-10^6 daltons for the unique DNA, the
1000-fold repetition of the polytene chromosomes in
Drosophila gives a localized concentration of 10^8-10^9
daltons of the DNA to be hybridized. This may be compared with the amounts of DNA which have been detected
in our previous experiments using cRNAs of the highest
specific activity obtainable. The diploid mouse genome
contains 4×10^{11} daltons of double stranded satellite
DNA, an average of 10^{10} daltons per chromosome. We
were able to detect this easily after a 5 day autoradi-

ographic exposure. In fact, because of the compactness
of the metaphase chromosomes, the film over the centromeres had already become saturated in some cases. The
nucleolus organizer of *Xenopus* contains some $5-7 \times 10^9$
daltons of double stranded rDNA and can be detected

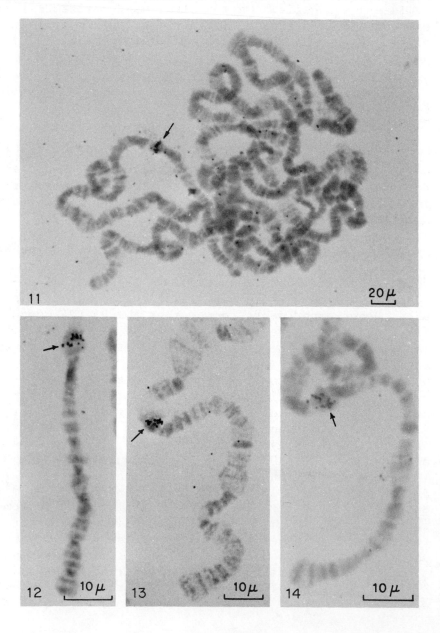

after a 4 day autoradiographic exposure. In neither
experiment were the conditions of hybridization carefully optimized for the sequences being studied so it
is probable that saturation was not achieved. Thus the
localization of unique DNA sequences of 10^5-10^6 daltons
should be feasible on Dipteran polytene chromosomes.

SUMMARY

In all published work, cytological hybridization has
been limited to the study of reiterated DNA sequences.
However, in favorable cases, unique sequences might be
detected in polytene chromosomes where the close appo-
sition of identical regions of many chromatids provides
a locally high degree of DNA sequence repetition.
In spite of the requirement for repetitious DNA sequences cytological hybridization has been used to investigate a variety of problems: 1) Hybridization of
the mysterious families of highly repetitive DNA sequences has been used to study their localization and relation to the rest of the cellular DNA. 2) Hybridization
of the DNA coding for ribosomal RNA and of sequences
present in centromeric heterochromatin has given new
information on the organization of polytene chromosomes.
3) Genes, such as those coding for 5S RNA, which have
not been localized by conventional genetic studies can
be localized by cytological hybridization. 4) The an-

Figure 11-14. (Opposite page.) Autoradiographs of
salivary chromosomes from *Drosophila melanogaster* hybridized with ^{3}H-5S RNA from *Xenopus laevis*. The hybridization is localized over region 56E-F on the right
arm of chromosome 2. The DNA of the chromosomes was
denatured with 0.07 N NaOH for 3 min and hybridized for
12 hr with 1 µg/ml of 5S RNA in 6 x SSC at 66°. The
RNA had a specific activity of 7.7 x 10^5 cpm/µg. Slide
stained with Giemsa. Exposure 149 days. (11) The entire chromosome complement of a polytene nucleus showing hybridization of ^{3}H-5S RNA only to region 56E-F
(arrow). (12-14) The distal end of the right arm of
chromosome 2 showing localization of the cistrons for
5S RNA in region 56E-F.

nealing of sequences specific for particular chromosome regions, such as centromeres, gives information on the arrangement of chromosomes within interphase nuclei. 5) Cytological hybridization permits the study of under- and over-replication of particular DNA sequences within individual cells which have not been fractionated from the surrounding tissue.

REFERENCES

Amaldi, F., and Buongiorno-Nardelli, M. (1971). *Exptl.. Cell Res.* *65*:329.

Arrighi, F. E., Getz, M. J., Saunders, G. F., Saunders, P., and Hsu, T. C. (1971). *Abstr. Intern. Cong. Human Gen., Paris, 1971.*

Arrighi, F. E., Hsu, T. C., Saunders, P., and Saunders, G. F. (1970). *Chromosoma 32*:224.

Arrighi, F. E., Saunders, P. P., Saunders, G. F., and Hsu, T. C. (1971). *Experientia* (in press).

Baltimore, D. (1970). *Nature 226*:1209.

Beerman, W. (1960). *Chromosoma 11*:263.

Berendes, H. D., and Keyl, H. G. (1967). *Genetics 57*:1.

Birnstiel, M., Sells, B., and Purdom, I. (1971). *J. M Mol. Biol.* (in press).

Birnstiel, M., Spiers, J., Purdom, I., Jones, K., and Loening, U. E. (1968). *Nature 219*:454.

Bostock, C. J., and Prescott, D. M. (1971). *Exptl. Cell Res. 64*:267.

Britten, R. J., and Kohne, D. E. (1968). *Science 161*: 529.

Brown, D. D., and Blackler, A. W. (1971). *J. Mol. Biol.* (in press).

Brown, D. D. and Weber, C. S. (1968). *J. Mol. Biol. 34*: 661.

Buongiorno-Nardelli, M., and Amaldi, F. (1970). *Nature 225*:946.

Chang, C. Y., and Witschi, E. (1955). *Proc. Soc. Exptl. Biol. Med. 89*:150.

Cooper, K. W. (1959). *Chromosoma 10*:535.

Corneo, G., Ginelli, E., and Polli, E. (1970). *J. Mol. Biol. 48*:319.

Corneo, G., Ginelli, E., Soave, C., and Bernardi, G. (1968). *Biochem. 7*:4373.

Davidson, E. H., and Hough, B. R. (1969). *Proc. Nat.. Acad. Sci. USA* 63:342.

Eckhardt, R. A., and Gall, J. G. (1971). *Chromosoma 32:* 407.

Flamm, W. G., Bernheim, N. J., and Brubaker, P. E. (1971). *Exptl. Cell Res. 64*:97.

Flamm, W. G., McCallum, M., and Walker, P. M. B. (1967). *Proc. Nat. Acad. Sci. USA 57*:1729.

Flamm, W. G., Walker, P. M. B., and McCallum, M. (1969). *J. Mol. Biol. 40*:423.

Gall, J. G. (1969). *Genetics 61* (suppl.):121.

Gall, J. G., Cohen, E. H., and Polan, M. L. (1971). *Chromosoma 33*:319.

Gall, J. G., and Pardue, M. L. (1969). *Proc. Nat. Acad. Sci. USA 63*:378.

Gall, J. G., and Pardue, M. L. (1971). In *Methods in Enzymology*, ed. L. Grossman and K. Moldave, *Vol. 21D* (New York: Academic Press), p. 470.

Gelderman, A. H., Rake, A. V., and Britten, R. J. (1971). *Proc. Nat. Acad. Sci. USA 68*:172.

Gerbi, S. A. (1971). *J. Mol. Biol. 58*:499.

Giardina, A. (1901). *Int. Mschr. Anat. Physiol. 18*:418.

Harel, J., Hanania, N., Tapiero, H., and Harel, L. (1968). *Biochem. Biophys. Res. Comm. 33*:696.

Hegner, R. W., and Russell, C. P. (1916). *Proc. Nat. Acad. Sci. USA* 2:356.

Heitz, E. (1934a). *Z. Zellforsch. 20*:237.

Heitz, E. (1934b). *Biol. Zbl. 54*:588.

Hennig, W., Hennig, I., and Stein, H. (1970). *Chromosoma 32*:31.

Jacob, J., Todd, K., Birnstiel, M. L., and Bird, A. (1971). *Biochim. Biophys. Acta 228*:761.

John, H., Birnstiel, M., and Jones, K. (1969). *Nature 223*:582.

Jones, K. W. (1970). *Nature 225*:912.

Jones, K. W., and Robertson, F. W. (1970). *Chromosoma 31*:331.

Kaufmann, B. P. (1934). *J. Morph. 56*:125.

Kaufmann, B. P., McDonald, M. R., Gay, H., Wilson, K., Wyman, R., and Okuda, N. (1948). *Carnegie Inst. Wash. Year Book 47*:144.

Kedes, L. H., and Birnstiel, M. L. (1971). *Nature New Biol. 230*:165.

Kedes, L. H., and Gross, P. R. (1969). *Nature* 223:1335.
Lima-de-Faria, A. (1962). *Chromosoma* 13:47.
Lima-de-Faria, A., and Jaworska, H. (1968). *Nature* 217: 138.
Lindsley, D. L., and Novitski, E. (1958). *Genetics* 43: 790.
Macgregor, H. C. (1968). *J. Cell Sci.* 3:437.
Macgregor, H. C., and Kezer, J. (1971). *Chromosoma* 33: 167.
Maio, J. J., and Schildkraut, C. L. (1969). *J. Mol. Biol.* 40:203.
Melli, M., and Bishop, J. O. (1969). *J. Mol. Biol.* 40: 117.
Melli, M., Whitfield, C., Rao, K. V., Richardson, M., and Bishop, J. O. (1971). *Nature New Biol.* 231:8.
Mulder, M. P., van Duijn, P., and Gloor, H. J. (1968). *Genetica* 39:385.
Nicklas, R. B., and Jaqua, R. A. (1965). *Science* 147: 1041.
Ohno, S., Christian, L. C., and Stenius, C. (1963). *Exptl. Cell Res.* 32:590.
Pardue, M. L. (1970). Ph.D. Thesis, Yale University.
Pardue, M. L., and Gall, J. G. (1969). *Proc. Nat. Acad. Sci. USA* 64:600.
Pardue, M. L., and Gall, J. G. (1970). *Science* 168:1356.
Pardue, M. L., and Gall, J. G. (1971). In *Chromosomes Today*, Vol. 3.
Pardue, M. L., Gerbi, S. A., Eckhardt, R. A., and Gall, J. G. (1970). *Chromosoma* 29:268.
Pelling, C. (1959). *Nature* 184:655.
Pelling, C. (1964). *Chromosoma* 15:71.
Priest, J. H., Heady, J. E., and Priest, R. E. (1967). *J. Cell Biol.* 35:483.
Prokofyeva-Belgovskaya, A. A. (1935). *Cytologia* 6:438.
Rae, P. M. M. (1970). *Proc. Nat. Acad. Sci. USA* 67:1018.
Ritossa, F. M., Atwood, K. C., Lindsley, D. L., and Spiegelman, S. (1966). *Nat. Cancer Inst. Monogr.* 23: 449.
Ritossa, F. M., and Spiegelman, S. (1965). *Proc. Nat. Acad. Sci. USA* 53:737.
Rudkin, G. T. (1964). In *Genetics Today*, Proc. XI Int. Congr. Genetics (The Hague: Pergamon), p. 359.

Rudkin, G. T. (1969). *Genetics 61* (suppl.):227.
Schultz, J. (1939). *Proc. VII Int. Genetics Congr.*
Slizynski, B. M. (1945). *Proc. Royal Soc. Edin. 62*:114.
Smith, B. J. (1970). *J. Mol. Biol. 47*:101.
Southern, E. M. (1970). *Nature 227*:794.
Sutton, W. D., and McCallumn, M. (1971). *Nature New Biol. 232*:83.
Tartof, K. D., and Perry, R. P. (1970). *J. Mol. Biol. 51*:171.
Temin, H. M., and Mizutani, S. (1970). *Nature 226*:1211.
Tobia, A. M., Schildkraut, C. L., and Maio, J. J. (1970). *J. Mol. Biol. 54*:499.
Walker, P. M. B. (1971a). *Nature 229*:306.
Walker, P. M. B. (1971b). In *Progress in Biophysics and Molecular Biology, Vol. 23,* eds. J. A. V. Butler, and D. Noble (London: Pergamon Press).
Wallace, H., and Birnstiel, M. L. (1966). *Biochim. Biophys. Acta 114*:296.
Waring, M., and Britten, R. J. (1966). *Science 154*:791.
Wimber, D. E., and Steffenson, D. M. (1970). *Science 170*:639.
Yasmineh, W. G., and Yunis, J. J. (1970). *Exptl. Cell Res. 59*:69.

The Evolutionary Solution
to the Antibody Dilemma

Donald D. Brown

Department of Embryology
Carnegie Institution of Washington
Baltimore

Most immunologists now assume that at least two separate germ line genes code for each immunoglobulin polypeptide (Dreyer and Bennett, 1965; reviewed in the Symposium on *Two Genes, One Polypeptide, Fed. Proc. 31,* 1972). The two genes are thought to be joined during the development of each immunocyte. Other genetic events have been proposed to occur during development to account for antibody variability. These include hypermutation of one or a limited number of germ line genes (Brenner and Milstein, 1966) and specific recombination of genes during development (Gally and Edelman, 1970; Smithies, 1970). If a specific genetic event occurs during development of an immunocyte, this event should also be involved in its stable commitment to express these rearranged or mutated genes. It seems economical to predict that these developmental genetic events determine which genes will be expressed and endow the cell with all of the classical features of a

committed or determined cell. If this is so, then the
immunologists have provided us with a novel molecular
mechanism for cell determination. Furthermore, this
mechanism involves "differential gene alteration"
rather than the time honored hypothesis for control of
transcription—"differential gene activation."

It is the purpose of this article to describe the
kind of DNA arrangement of antibody genes as well as
the evolutionary restrictions which would have to be
placed upon these genes if no such genetic change occurred during differentiation of the immunocyte, and
control of gene expression was exercised exclusively
by differential gene activation. Whereas the *developmental genetic* models require "heretical" genetic
events in somatic cells, the *evolutionary* model requires "heretical" mechanisms in evolution. Most of
these ideas have been set forth clearly many times but
particularly well by Edelman and Gally (1971) and Hood
(1972). The only difference is that one man's heresy
is another man's theory.

*Some Facts About Antibody Structure
and Genetics and the Hypotheses
Invoked to Explain Them*

Each antibody molecule is basically a tetramer
(L_2H_2) consisting of two identical light (L) chains
and two identical heavy (H) chains joined by disulfide
bonds (reviewed by Edelman and Gall, 1969). The
marked heterogeneity of gamma globulin became analyz-
able with the realization that individual myeloma tumors of mice and humans result in the production of
homogeneous immunoglobulins or L or H chains whose
amino acid sequence could be determined (Potter, 1967).
This has led to the partial or complete sequencing of
literally hundreds of different immunoglobulins (reviewed by Edelman and Gall, 1969; Smith, Hood and
Fitch, 1971; Pink, Wang and Fudenberg, 1971).

The H chains of immunoglobulin usually consist of
about 400 amino acids while the L chains are half that
size. The dilemma which led to heretical developmental
genetic models was the realization that the variability
in amino acid sequence between molecules is confined

mostly to one region of each polypeptide termed the
variable (V) region. This comprises about 110 amino
acid residues at the amino terminal end of both L and H
polypeptides. The remainder of each protein is called
the constant (C) region. Whereas individual polypeptides within a class usually have different amino acid
sequences in their V region, they have identical C
regions.

Glossary of Terms

The *class* of an H chain and the *type* of L chain depend upon their C region. Developmental genetic models
propose that the C regions of all immunoglobulins of
the same class or type are coded for by a single gene
in the germ line. L chain types are called K and λ.
H chain classes are γ, α, μ, δ and ε. Antibody genes
within a linkage group comprise a *family*. The genes
for λ, K and H chains each are thought to make up a
separate gene family.

Subgroups of immunglobulins refer to related V regions. Sequences of many V regions fall into certain
groups. Although individual members of a subgroup may
differ by several amino acid residues, they are closer
to each other in sequence than they are to members of
other subgroups (reviewed in Pink, Wang and Fudenberg,
1971). Some workers believe that each subgroup is encoded by a single or limited number of germ line genes
which mutate during development (reviewed by Milstein
and Pink, 1970); others believe that each V region is
encoded by a separate gene and that those coding for
the members of a single subgroup are distinct but have
diverged less from an ancestral copy (reviewed by Hood
and Talmage, 1970).

Alleles are different forms of the same gene which
segregate according to Mendelian laws. Some classes
(C region) and some subgroups (V region) behave as alleles. This is part of the reason that individual
classes and even some subgroups have been thought to
be encoded by a single germ line gene. Immunologists
apply the term allele because of the way these classes
or subgroups segregate in breeding experiments irrespective of whether one gene or a group of linked genes

is thought to be involved. Presumably multiple linked genes can behave like an allele if they do not undergo recombination at meiosis.

I should now like to summarize briefly the structural and genetic information which has to be accounted for by any hypothesis. I will point out the "heresies" required for the developmental genetic models on the one hand and those for the evolutionary model on the other. Experiments which have been carried out with human, mouse and rabbit immunoglobulins will be interchanged in the discussion. Each of these observations has recently been extensively reviewed.

Different Subgroups of V Regions can be Associated with Allelic Forms of a C Region.

TABLE 1

The proposed arrangement of genes in the germ line

One *developmental genetic* model

$$(V_1)_n (V_2)_m (V_3)_p - C_a$$

An *evolutionary* model

$$(V_1 C_a)_n (V_2 C_a)_m (V_3 C_a)_p$$

Over 100 human K chains have been sequenced. No two polypeptides have been found to be identical in the V region (reviewed by Quattrocchi, Cioli and Baglioni, 1969). In contrast all of them share one of two allelic forms in the C region (INV locus). This is a single amino acid interchange at position 191 (valine-leucine; Milstein, 1966). The amino acid sequences of human V_K fall into three subgroups. Individual K proteins have been found which have any of the three subgroups in combination with the same C polypeptide (Milstein, Milstein and Jarvis, 1969). Several developmental genetic hypotheses have been invoked to explain this finding. First, multiple germ line genes exist for each of the many V regions, but there is only one C region gene. In each immunocyte a translocation event in the DNA

occurs so that one of the V genes becomes adjacent to
the single C region gene. Variability is produced in
an individual by different cells which have translocated different V genes. Second, there are a limited
number of V genes in the germ line, perhaps one for
each subgroup of V sequences. During differentiation
the V gene(s) in each immunocyte undergoes genetic
change so that each mature immunocyte contains a V
gene(s) with a different nucleotide sequence. Those
who believe in this latter method for generating diversity also invoke a translocation event since the existence of subgroups appears to preclude a single gene
which codes for all of the possible V regions that are
found joined to any single C region. The heresies
involved in these developmental genetic models are
translocation of genes, hypermutation or recombination
of specific nucleotide sequences; these events are supposed to occur in somatic cells during differentiation.

The evolutionary model proposes that the genome in
all immunocytes is identical to the germ line genome.
Each gene for a V sequence is adjacent to one for a C
sequence. The heresies involved here are evolutionary.
In some manner (to be discussed later) evolutionary
mechanisms have allowed for the divergence of genes for
V regions while keeping multiple genes for C regions
identical. There would have to be thousands of such
V-C genes coding for human or mouse K chains.

Different Subgroups of V Regions can
be Associated with Nonallelic Classes
of the C Region.

TABLE 2

The proposed arrangement of genes in the germ line

One *developmental genetic* model

$$(V_1)_n(V_2)_m(V_3)_p - C_aC_b$$

An *evolutionary* model

$$(V_1C_a)_n(V_2C_a)_m(V_3C_a)_p(V_1C_b)_q(V_2C_b)_r(V_3C_b)_s$$

In human λ and H chains, nonallelic C regions appear
to share a common pool of V regions. For example,
human C_λ regions exist in two nonallelic forms (OZ^+ and
OZ^-) (Hood and Ein, 1968). These differ by a single
amino acid interchange at position 190 (arg-lys). Unlike the INV difference described above for human K
chains, these are not alleles since sera from all individuals contain λ chains which are OZ^+ and OZ^-. Sequence studies have demonstrated that individual λ
polypeptides which are OZ^+ or OZ^- can share V regions
which are drawn from the same subgroups (Tischendorf
and Tischendorf, 1970). Developmental genetic models
propose that the translocation event can recombine any
V gene with either C gene. The evolutionary model proposes that several kinds of V region genes have evolved
next to either of the two kinds of C region genes.
This *parallel* evolution must be explained by either
natural selection or by a new mechanism of evolution.

*One V Subgroup has Allelic forms
that Associate with Nonallelic
C Regions.*

TABLE 3

The proposed arrangement of genes in the germ line

One *developmental genetic* model

   $(V_1)_n C_a C_b C_c$

An *evolutionary* model

   $(V_1 C_a)_n (V_1 C_b)_m (V_1 C_c)_p$

The A_1, A_2, and A_3 allelic subgroups of rabbit H
chains are localized in the V region (reviewed by Mage,
1971). These subgroups associate with at least four
classes of H chain C regions (γ, α, μ, and δ) (Prahl
et al., 1970). According to the evolutionary model,
multiple DNA sequences for V regions which are on the
same chromosome (cis) and are adjacent to different C
region genes have retained sufficient homology to code
for polypeptides which react with the same antiserum.

Once again this means parallel evolution of genes in
cis configuration. This parallel evolution has not
prevented divergence of the C genes for each nonallelic
class. It has prevented divergence of multiple C genes
within each allelic class.

One developmental genetic explanation to account for
the rabbit subgroup genetics supposes that there is a
single germ line V gene and multiple germ line C genes.
During differentiation the V gene changes in each immunocyte to generate the variability, and the altered
V gene is translocated to be joined with one of the
different C genes. Two heresies are required—hypermutation and translocation. Another developmental
genetic explanation actually requires a component of
evolutionary heresy. It is proposed that all possible
V genes are present in the germ line in a cluster and
closely linked with the four (or more) C genes (one
for each class). Translocation places one of the V
genes next to one of the C genes during differentiation. A different combination occurs in each immunocyte. However, this explanation requires the parallel
evolution of multiple V genes so that those arranged
in cis configuration at least contain enough similarity
to give rise to proteins of the same subgroup. The
difference between this parallel evolution and that
proposed by the evolutionary model is that the multiple
V genes are adjacent while in the evolutionary model
they are interspersed with C genes in the germ line.
Clearly, these two models can be distinguished by finding individual recombinants, and these investigations
are being pursued (Mage, Young-Cooper and Alexander,
1971).

*A Human Myeloma Patient (Til) Produces
Homogeneous IgM and IgG whose V Regions
of the H Chains may be Identical*
(Wang *et al.*, 1970).

More than 35 residues of the homogeneous μ and γ
proteins produced by this patient have been sequenced
in their V regions and have been found to be identical.
Assuming that their entire V regions will be found to
be identical, not only can two C regions share the same

TABLE 4

The germ line of patient Til

One *developmental genetic* model

$$V_{Til}C_\gamma C_\mu$$

An *evolutionary* model

$$V_{Til}C_\gamma V_{Til}C_\mu$$

V subgroup but two C regions can share an identical V sequence. This result imposes an additional restriction on the evolutionary model by requiring the *exact* parallel evolution of at least two V region gene sequences—one next to a C_γ gene sequence and one next to a C_μ gene sequence.

Til^μ raises additional questions for both kinds of models, questions which we have not discussed which are of general importance for the antibody dilemma. These are problems of control mechanisms. What role do control mechanisms play in the decision by a cell to express one H and one L gene? Different cells of Til produce the IgM and IgG. If a single cell can express both H chain genes but at different steps of its differentiation, the developmental genetic model requires that the same V gene (or a replica of it) had to have been translocated next to two C genes. If separate progenitor cells exist to account for the formation of the two kinds of immunoglobulin, then they must share some kind of common control mechanism to cause translocation of the same V gene to different C genes. The evolutionary model also needs an additional control mechanism to cause the sequential expression of two genes with identical nucleotide sequences coding for the V region in the same or different cells.

The importance of transcriptional control mechanisms in the antibody dilemma is demonstrated by the phenomenon referred to as "allelic exclusion" (Cebra, Colberg and Dray, 1966). Allelic exclusion refers to the fact that a single cell usually only synthesizes one H and one L chain. The term is a misnomer since the exclusion is much more than allelic exclusion. A single cell synthesizes either a K or a λ light chain, and these are not alleles. The phenomenon is best envisioned as the switching on of one L and one H gene in each cell. The developmental genetic model by definition must predict that the H and L chains which are expressed are the ones which underwent the translocation event. However, immunoglobulin genes in each cell are capable theoretically of undergoing at least six translocation events, two for each of the three gene families—H, K and λ. The developmental genetic model is not sufficient to account for the phenomenon of allelic exclusion. Either it must be proposed that only two of the six gene clusters undergo translocation in a cell or control mechanisms under the broad category of "differential gene activation" select two of the six translocated genes for expression. The evolutionary model simply relies on differential gene activation. It says that there are hundreds or thousands of genes in each of the three gene clusters. Only one H and one L gene is switched on in any cell by the same process which singles out a limited number of genes for expression in other differentiated cell types.

V and C Regions are Coded for by
Genes that are Closely Linked
in the Germ Line.

This fact is required by an evolutionary model but need not be true for developmental genetic models. The rabbit γ chains have allelic markers in the V region ($A_{1,2}$) and in the C region ($A_{14,15}$). When heterozygote rabbits are produced from a mating of two rabbits which are homozygous for both V and C markers and their H chains isolated, more than 95% of individual polypeptides contain both markers derived from the same parent (Tosi, Mage and Dubiski, 1970).

	Mother	*Father*	F_1
	A_1A_{14}	A_2A_{15}	A_1A_{14}
	x		→
	A_1A_{14}	A_2A_{15}	A_2A_{15}

This demonstrates that if a translocation event occurs
it mainly links up two genes which are on the same
chromosome (cis).

The Evolution of Linked and Nonlinked
Genes and the Role of Natural
Selection

All immunoglobulin genes appear to have evolved from
a common ancestor whose length was about 110 amino acid
residues (Singer and Doolittle, 1966; Hill *et al.*,
1966). Thus each L chain is encoded by two such gene
regions in tandem while each H chain is the product of
four (or five) of these basic genes. The most parsi-
monious developmental genetic model admits to the presence of several germ line genes for immunoglobulins if
only one for each class and subgroup. The existence of
these related genes in a single genome demonstrates
that gene duplication occurred during evolution. From
subclass and subgroup sequence determinations it is
clear that unlinked genes have diverged more than
linked genes. Thus, if one supposes that each of the
V_K subgroups and each of the V_λ subgroups are encoded
by separate genes, the K subgroup sequences are closer
to each other than they are to any of the λ subgroup
sequences. The V_H subgroups cannot be mistaken for V_λ
or V_K sequences. Likewise, the sequences of the different H chain C classes are closer to each other than
they are to C_λ or C_K sequences. Thus linked genes have
not diverged to the extent that unlinked genes have in
the same organism.

Having made this generalization, it must be emphasized that some linked C regions have diverged considerably such as human C_μ and C_γ (Shimizu *et al.*, 1971).
In addition, considerable divergence in C regions has
occurred with speciation. It has been pointed out that
V_K sequences of mouse and human have fewer differences
than their C regions (Kabat, 1967). These observations
show that irrespective of which model explains the antibody dilemma there must be complex and *different* selective forces involved in the evolution of DNA coding
for V regions compared to that coding for C regions.
This difference is not surprising in view of the sepa-

rate functions of V and C regions in antibodies nor
does it require the two DNA sequences to be separate in
the germ line.

The DNA Predicted by the
Evolutionary Model

The evolutionary model for the antibody dilemma
predicts thousands of antibody genes. These genes
should be clustered in at least three linkage groups.
Each class of antibodies within each group will itself
be coded for by many genes. Genes within a class will
comprise a highly repetitive DNA in which V and C regions will alternate. The DNA coding for the V regions
will behave as a related but not identical group of
sequences, while those for the multiple C regions will
be more homogeneous. Despite the identity of multiple
C region proteins, the genes themselves could have diverged considerably in the third nucleotide position.
The evolutionary model would be supported if any component of eukaryotic DNA had this unusual structure.

We have studied the structure of two kinds of repetitious DNAs in the genome of the amphibians *Xenopus*
laevis and *Xenopus mulleri*. I wish to point out some
features of their structure which resemble the repetitious DNA that the evolutionary model predicts for
multiple immunoglobulin genes.

The Ribosomal DNA of X. laevis and
X. mulleri—Multiple, Adjacent
Homogeneous Genes

Ribosomal DNA contains the genes for the 18S and
28S rRNAs (Wallace and Birnstiel, 1966; Brown and
Weber, 1968). In *Xenopus* about 450 of these genes are
clustered together on one chromosome at the "nucleolar
organizer" locus. The multiple genes on one chromosome
behave superficially as an allele since they segregate
in a Mendelian fashion when paired with a chromosome
which is partly or completely deleted of these genes
(reviewed by Brown, 1967).

The arrangement of ribosomal genes is regular within
the cluster (Wensink and Brown, 1971). Individual 18S

and 28S genes are either adjacent or separated by no
more than a few hundred nucleotides (Dawid, Brown and
Reeder, 1970). Each pair of 18S and 28S genes alternates with a "spacer" region. Together these are
referred to as a single repeating unit (Fig. 1). Some-

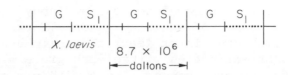

Figure 1. A schematic diagram of the arrangement of repeating sequences within the
ribosomal DNAs of *X. laevis* and *X. mulleri*.
The gene region (G) which contains the sequences for 18S and 28S rRNA alternates
with a spacer sequence (S) of about the
same length. Whereas the gene regions are
very similar if not identical between the
two species, the spacer sequences are very
different. Three repeats of each DNA are
shown; there are about 400 to 500 tandem
repeats clustered on a single chromosome.

what more than half of each repeat is transcribed *in
vivo* as a 40S RNA molecule (about 2.5×10^6 daltons).
This molecule is cleaved to form a 28S and an 18S rRNA
molecule (1.5×10^6 and 0.7×10^6 daltons, respectively)
(Loening, Jones and Birnstiel, 1969); the additional
nucleotide sequences (about 0.3×10^6 daltons) are discarded. The mass of one repeat of native rDNA is
8.7×10^6 daltons (Wensink and Brown, 1971). The DNA
coding for the 40S precursor is therefore 5.0×10^6
daltons (twice the mass of the 40S RNA since it is
double stranded) and the remainder, the spacer DNA, is
3.7×10^6 daltons. The rDNA within *X. mulleri* and *X.*

laevis is exceedingly homogeneous with respect to the lengths of the repeats and their nucleotide sequences. The former has been analyzed by electron microscopy (Wensink and Brown, 1971; Brown, Wensink and Jordan, 1972), the latter by molecular hybridization (Dawid, Brown and Reeder, 1970; Brown, Wensink and Jordan, 1972). This homogeneity of rDNA is as true for the spacer sequences as it is for the "gene" sequences (those which code for the 40S rRNA). These multiple, adjacent DNA repeats have evolved together. Whereas the multiple 18S and 28S genes of *X. mulleri* and *X. laevis* are indistinguishable, the spacers are very different in the two animals (Brown, Wensink and Jordan, 1972). The structure of these two rDNAs is relevant to the antibody dilemma in several respects.

(1) *Hundreds of clustered identical genes can behave as a single allele.* If the number of ribosomal genes in the germ line of *X. laevis* were not known by direct hybridization measurements, classical genetics would indicate that only a single 18S and 28S gene existed in the germ line. We know that there are hundreds.

(2) *Multiple genes can be maintained identical or nearly identical in a single genome without substantial divergence.*

(3) *Multiple genes can consist of two parts which have different evolutionary restrictions.* Thus the spacer sequences but not the gene sequences have evolved since speciation of *X. mulleri* and *X. laevis*. The evolution of multiple ribosomal genes is an example of parallel evolution since the multiple spacers have evolved together. The maintenance of many similar or identical genes in a genome may apply only to linked genes. The spacer sequences of rDNA are analogous to multiple identical C region genes (if they exist) within a single immunoglobulin class. These genes are homogeneous within a species; they are transmitted as a single allele, but they have diverged in the two *Xenopus* species.

The 5S DNA—Multiple, Adjacent,
Heterogeneous Genes

Whereas rDNA provides a model for certain features
of the antibody dilemma, it is clearly deficient in one
respect. It cannot account for a repeating DNA in
which half (L) or one-fourth (H) of each repeat varies
while the remainder is constant. Are there genes in the
animal genome which are constructed in this manner?
Analyses of eukaryotic DNA have revealed families
of DNA which contain poorly matched and well matched
sequences as measured by the fidelity with which the
base pairs reassociated after having been denatured
(Britten and Kohne, 1968). Not much is known about the
clustering of this repetitious DNA within the genome.
A repetitious DNA which has some similarities to the
proposed genes for antibodies has been isolated recently in our laboratory from the nuclear DNA of *X. laevis*
(Brown, Wensink and Jordan, 1971). This DNA contains
the genes which code for the low molecular weight RNA
of ribosomes referred to as 5S RNA. The 5S DNA comprises about 0.7% of *X. laevis* DNA. The sequences for
5S RNA are coded for by about 14% of the base pairs of
5S DNA, and we calculate that there are about 24,000
of the 5S DNA repeats in each haploid complement of DNA.
There is no known function for the remaining 86% of 5S
DNA, and it is not known whether this DNA is transcribed. We again refer to it as spacer DNA. The
genes for 5S RNA, like those for rRNA, are regularly
interspersed between the spacer sequences. This alternating arrangement was discovered because of the remarkable base composition heterogeneity within 5S DNA.
The thermal denaturation profile of 5S DNA demonstrates
this heterogeneity (Fig. 2). The overall base composition of 5S DNA is about 34% GC, but it consists of
two regions that denature with T_m values which differ
by 10°, a difference corresponding to at least 20% GC.
The region of 5S DNA which codes for 5S RNA is probably
the last region of 5S DNA which denatures because 5S
RNA is known to have a high GC content of about 57%
(Mairy and Denis, 1971). Because of the heterogeneity
in GC content it has been possible to partly denature
this DNA and examine it under the electron microscope.

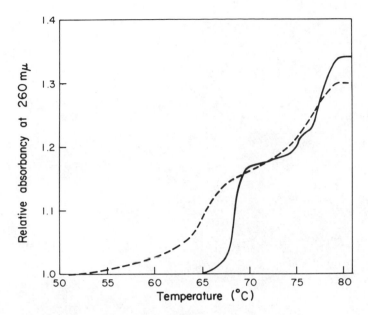

Figure 2.   Thermal denaturation profile of
native and reassociated 5S DNA.   The 5S DNA
(average molecular weight of about 10^7 daltons) was denatured in 0.1 SSC (solid line)
and then allowed to reassociate for four
days at 47°.   The reassociated DNA was re-
melted (dashed line).   Similar first and
second melt profiles have been obtained for
sonicated 5S DNA (about 0.5×10^6 daltons)
as well as the 5S DNA isolated from a single
animal.   A displacement of 1.6° is equivalent to about 1% mismatching of DNA (B. J.
McCarthy, personal communication).

The "denaturation map" (Fig. 3) is exceedingly regular
with alternating denatured and native regions.   One
repeat length (from the beginning of one denatured region to the next) contains 500,000 daltons of DNA.
This agrees well with the predicted mass of one repeat
length if each contains a single 5S gene (84,000 daltons of native DNA).
     One advantage in analyzing 5S DNA is that heterogen-
eity within the high and the low GC regions of the DNA

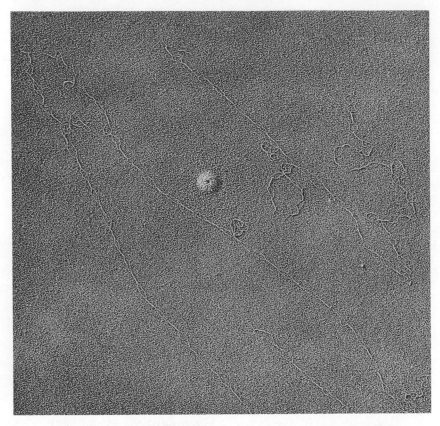

Figure 3. Electron micrograph of partially
denatured molecules of 5S DNA. Purified 5S
DNA was partly denatured in alkali by the
method of Inman and Schnös (1970) as described by Wensink and Brown (1971) and
Brown, Wensink and Jordan (1971). The high
AT regions denature first and are fixed
with formaldehyde. The contour length of
DNA was measured by spreading intact molecules of native T7 DNA on the same grids.
Their known length is 25 x 10^6 daltons
(Freifelder, 1970), and in these preparations 1 µ equals 1.7 x 10^6 daltons of DNA.
The line is 0.5 µ. The electron micrograph
was prepared by P. C. Wensink.

can be assessed separately. This heterogeneity has been analyzed by reassociating the denatured 5S DNA and carrying out a second melt. DNA with a unique sequence such as phage DNAs gives a second melt whose T_m is not more than 1° below that of the native DNA. Reassociated rDNA melts with this degree of fidelity (Dawid, Brown and Reeder, 1970; Brown, Wensink and Jordan, 1972). However, 5S DNA shows considerable heterogeneity. The region which denatures first (low GC) renatures poorly (Fig. 2). The reduction of the T_m ranges from 2° to 8° which corresponds to mismatching of about 1.4 to 5.5% of the base pairs throughout this part of the DNA (B. J. McCarthy, personal communication). The DNA region which denatures at higher temperatures is better matched. The best matched part of 5S DNA is the last region which is denatured, the DNA sequences which we presume code for 5S RNA. This agrees with what we know about 5S RNA. This RNA is highly conserved in evolution. Partial sequence studies of X. *laevis* 5S RNA estimate that only about 10 of its 120 nucleotide residues differ from human 5S RNA (Williamson, 1970). Furthermore, the simple oligonucleotide pattern produced by enzymatic digestion is compatible with a single nucleotide sequence for 5S RNA of X. *laevis*. Yet we know that it is coded for by about 24,000 germ line genes.

The electron microscope picture shows that the repeating units within 5S DNA are highly clustered. As many as 86 repeats have been visualized on a single DNA molecule. However, we do not know whether all 24,000 repeats are clustered on a single chromosome like the repeats which comprise rDNA. We already know that the sequence heterogeneity in the low GC region of 5S DNA exists in the 5S DNA isolated from a single animal and therefore was not introduced by pooling the DNA of many animals.

The Evolutionary Heresies—Some Facts and Hypotheses

The structure of 5S DNA requires some of the same heresies for its evolution and maintenance as those proposed for the antibody dilemma. How can multiple tandem genes be maintained in which one region within

each repeat is related but not identical (the low GC
spacer sequence of 5S DNA and the multiple V regions of
an antibody subgroup), this region alternating along
the DNA with another sequence which is identical or
nearly so in each repeat (the sequence of 5S DNA coding
for 5S RNA and the multiple C regions within a class)?

Thousands of antibody genes.   If there is one
gene in the germ line for each immunoglobulin polypeptide then they must number in the thousands.   There
must be at least a few thousand K genes in the human
since over 100 K polypeptides have been sequenced, and
no two sequences are the same (Quattrocchi, Cioli and
Baglioni, 1969).   Ten thousand K, λ, and H genes would
comprise together about 1% of the human genome.   Since
about 40% of human DNA is known to be repetitious
(Britten, 1969), there is ample DNA of the appropriate
complexity for even 10 times this number of genes.

The role of natural selection.   The homogeneity of
rDNA and presumably 5S genes can be explained at least
in part by natural selection.   Multiple rDNA genes are
required to produce enough rRNA for animal cells.
Xenopus cannot survive with fewer than about 230 ribosomal genes in each cell (Miller and Knowland, 1970);
the minimal requirement for 5S genes is unknown.   The
genes for 18S and 28S as well as 5S RNA have evolved
very slowly, no doubt reflecting limitations imposed by
strict requirements for the ribosomal structure.   Therefore both the quantity and nucleotide sequence of these
genes have selective importance for survival of the
animal.   Evolutionary pressures on the spacer sequences
of rDNA and 5S DNA are unknown.   In the case of rDNA it
appears as though one particular nucleotide sequence
may not be required but the length of the spacer may be
important (Brown, Wensink and Jordan, 1972), perhaps
playing a role in the structure and folding of the
chromosome.   Antibody genes resemble 5S and rDNA since
they consist of two regions (V and C) under different
evolutionary pressures.   Presumably an animal needs
multiple V region genes not for quantitative reasons
as is the case for 5S and rDNA but for qualitative
reasons, i.e., to generate antibody diversity.   This

obvious reason to maintain multiple different V genes
does not account for the selective pressures required
to maintain the homogeneity of multiple C genes of a
class. A variety of functions has been assigned to the
C region of immunoglobulins (Cohen and Porter, 1964),
but none of these satisfactorily accounts for the lack
of genetic drift which the evolutionary model predicts
for C gene regions of one class. Certainly one of the
least satisfactory features of the evolutionary model
is the absence of a good reason to maintain multiple C
sequences constant in a single species while permitting
their extensive evolution between species. If such a
reason does exist, then natural selection could solve
the antibody dilemma by providing selective advantage
to an animal which can maintain thousands of antibody
genes with a variable part for antibody specificity and
a constant part for some as yet undefined but important
function.

   Restrictions on recombination. The crossover rate
of linked C region classes is exceedingly low. No
crossovers between markers on the H chain of different
classes were found in more than 2,000 backcrosses in
mice (reviewed by Herzenberg, McDevitt, and Herzenberg,
1968). Rare crossovers have been described between
linked H chain markers in humans (Natvig, Kunkel and
Gedde Dale, 1968) and rabbits (Mage, Young-Cooper,and
Alexander, 1971). The low crossover rate has been interpreted as additional evidence that each C region is
encoded by a single gene and the C genes are tightly
linked. Alternatively, there may be restrictions on
crossing over in the multigene loci predicted by the
evolutionary model. Some restrictions on recombination
must be true for these linked genes; otherwise extensive divergence of linked DNA sequences coding for C
and C could not have occurred. Therefore, crossing
over of linked genes within the same class is rare as
is nonhomologous pairing and crossing over between
linked genes of different classes.

   The master-slave hypothesis. This hypothesis (Callan and Lloyd, 1960; Callan, 1967) proposes that a
single repeat (or gene) corrects adjacent gene copies

at meiosis (or even during mitosis) so that all repeats
have the identical nucleotide sequence as the master
gene. This hypothesis would have to be modified for
the antibody genes. Each master gene would faithfully
correct the C region but not the V region. Each immunoglobulin class or even subgroup could comprise a
self-correcting group of genes.

   Expansion-contraction model. This model has been
proposed to explain the maintenance of homogeneous rDNA
repeats (Brown, Wensink and Jordan, 1972) and also to
account for some features of the evolution of antibody
genes (Smith, Hood and Fitch, 1971). Once again we can
assume that each class and perhaps each subgroup self-
corrects independently from the others. An occasional
"contraction" occurs by deletion or unequal crossing
over which drastically reduces the number of genes in
a class from thousands to a few genes. One or more of
these genes is expanded linearly, perhaps by a rotating
circle type of replication, to produce once again thousands of tandem genes. At this point, however, the
genes are identical and of limited value for antibody
diversity. This corrected group of genes must now
evolve to diverge the different V regions.
   This kind of correction and divergence makes certain
predictions.

   (1) Some classes of antibodies in individual animals
       or species will have limited heterogeneity if
       they are derived from a family of genes which
       has been contracted or corrected recently. This
       could be the case with chains of the mouse
       (Appella, 1971). Another example might be the
       "Lepore" type immunoglobulin described by Kunkel,
       Natvig and Joslin (1969). An individual has
       been found who contains no ordinary γ_1 or γ_3 H
       chain proteins but only hybrid molecules. If
       this were due to a deletion or unequal crossover
       so that one part of a γ_1 gene was fused with
       part of a γ_3 gene and then this hybrid gene was
       reexpanded to form many copies, we should pre-
       dict that the V regions associated with this new
       C region are more homogeneous than those associ-

ated with normal human $C_{\gamma 1}$ or $C_{\gamma 3}$. As far as I
know this information is not now available.

(2) The population of natural antibodies will not
    reflect accurately the variety of antibody genes
    in the genome. This is due to the fact that a
    variety of mutations and deletions will have oc-
    curred which have rendered individual genes non-
    functional or unable to produce a specific anti-
    body. The cells committed to the expression of
    these defective immunoglobulins will not normal-
    ly be called upon to proliferate and contribute
    substantially to the antibody population. The
    array of myelomas which an animal can produce
    may more faithfully reflect the variety of im-
    munoglobulin genes in the animal's germ line.
    Many examples of defective antibody synthesis
    which are produced by myeloma cells have been
    recorded, including production of only one kind
    of chain (H or L), and production of shortened
    chains which appear to be coded for by genes
    with internal deletions (reviewed by Pink, Wang
    and Frudenberg, 1971).

*Limitations on mutations within the C region and
much of the V regions*. A simple heresy to invoke would
be that the C portion of each gene is somehow protected
against mutation once it has been expanded to multiple
copies. Alternatively, a mutation within the C region
renders that gene totally inactive so it can never
again be expressed until it is "corrected."

SUMMARY

The question has been raised whether new information
gained from the arrangement and evolution of DNA in
eukaryotes might account for the dilemma of antibody
diversity, genetics and structure. Most immunologists
propose one or more developmental genetic mechanisms to
explain these unusual facts. However, a germ line hypothesis can be proposed in which there is a single
gene (with a variable and a constant region) for each
antibody polypeptide. This evolutionary model requires

one or more evolutionary "heresies," that is, mechanisms for parallel evolution of multiple tandem genes for which we have no proven mechanisms. The ribosomal DNA and the 5S DNA purified from the genome of *Xenopus laevis* have certain characteristics which resemble the proposed complexity of antibody genes, and this fact in itself suggests that evolutionary mechanisms must exist which allow the creation and maintenance of such gene clusters. The driving force to maintain these multiple genes in the genome of eukaryotes is presumed to be natural selection. In the case of 5S DNA and rDNA multiple genes are maintained to provide a sufficient number of templates for RNA synthesis—a quantitative reason; if multiple antibody genes exist, it will be to generate diversity—a qualitative reason.

Control of gene expression in immunocytes according to developmental genetic methods proceeds by gene alteration as well as by gene activation. According to the evolutionary model, control of gene expression is by differential gene activation alone.

Analysis of the number of antibody genes by molecular hybridization using purified radioactive messenger RNA should decide whether vast numbers of these genes are present, the extent to which the gene family is related, and how they are linked. If the number of genes is very small in the germ cells, the evolutionary model will be ruled out.

ACKNOWLEDGMENTS

I am grateful to Dr. I. B. Dawid for many stimulating discussions which helped to focus these ideas. I thank Drs. R. Mage, L. Hood and G. Edelman for the opportunity to see their unpublished manuscripts and the chance to try these ideas out on them. It became clear that none of these ideas are new to immunologists except perhaps their emphasis. The manuscript was improved by the suggestions of Drs. I. Dawid, M. Edidin, L. Gage, R. Mage and R. Roeder, Mrs. S. Craig and Mr. R. Stern. The studies on the fine structure and homogeneity of *Xenopus* ribosomal and 5S DNAs were done in collaboration with Dr. Pieter C. Wensink. Mrs. E. Jordan provided expert technical assistance.

REFERENCES

Appella, E. (1971). *Proc. Nat. Acad. Sci. USA 68*, 590.

Brenner, S. and Milstein, C. (1966). *Nature 211*, 242.

Britten, R. J. (1969). *Carnegie Inst. Wash. Year Book 67*, 327.

Britten, R. J. and Kohne, D. E. (1968). *Science 161*, 592.

Brown, D. D. (1967). In *Current Topics in Developmental Biology*, A. Monroy and A. Moscona, eds. *Vol. 2*, (New York: Academic Press), p. 48.

Brown, D. D. and Weber, C. S. (1968). *J. Mol. Biol. 34*, 661.

Brown, D. D., Wensink, P. C., and Jordan, E. (1972). *J. Mol. Biol. 63*, 57.

Brown, D. D., Wensink, P. C., and Jordan, E. (1971). *Proc. Nat. Acad. Sci. USA 68*, 3175.

Callan, H. G. (1967). *J. Cell Sci. 2*, 1.

Callan, H. G. and Lloyd, L. (1960). *Phil. Trans. Roy. Soc. B 243*, 135.

Cebra, J. J., Colberg, J. E., and Dray, S. (1966). *J. Exp. Med. 123*, 547.

Cohen, S. and Porter, R. R. (1964). *Adv. Immunol. 4*, 287.

Dawid, I. B., Brown, D. D., and Reeder, R. H. (1970). *J. Mol. Biol. 51*, 341.

Dreyer, W. J. and Bennett, C. J. (1965). *Proc. Nat. Acad. Sci. USA 54*, 864.

Edelman, G. M. and Gall, W. E. (1969). *Ann. Rev. Biochem. 38*, 415.

Edelman, G. M. and Gally, J. A. (1971). In *The Neurosciences; 2nd Study Program*, S. O. Schmitt, ed. (New York: Rockefeller University Press).

Ein, D. (1968). *Proc. Nat. Acad. Sci. USA 60*, 982.

Freifelder, D. (1970). *J. Mol. Biol. 54*, 567.

Gally, J. A. and Edelman, G. M. (1970). *Nature 227*, 341.

Herzenberg, L. A., McDevitt, H. O., and Herzenberg, L. A. (1968). *Ann. Rev. Gen. 2*, 209.

Hill, R. L., Delaney, R., Fellows, R. E., Jr., and Lebovitz, H. E. (1966). *Proc. Nat. Acad. Sci. USA 56*, 1762.

Hood, L. E. (1972). *Fed. Proc. 31*, 177.

Hood, L. E. and Ein, D. (1968). *Nature 220*, 764.

Hood, L. E. and Talmage, D. (1970). *Science 168*, 325.

Inman, R. B. and Schnös, M. (1970). *J. Mol. Biol. 49*, 93.

Kabat, E. A. (1967). *Proc. Nat. Acad. Sci. USA 57*, 1345.

Kunkel, H. G., Natvig, J. B., and Joslin, F. G. (1969). *Proc. Nat. Acad. Sci. USA 62*, 144.

Loening, U. S., Jones, K., and Birnstiel, M. L. (1969). *J. Mol. Biol. 45*, 353.

Mage, R. G. (1971). In *Prog. in Immunol.*, B. Amos, ed. (New York: Academic Press), p. 47.

Mage, R. G., Young-Cooper, G. O., and Alexander, C. (1971). *Nature New Biology 230*, 63.

Mairy, M. and Denis, H. (1971). *Devel. Biol. 24*, 143.

Miller, L. and Knowland, J. (1970). *J. Mol. Biol. 53*, 329.

Milstein, C. (1966). *Nature 209*, 370.

Milstein, C., Milstein, C. P., and Jarvis, J. M. (1969). *J. Mol. Biol. 46*, 599.

Milstein, C. and Pink, J. R. L. (1970). *Prog. Biophys. Mol. Biol. 21*, 208.

Natvig, J. B., Kunkel, H. G., and Gedde Dale, T., Jr. (1968). *Cold Spring Harbor Symp. Quant. Biol. 32*, 313.

Pink, R., Wang, A. C., and Fudenberg, H. H. (1971). *Ann. Rev. Med. 22*, 145.

Potter, M. (1967). *Meth. Cancer Res. 2*, 105.

Prahl, J. W., Mandy, W. J., David, G. S., Steward, M. W., and Todd, C. W. (1970). In *Protides of the Biological Fluids*, H. Peeters, ed., *Vol. 17* (Oxford: Pergamon Press), pp. 125-130.

Quattrocchi, R., Cioli, D., and Baglioni, C. (1969). *J. Exp. Med. 130*, 401.

Shimizu, A., Paul, C., Kohler, H., Shinoda, T., and Putnam, F. W. (1971). *Science 173*, 629.

Singerm S. J. and Doolittle, R. F. (1966). *Science 153*, 13.

Smith, G. P., Hood, L., and Fitch, W. M. (1971). *Ann. Rev. Biochem. 40*, 968.

Smithies, O. (1970). *Science 169*, 882.

Tischendorf, F. W. and Tischendorf, M. M. (1970). *Europ. J. Biochem. 13*, 398.

Tosi, S. L., Mage, R. G., and Dubiski, S. (1970). *J. Immunol. 104*, 641.

Wallace, H. and Birnstiel, M. L. (1966). *Biochim. Biophys. Acta 114*, 296.

Wang, A. C., Wilson, S. K., Hopper, J. E., Fudenberg, H. H., and Nisonoff, A. (1970). *Proc. Nat. Acad. Sci. USA 66*, 337.

Wensink, P. C. and Brown, D. D. (1971). *J. Mol. Biol. 60*, 235.

Williamson, R. (1970). *Eighth International Congress of Biochemistry*.

Specific Hybridization of the Silk Fibroin Genes in Bombyx mori

L. Patrick Gage, Yoshiaki Suzuki, and Donald D. Brown

Department of Embryology
Carnegie Institution of Washington
Baltimore

INTRODUCTION

The posterior silk gland of *Bombyx mori*, the commercial silkworm, is highly specialized for the production of a single protein, silk fibroin. During larval development the silkworm grows about ten-thousandfold in weight. Cells of the silk gland grow in proportion with the larva and undergo DNA synthesis without cell division, a process termed 'polyploidization' (Gillot and Daille, 1968). Chromosomes are never seen in these enormous polyploid cells which eventually contain 0.1 to 0.2 µg of DNA. Each of the giant cells produces about 300 µg of silk fibroin during the last two to three days of larval life (Tashiro *et al.*, 1968). Most of this massive protein synthesis occurs after DNA synthesis has stopped, a property common to cells which synthesize highly specialized proteins. One reason for this characteristic order of events could be that the

127

terminal DNA synthesis which occurs before cell specific protein synthesis not a uniform replication of the
genome but a specific amplification of those genes is
which the cell expresses in its differentiated state.
Specific amplification of the genes for rRNA occurs in
developing oocytes (Brown and Dawid, 1968; Gall, 1968)
but has not been described for other cell types.

Experiments are described here that indicate that
polyploidization in the silk gland occurs by uniform
replication of the entire genome. In addition specific
hybridization of *B. mori* DNA with purified mRNA for
fibroin (Suzuki and Brown, 1972) has determined the
relative abundance of fibroin gene sequences in highly
polyploid silk gland cells and carcass cells which contain several orders of magnitude less DNA.

MATERIALS AND METHODS

Larvae of *B. mori* were raised as described by Suzuki
and Brown (1972). DNA was prepared from silk glands by
homogenization and incubation for 1 hr at 37° in 1 X
SSC, 0.05 M Tris, pH 7.5, and 0.5% SDS (sodium dodecyl
sulfate) to which pronase, predigested for 30 min at
25°, had been added to 1 mg/ml. About 1 to 2 ml/gland
was used. The lysate was extracted with phenol twice,
NaCl was added to 0.3 M. and the DNA wound-out after
the addition of one to two volumes of ethanol. The DNA
was dissolved in SSC-Tris and incubated at 37° for 3 hr
with 50 µg/ml pancreatic RNase and 25 units/ml T1 RNase
(each pretreated at 80° for 10 min), followed by digestion for 1 hr with 500 µg/ml of predigested pronase
and 0.5% SDS. The DNA was purified by two additional
phenol extractions and two to three ethanol precipita-
tions. DNA was prepared in the same manner from car-
casses which had first been shredded in a blendor in
SSC-Tris.

All DNA was sheared in an Omni-Mixer at full speed
for 3 min in SSC, and its weight average molecular
weight was 2.5×10^6 daltons as measured by analytical
band sedimentation (Studier, 1965).

DNA was centrifuged to equilibrium in CsCl or Cs_2SO_4
at 25° in either the #65 fixed angle rotor (5 ml/tube)

or the #50.1 rotor (20 ml/tube) at 33,000 rpm for 2 to
3 days. Silver was complexed to DNA in the ratio 0.3
M Ag/M of DNA-PO₄ (0.138 μg Ag₂SO₄/μg DNA) (Jensen and
Davidson, 1966; Corneo *et al.*, 1968; Brown *et al.*, 1971).

Fibroin mRNA was prepared by two sucrose gradient
fractionations, as described by Suzuki and Brown (1972).
This mRNA was purified further by centrifugation in the
SW 41 rotor at 38,000 rpm for 24 hr on a 13 ml 5-20%
(w/w) sucrose gradient in the presence of 70% formamide,
3 mM EDTA and 3 mM Tris, pH 7.5 at 25°. Only RNA sedimenting faster than intact *Xenopus laevis* 28S rRNA was
saved (about a 50% yield). *B. mori* 18S and 28S rRNA and
most of their 40S precursor cosediment with intact *X.
laevis* 18S rRNA when they are denatured with formamide.

Hybridizations were performed at 50° in 50% formamide and 4 X SSC (McConaughy, Laird and McCarthy, 1969)
for two days, and filters were assayed for pancreatic
RNase resistant hybrid as described by Brown and
Weber (1968).

RESULTS

*Polyploidization is a Uniform
Replication of the Genome*

Polyploidization was analyzed by comparing DNAs
isolated from the posterior and middle silk gland, and
cells of the carcass which contain orders of magnitude
less DNA. Comparative thermal denaturation and analytical CsCl equilibrium centrifugation could not distinguish these DNAs (data not shown). Ribosomal RNA (rRNA)
hybridizes with about 0.13% of each kind of DNA demonstrating that rDNA is replicated in proportion with the
bulk DNA during polyploidization (Gage, 1971). The
DNAs also hybridize equally well with each of the complementary RNAs made from them *in vitro* with *Escherichia coli* RNA polymerase (Gage, 1971). Therefore the
polyploid silk gland cells contain on the whole the
same relative abundance of these sequences as do the
carcass cells. However, substantial change in the
abundance of a single sequence, like specific amplification of the fibroin gene(s), might have gone unde-

tected.   Accurate fibroin gene concentrations could
only be determined by specific hybridization with pure
fibroin mRNA.

Detection and Quantitation of
Fibroin Genes

Sixty percent of each fibroin molecule consists
largely of repeats of the hexapeptide sequence shown
in Figure 1 (Lucas and Rudall, 1968).   Fibroin is 45%

$$gly - ala - gly - ala - gly - ser$$

$$GG_A^U \cdot GCU \cdot GG_A^U \cdot GCU \cdot GG_A^U \cdot UCA$$

Figure 1.   The predominant polypeptide
(Lucas and Rudall, 1968) and oligonucleotide sequences of silk fibroin and its mRNA.
The major codons of glycine, alanine and
serine found by partial sequence determination are included (Suzuki and Brown,
1972).

glycine, 29% alanine and 12% serine; glycylglycine di-
peptides are completely absent.   The proposed size of
one gene for fibroin should be about 4.4×10^6 daltons
(native DNA) if we accept the molecular weight of the
protein as 1.7×10^5 (Tashiro and Otsuki, 1970a, b).
Assignment of codons for the principal amino acids
predicts that the fibroin mRNA should be 57 to 65% GC
with 40% G content (Fig. 1).   An RNA fraction has been
isolated from the mature posterior silk gland which is
40% G, 20% C and which is not found in the adjacent
middle silk gland or carcass (Suzuki and Brown, 1972).
This RNA was identified as the mRNA for fibroin by
partial sequence determination.

Detection of a specific gene by hybridization requires that the RNA be pure and highly labeled.   The
fibroin mRNA was labeled *in vivo* at high specific activity, and partial sequencing had indicated that it
was at least 80% pure (Suzuki and Brown, 1972).   The

fibroin genes should be separable from the bulk of *B.
mori* DNA by prefractionation of the DNA by equilibrium
density gradient centrifugation. The bulk DNA of *B.
mori* is 39% GC (Neulat, 1967) while the rRNA genes are
about 48% GC (Suzuki and Brown, 1972). Carcass DNA was
sheared to about half fibroin gene size, fractionated
in CsCl and hybridized with (^{32}P) fibroin mRNA (Fig.
2a). The (^{32}P) RNA hybridized coincidentally with

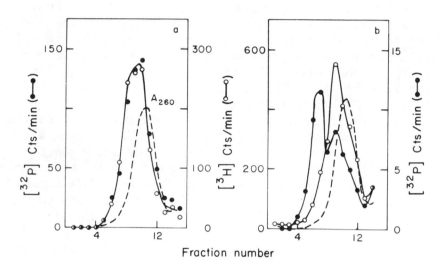

Figure 2. Hybridization of sheared CsCl
fractionated DNA with fibroin mRNA preparations before (a) and after (b) formamide-
sucrose gradient centrifugation. 100 µg
of *B. mori* carcass DNA, fractionated on
CsCl, was hybridized with either a) 1 µg/ml
(^{32}P) fibroin mRNA (2×10^4 cpm/µg) mixed
with 0.6 µg/ml of (^{3}H) *Xenopus laevis* rRNA
(3.5×10^5 cpm/µg); or b) 1 µg/ml (^{32}P) fibroin mRNA (2×10^4 cpm/µg), after fractionation in formamide-sucrose mixed with 2.4
µg/ml (^{3}H) *X. laevis* rRNA. Preparation and
centrifugation of fibroin mRNA and DNA are
described in the Materials and Methods
section.

(^{3}H) *X. laevis* rRNA across the gradient and the base
composition of the hybridized (^{32}P) RNA was similar to
that of *B. mori* rRNA. Ribosomal RNA is only a low-
level contaminant of the mRNA preparation so its dom-
ination of the hybridization reaction demonstrated that
the genes for rRNA must be much more abundant than
those for fibroin in the DNA.

Sedimentation of the (^{32}P) mRNA through a 70% formamide-sucrose gradient released most of the rRNA contaminant from aggregates with the larger fibroin mRNA.
Hybridization of the same mRNA preparation which was
used in the experiment of Figure 2a, but after purification by formamide-sucrose centrifugation, is shown
in Figure 2b. The formamide-sucrose centrifugation
step lowered the rRNA contamination to the point where
fibroin mRNA hybridization was detectable.

The fibroin genes can be separated further from the
rRNA genes and the bulk DNA when the DNA is complexed
with silver and centrifuged to equilibrium in Cs_2SO_4
(Fig. 3). Besides the greater separation of the fibroin genes from rDNA and main band DNA, more DNA can be
centrifuged in the steeper Cs_2SO_4 gradient. By using
the enriched fibroin gene fractions from 9 mg of carcass DNA, enough (^{32}P) mRNA was hybridized to prove its
identity as fibroin mRNA by base composition and partial sequencing (Suzuki, Gage and Brown, in preparation).

If fibroin genes are amplified during silk gland
polyploidization, its DNA will hybridize with more fibroin mRNA than the DNA from a control tissue (carcass
or middle gland). Fibroin gene levels were determined
in DNA from carcass cells, from highly polyploid middle
silk gland cells which make silk sericin rather than
fibroin, and from cells of the posterior silk gland
where the fibroin gene is expressed so actively. About
1 mg of each DNA was fractionated by Ag-Cs_2SO_4, bound
to filters and hybridized together with saturating
quantities of formamide purified (^{32}P) fibroin mRNA
mixed with (^{3}H) rRNA (Fig. 4). The fibroin mRNA hybridized with an exceedingly small but equal fraction
(0.002%) of each DNA.

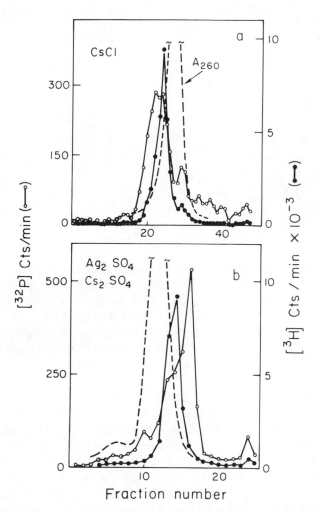

Figure 3. Hybridization of *B. mori* carcass DNA after it was fractionated by equilibrium density centrifugation. Hybridization with 0.7 μg/ml of (^{32}P) fibroin mRNA (purified through a formamide-sucrose gradient) and 0.6 μg/ml (^{3}H) *X. laevis* rRNA with: a) 2.3 mg *B. mori* carcass DNA centrifuged in neutral CsCl; b) 2.0 mg *B. mori* carcass DNA complexed with silver and centrifuged in Cs_2SO_4.

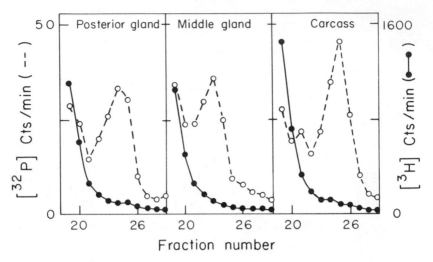

Figure 4. Specific hybridization of fibroin
genes in *B. mori* silk gland and carcass DNA.
About 1 mg of each DNA was sheared, complexed
with silver and centrifuged in Cs_2SO_4. Filter bound DNA from the fibroin gene regions
(11 of 35 fractions) were hybridized together
for 48 hr with 0.9 µg/ml formamide purified
(^{32}P) fibroin mRNA (1.4×10^4 cpm/µg), 6.6
µg/ml (^3H) *B. mori* rRNA (1.55×10^5 cpm/µg),
and 130 µg/ml unlabeled total carcass RNA.
After the DNA-filters were counted, the DNA
on each filter was measured by hydrolyzing
it in 1 N HCl (Brown and Weber, 1968). In
this experiment the fibroin genes from 550
µg of posterior gland DNA, 620 µg of middle
gland DNA, and 690 µg of carcass DNA was hybridized. The amount of contaminating (^{32}P)
rRNA which bound to DNA was corrected from
the amount of (^3H) rRNA which hybridized to
each filter. Backgrounds of 4 cpm (^{32}P) and
32 cpm (^3H) were subtracted from each fraction before correction. In this way, each
DNA was shown to have hybridized only 2.0 ±
0.1×10^{-5}µg RNA per µg DNA.

DISCUSSION

These conclusions only refer to the comparison between two highly specialized polyploid cell types—the middle and posterior silk gland cells and the carcass cells which are orders of magnitude lower in DNA content than the silk gland cells but of unknown ploidy. Since germ line DNA of *B. mori* has not been analyzed by these methods we cannot rule out some selective change in the genome of somatic cells which would be common to the DNA of all three of these somatic cell types.

Polyploidization of the silk gland results in an increase in cellular DNA content of 10^4 to 10^6 times that of a diploid cell (Gillot and Daillie, 1968). The DNAs from posterior and middle silk glands and the carcass are identical in their physical properties, in their content of rRNA genes, and in at least some of their repetitive sequences. Neulat (1967) showed that DNA isolated from mature silk glands and from embryos, which are composed primarily of diploid cells, are identical in base composition and the distribution of their pyrimidine isostiches. We conclude that most if not all of the cellular DNA increase during polyploidization of the silk gland results from uniform replication of the same DNA found in the carcass cells. In particular, the fibroin genes are not amplified during silk gland polyploidization since they are equally abundant in silk gland and carcass DNA.

Fibroin mRNA (presumed molecular weight of about 2.3×10^6 daltons) hybridizes with about 0.002% of total *B. mori* DNA. The genome size has been estimated to be 0.15 to 0.3 pg of DNA (Gage, unpublished data). From these values we calculate that there are less than three genes per haploid complement of *B. mori* DNA, and there may be only a single gene. The exact number of fibroin genes cannot be determined until more accurate numbers have been obtained for the molecular weight of the mRNA and the genome size of *B. mori*.

We can estimate the rates of fibroin mRNA transcription and fibroin translation. Each posterior silk gland cell synthesizes about 10^{15} molecules of fibroin in about two to three days (Tashiro *et al.*, 1968).

This is accomplished with about 10^{10} molecules of fibroin mRNA and 10^{12} ribosomes in each cell (Suzuki and
Brown, 1972), so that each mRNA molecule supports the
synthesis of about 10^5 fibroin molecules. Therefore
the average rate of fibroin synthesis during this period is about 10 amino acids polymerized/ribosome/second.

Fibroin mRNA is about the size of the two rRNAs together, and ribosomal RNA is approximately 80-fold more
abundant than fibroin mRNA in the posterior silk gland
(Suzuki and Brown, 1972), a ratio which approximates
the relative abundance of their genes. If the rate of
transcription of fibroin genes and ribosomal genes is
the same and their RNA products are equally stable, the
fibroin mRNA would be expected to accumulate in the
cell to the level which has been found—about 1 to 2%
of the rRNA content (Suzuki and Brown, 1972).

We conclude from these experiments that synthesis of
massive quantities of fibroin by these very specialized
cells proceeds by differential gene activation and the
efficient stabilization and utilization of the mRNA for
translation. Amplification of the fibroin genes has
not occurred in the cells of the posterior gland. Gene
amplification may only be required for genes which code
for RNAs (such as rRNA) which are not translated into
proteins and are needed in large amounts by cells.

SUMMARY

The extensive polyploidization which occurs in cells
of the posterior silk gland, where silk fibroin is synthesized, and the middle silk gland of *Bombyx mori* results in a uniform replication of most if not all of
the DNA sequences found in carcass DNA. Purified messenger RNA for silk fibroin hybridizes to the same
fraction (0.002%) of carcass and middle gland DNA as it
does to posterior gland DNA. We estimate that this
amount of DNA includes from one to three fibroin genes
for each haploid equivalent of DNA. Therefore the
specific expression of fibroin genes in the posterior
silk gland is not a result of gene amplification but
presumably involves differential gene activation; the
massive yield of fibroin can be attributed to the great
translational potential of its stable messenger RNA.

REFERENCES

Brown, D. D., and Dawid, I. B. (1968). *Science 160*, 272.
Brown, D. D., and Weber, C. S. (1968). *J. Mol. Biol. 34*, 661.
Brown, D. D., Wensink, P. C., and Jordan, E. (1971). *Proc. Nat. Acad. Sci. USA 68*, 3175.
Corneo, G., Ginelli, E., Soave, C., and Bernardi, G. (1968). *Biochemistry 7*, 4373.
Gage, L. P. (1971). *Carnegie Inst. Wash. Year Book 70*, p. 39.
Gall, J. G. (1968). *Proc. Nat. Acad. Sci. USA 60*, 553.
Gillot, S., and Daillie, M. J. (1968). *C. R. Acad. Sci. Paris 266*, 2295.
Jensen, R. H., and Davidson, N. (1966). *Biopolymers 4*, 17.
Lucas, F., and Rudall, K. M. (1968). In *Comprehensive Biochemistry*, *Vol. 26*, Part B (New York: American Elsevier), p. 475.
McConaughy, B. L., Laird, C. D., and McCarthy, B. J. (1969). *Biochemistry 8*, 3289.
Neulat, M. (1967). *Biochim. Biophys. Acta 149*, 422.
Studier, F. W. (1965). *J. Mol. Biol. 11*, 373.
Suzuki, Y., and Brown, D. D. (1972). *J. Mol. Biol. 63*, 409.
Tashiro, Y., Morimoto, T., Matsuura, S., and Nagata, S. (1968). *J. Cell Biol. 38*, 574.
Tashiro, Y., and Otsuki, E. (1970a). *J. Cell Biol. 46*, 1.
Tashiro, Y., and Otsuki, E. (1970b). *Biochim. Biophys. Acta 214*, 265r.

Information Flow from the Genome to the Cytoplasm and Back

Introduction

Experiments with Procaryotes have shown that the transcription of DNA into RNA is a primary site of the regulation of gene expression. This implies that at a given time a particular RNA polymerase molecule can recognize a specific initiation site on the template, bind to it, transcribe, reach a termination site, and stop. What this means in molecular terms with respect to the composition of the polymerase and the physical or stereochemical state of the template are central questions currently under attack. Marked successes have been attained by purifying and characterizing polymerases, first from Procaryotes and now from Eucaryotes as well, and by the development of elegant methods for visualizing both the template and the template-polymerase complex in the act of transcription.

That transcripts have to be processed and parts discarded prior to use became obvious some years ago in studying the ontogeny of ribosomal RNA in Eucaryotes.

It was not obvious in the case of mRNA and only during
1971 have the first hints of such a process begun to
be appreciated. These preliminary results also have
profound implications with respect to how mRNA is
transported from nucleus to cytoplasm and perhaps even
how translation may be regulated.

Many studies including some to be described in the
next section have demonstrated clearly that information
about morphogenetic or metabolic events occurring in
the cytoplasm is fed back and helps to regulate the
differential activation of the genome. The nature of
this information and the molecular means for its conveyance are, except for several notable examples involving physiological modulations in bacteria, still
mysterious. In the case of Eucaryotes several potentially fruitful systems have recently been developed
wherein combinations of nuclei and cytoplasms from
cells in different states of metabolic and morphogenetic activity can be experimentally created. These include nuclear transplantations and induced heterocaryons of Metazoan cells and of Metaphytic cells. Two
elegant examples are included in this volume but were
placed in Section IV on grounds of comparative relevance.

Since 1970 it has become clear that at least in the
case of some RNA viruses, RNA initially present in the
cytoplasm can, by the action of a reverse transcrip-
tase, be debircsnart into DNA and the product can be
integrated into the genome. This reverse flow of information has exciting implications not only for the
understanding of virally induced developmental abnor-
malities but might conceivably play a role in normal
developmental processes, for instance in the creation
of a competent antibody producing cell via permanent
modification of its DNA.

RNA Polymerases and Transcriptive Specificity in Eukaryotes

William J. Rutter, C. James Ingles,
Robert F. Weaver, Stanley P. Blatti,
and Paul W. Morris

Department of Biochemistry and Biophysics
University of California
San Francisco

One of the major questions, if not the major question, posed by developmental biology concerns the mechanism by which genes are selectively expressed in differentiated cells. Until recently, the regulation of gene expression has been equated with transcriptional specificity. Subsequent papers in this volume will show post-transcriptional and reverse-transcriptional mechanisms may also be involved. This paper is concerned with transcription itself. *A priori*, there are several plausible mechanisms for achieving specificity. These include (Fig. 1) the frequently hypothesized template restriction model by which the RNA polymerase simply is prevented from acting on certain regions because of physical sequestration of the DNA by histones or other chromosomal proteins. This mechanism is analogous to negative regulation by repressors. Alternative positive regulatory mechanisms may be visualized by which certain genes are selected for reading, either

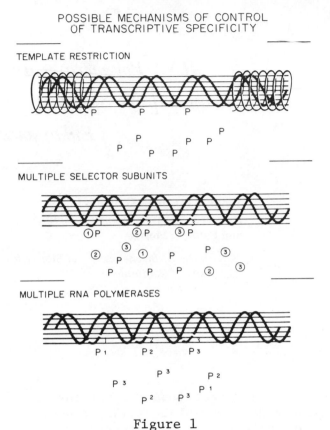

Figure 1

by the stringent specificity of the RNA polymerases or by the action of "specifier subunits" which confer specificity on an indiscriminate polymerase. The former mechanism would require that each transcriptive system include a large number of polymerases with discrete specificities while the latter (in the limiting case) would require only a single core polymerase molecule and multiple specifier subunits.

The presently available experimental information suggests that all three mechanisms may be operative in eukaryotic organisms. Whereas the reading of naked DNA templates by RNA polymerase appears to be largely indiscriminate (Roeder and Rutter, 1969), isolated chromatin retains some degree of specificity, even when

prokaryotic polymerase is employed as a transcribing
agent (Bonner *et al.*, 1968; Paul and Gilmour, 1968;
Smith, Church and McCarthy, 1969). This specificity
is lost by dissociation of the chromosomal proteins
from the DNA, and is regained on reconstitution of
the chromatin (Paul and Gilmour, 1968; Bekhor, 1969;
Huang and Huang, 1969). These experimental results
indicate at least some transcriptive specificity re-
sides in the chromosomal matrix. The experiments,
however, do not discriminate between positive and negative mechanisms of control.

Multiple RNA Polymerases

My colleagues and I have been concerned with possible transcriptive regulation by multiple RNA polymerases. It had been known since the experiments of
Widnell and Tata (1966) that changes in the divalent
metal ion and ionic strength of the incubation medium
can produce a qualitative difference in the RNA synthesized by various nuclear preparations. For example,
in Mg^{++} and low salt, a high GC RNA is produced, while
in Mn^{++} and higher salt, the base composition of the
synthesized RNA resembles that of the DNA. These observations could imply the presence of polymerases with
different ionic requirements or simply could be the
result of effects on the template or on other regulatory components of the system.
After solubilizing and stabilizing the polymerase
activity in crude extracts, Robert Roeder and I (1969)
demonstrated that three major, but not necessarily
homogeneous, peaks of RNA polymerase activity could be
resolved by DEAE-Sephadex chromatography (see Table 1).
These enzyme activities were completely dependent upon
added DNA as a template, and rechromatographed as
distinct entities. They thus appeared to be different
molecular species rather than simply nonspecific aggregates. The peaks of enzymatic activity furthermore exhibited different catalytic characteristics, such as
salt activity profiles and divalent metal ion optima,
and activity on native (double-stranded) and denatured
(single-stranded) DNA. It was later shown that RNA
Polymerase II activity is specifically inhibited by

TABLE 1

Properties of nuclear RNA polymerase*

	I	II	III
DEAE elution [~M(NH₄)₂SO₄]	.15	.25	.35
$\frac{Mn^{++}}{Mg^{++}}$ activity ratio	1	5-50	2-3
Salt optimum [M(NH₄)₂SO₄]	.05	.10-.15	0-0.2
$\frac{Denatured\ DNA}{Native\ DNA}$ activity ratio	1-1.2	~2	~1
α-Amanitin inhibition (%)	0	100	0
Rifamycin inhibition (%)	0	0	0

*Calf thymus, Rat liver, Sea urchin, Yeast.

α-amanitin, the toxin produced by the mushroom *Amanita phalloides* (Lindell *et al.*, 1970; Kedlinger *et al.*, 1970; Jacob, Sajdel and Munro, 1970). These collective observations argue that the molecular structures of Polymerases I, II and III are each somehow different, but the question whether they are unique species is still open. It was of particular interest that the properties of the various polymerases from sea urchins resembled their corresponding forms from rat liver. More recent studies with yeast confirm the general similarity of the various cognate forms of RNA polymerase. This suggests that the molecular characteristics of each polymerase species are highly conserved and, hence, presumably tailored for some specific biological function.

Polymerases I and II are the dominant activities in all eukaryotic systems examined (Roeder and Rutter, 1969). Polymerase II is less readily detected in most mammalian systems studied. In rat liver, for example, it is usually no more than a few percent of the total activity and is very labile in crude extracts (Blatti *et al.*, 1970). In contrast, in nuclei from sea urchin embryos, Polymerase III represents about 20% of the total enzymatic activity (Roeder and Rutter, 1970).

Polymerase III is also present in significant levels
in yeast nuclei (R. G. Benson and W. J. Rutter, unpublished data).

The limited available evidence suggests separate
transcriptive roles for the polymerases. However, the
evidence is indirect and incomplete. As isolated, the
enzymes exhibit little transcriptive specificity on
DNA templates (Roeder and Rutter, 1969; Bonner *et al.*,
1968; Bekhor, 1969). There are significant, but not
dramatic, differences in their relative transcriptive
rates on various homopolymers (Blatti *et al.*, 1970).
Inherent transcriptive specificity, however, may not
be apparent on these templates, since they are highly
nicked and fragmented, and thus, may contain a large
number of artificial initiation sites.

The evidence for specificity is derived from studies on isolated nuclei, employing variations in the
incubation conditions which selectively favor the activity of one or the other of the polymerases in ionic
strength. For example, under conditions favoring
Polymerase I activity (low ionic strength, magnesium
ions, and α-amanitin to block Polymerase II activity),
the RNA produced has the characteristics of ribosomal
RNA (Blatti *et al.*, 1970; Ingles and Rutter, in press).
The base composition is very similar to that of ribosomal RNA precursor (see Table 2). Furthermore, the
hybridization of this RNA with DNA is competitively
inhibited by unlabeled ribosomal RNA (Fig. 2). Hybrid-
ization-saturation experiments also indicate that few
RNA species (<10% of total) are produced in the presence of α-amanitin (where Polymerase I and Polymerase
III are active) (Fig. 3). Thus the transcriptive
range of Polymerase I is limited. The synthesis of
the great majority of RNA species (>90%) is specifically inhibited by α-amanitin and, therefore, is a
result of Polymerase II activity. In accord with this
conclusion, Polymerase I is localized largely, if not
entirely, in nucleoli, whereas Polymerase II is found
in the nucleoplasm (Roeder and Rutter, 1970). Polymerase III activity is also detected in the nucleoplasm,
but the experiments do not preclude leaking of this
enzyme from the nucleolus during the experiment. The
collective data (Table 3) suggest that Polymerase I is

TABLE 2

Nucleotide base composition of RNAs synthesized in
isolated rat liver nuclei in the presence
and absence of α-amanitin

Base	Base composition (moles percent)		
	Rat liver* 45S RNA	Polymerase I(+III) (α-amanitin insensitive)	Polymerase I (α-amanitin sensitive)
U	20.8	18.7	28.5
A	14.5	15.6	22.5
G	35.3	34.0	23.1
C	29.4	31.6	25.2

In a series of experiments, isolated nuclei were incubated with one ^{3}H labeled, and the other three, unlabeled nucleotide triphosphates, all under assay conditions (Roeder and Rutter, 1969), but in the presence
of α-amanitin (largely Polymerase I) or in its absence
(largely Polymerase II).

*The nucleotide base composition of rat liver 45S RNA
is taken from Muramatsu, Hodnett and Busch (1966).

TABLE 3

Transcriptive specificity of nuclear RNA polymerases

	I	II	III
Nuclear localization	nucleolus	nucleoplasm	?
Transcriptive range	<10%	>90%	<10%
Major transcriptive product	ribosomal RNA	most species (hRNA, mRNA)	?

involved in a transcription of ribosomal RNA and perhaps a few other RNA species, but they do not preclude
the role of Polymerase III in these functions. The

enzyme responsible for transcribing 5S and transfer
RNA has yet to be determined. While these experimental results strongly suggest separate transcriptive
roles for Polymerases I (III) and II, they do not answer whether specificity lies with the enzyme itself,
with the template, or with other factors present in
the nuclei.

Recent experiments have demonstrated heterogeneity
in both Polymerases I and II. Two forms of Polymerase
I have been resolved by ion exchange chromatography
(Chesterton and Butterworth, 1971a). Two forms of
Polymerase II have also been isolated (Chesterton and
Butterworth, 1971b; Weaver, Blatti and Rutter, 1971;
Kedlinger, Nuret and Chambon, 1971). However, there
is yet no evidence for distinct specificity: these
species may be distinct molecules with separate functions, they may be different regulatory states of the
same molecule, or they may simply represent the first
stages of degradation (turnover). Whatever their
origin and role, the apparent degree of heterogeneity
of polymerases is not sufficient to account for a
broad range of transcriptive specificity. The regulation of transcription, therefore, must involve other
components as well.

Purification and Molecular Structure

Polymerases I and II have now been purified from
several sources: calf thymus (Chesterton and Butterworth, 1971b; Weaver, Blatti and Rutter, 1971; Kedlinger, Nuret and Chambon, 1971), rat liver (Weaver,
Blatti and Rutter, 1971; Mandel and Chambon, 1971),
sea urchin embryos (Roeder and Rutter, 1969; Roeder
and Rutter, 1970a), and mammalian cells in tissue
culture (Keller and Goor, 1970; Blatti, S., Weinberg,
F., and W. J. Rutter, unpublished observations).
Robert Weaver, Stanley Blatti and I (1971) have obtained Polymerase II in essentially homogeneous form
from both calf thymus and rat liver. The general out-
line of the purification procedure is presented in
Table IV. Calf thymus and rat liver enzymes were
purified 10,000- and 3,000-fold from nuclei respectively. Polymerase I is present in somewhat lower

quantities and is considerably more labile under the
conditions employed. Furthermore, it fractionates
with a larger portion of nuclear proteins. Thus, the

Figure 2. (Opposite page.) Competition by nuclear or
ribosomal RNA with the hybridization of *in vitro* labeled nuclear RNAs. (a) 39 µg of ^{3}H labeled nuclear
RNA, produced largely by Polymerase II (labeled in the
absence of α-amanitin) was hybridized to 8 µg DNA in
the presence of increasing amounts of unlabeled com-
peting rat liver nuclear (O) or 18 and 28S ribosomal
RNA (●). Similarly, (b) 33 µg of labeled nuclear RNA,
produced largely by Polymerase I (labeled in the presence of α-amanitin) was hybridized to DNA in the presence of increasing amounts of unlabeled nuclear or ribosomal RNA. Nuclei were isolated from rat liver by
the method of Blobel and Potter (1966) except that 1 mM
spermine was incorporated in the isolation media. The
nuclei were suspended in buffer containing 0.05 M Tris-
HCL, pH 7.9; 25% (v/v) glycerol; 5 mM $MgCl_2$; 0.1 mM
EDTA; 1.0 mM dithiothreitol; and 1.0 mM spermine. The
isolated nuclei were incubated at 30° for 20 min in a
medium containing ^{3}H-UTP (0.003 mM, 2.4 C/mM) together
with the unlabeled nucleoside triphosphates of adenosine, guanosine and cytosine at 0.6 mM concentration
in a medium containing 0.05 M Tris-HCl, pH 7.9; 10%
(v/v) glycerol; 0.04 mM EDTA; 2.0 mM $MgCl_2$; 1.6 $MnCl_2$;
6 mM NaF; 0.4 mM dithiothreitol; 1.6 mM 2-mercaptoeth-
anol; 0.04 M $(NH_4)_2SO_4$; and, when present, 6.8 x 10^{-6} M
α-amanitin. Nuclear RNA was prepared by ethanol extraction of the nuclear incubation mixture, ethanol
precipitation, and treatment with DNase, α-amylase and
pronase, re-extraction with phenol, and chromatography
on Sephadex G-50. The nuclear RNA labeled in the absence or presence of α-amanitin were hybridized to 8 µg
of alkali—denatured rat liver DNA immobilized on 5.0
mm diameter nitrocellulose filters. Hybridization was
conducted in 0.1 ml 50% formamide-4 x SSC for 20 hr at
37°, the filters were washed twice in formamide-SSC at
37°, twice in 2 x SSC at 25°, treated with 50 µg/ml
ribonuclease for 60 min at 25°, washed four times more
in 2 x SSC, air dried and counted in toluene based
scintillator fluid.

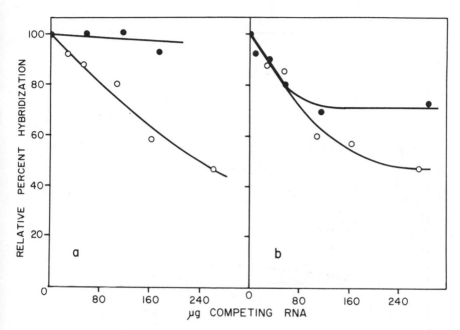

purification procedure is not so successful for this
enzyme. Polyacrylamide gel electrophoresis under conditions where the native structure is maintained re-
solves the highly purified Polymerase II preparation
into two components. A subsequent analysis of the
subunit structure has shown that there are two forms
of the enzyme. Electrophoresis of the highly purified
enzyme in polyacrylamide gels in the presence of sodium dodecyl sulfate (conditions under which protein
molecules are known to dissociate into their subunits),
produces five components as shown in Figure 4. There
is a stoichiometric correlation between the components
present and the RNA polymerase activity, as shown in
the lower part of the figure. The molecular weights of
the components were estimated to be about 190,000,
170,000, 150,000, 40,000 and 25,000. Quantitation of
the subunit profile indicates that the three smaller
subunits are present in stoichiometric amounts and the
two larger ones summate to a unit value (Table V).
Ordinarily the [170,000] component was present in
higher quantities than the [190,000] component, but
the proportion could be sharply reduced by including a
proteolytic inhibitor (phenylmethane sulfonyl fluo-

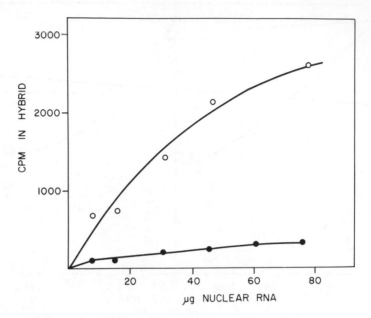

Figure 3. Hybridization of *in vitro*
labeled rat liver nuclear RNAs to rat DNA.
Increasing quantities of the nuclear RNAs
labeled in the absence (O) or presence (●)
of α-amanitin were hybridized to 8 µg alkali denatured rat liver DNA immobilized
on 5.0 mm diameter nitrocellulose filters.
Under the conditions employed, the highly
iterated sequences are selectively hybridized.   Thus the experiments do not provide
information of the transcription of the
unique sequences.

ride) during the purification procedure.   These results
suggest that the preparations contain two molecules—
one with the molecular structure, $[(190,000)_1 (150,000)_1$
$(40,000)_1 (25,000)_1]$, the other with the structure,
$[(170,000)_1 (150,000)_1 (40,000)_1 (25,000)_1]$.   Observations leading to a similar conclusion have been obtained with the Polymerase II isolated from calf
thymus.   In this instance, the isolated enzyme contains a preponderance of the enzyme containing the
190,000 subunit, but on "aging," the [190,000] and

TABLE 4

Purification of nuclear RNA polymerases

Enzyme	-Fold purified	Specific activity	Molecules/ nucleus
Calf thymus I	1-4,000	50-200	200
II	10-20,000	500-1000	800
Rat liver I	300	~30	10,000
II	3,000	~1000	10,000
E. coli	250	~500	500

Procedure: sonicated nuclei (high salt); minus chromatin (dilution); $(NH_4)_2SO_4$ precipitate; dialysis; DEAE-Sephadex (chromatography); phosphocellulose (chromatography); sucrose density gradient (centrifugation).

[170,000] subunits were present in about equal proportions.

These observations suggest that the 170,000 molecular weight subunit may be derived from the 190,000 subunit. If this were true, the proteolytic modification must be extraordinarily specific. (No evidence exists of other bonds being cleaved.) Furthermore, both forms of the enzyme are enzymatically active. Such reported conversion of one molecular form to the other does not preclude specific transcriptive roles for the two enzymes. Leighton, Frieze and Doi (1971), have shown that a specific serine protease is required for a functional modification of RNA polymerase during bacterial sporulation. Furthermore, a specific modification of the *E. coli* polymerase on infection with T4 bacteriophage has been reported (Seifert, Rabussay and Zillig, 1971). However, there is no evidence as yet to indicate whether the two forms of Polymerase II have different transcriptional roles.

These structural studies do emphasize the homology between the structures of the two forms of Polymerase II and those of the prokaryotic core polymerase (Table VI). The similarity in the two large and two small

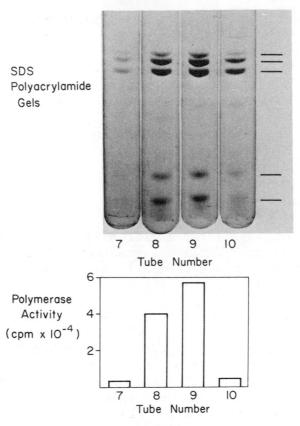

Figure 4. SDS Polyacrylamide gel electrophoresis patterns and polymerase activities
of sucrose gradient fractions of rat liver
Polymerase II. Fractions from sucrose density gradients were assayed for RNA polymerase activity, and protein, by standard methods (Roeder and Rutter, 1969); the specific activities are indicated by the bars in
the lower part of the figure. An aliquot of
each fraction was then subjected to SDS polyacrylamide gel electrophoresis as described
by Shapiro, Vinuela and Maizel,(1967). The
gels were then stained with a solution of
Coomassie blue (Lindell et al., 1970). Note
the correlation between the activity profile
and SDS gel electrophoresis patterns.

TABLE 5

'Subunit' composition of Polymerase II from rat liver

Subunit mass	(Relative proportions)	
	Control	+PMSF
190,000	0.1	0.5
170,000	0.9	0.5
150,000	1.0	1.0
35,000	1.0	1.0
25,000	1.0	1.0

Enzymes were isolated, in the absence (control) or in the presence of a proteolytic inhibitor, phenyl-methane sulfonyl fluoride (PMSF). The stoichiometric proportions of the subunits were determined by spectrophotometric scanning of Coomassie blue stained-SDS gels (after electrophoretic resolution of the subunits). The values obtained varied less than 10%.

subunits and the overall mass of the molecule is striking. If the transcriptive units in prokaryotes and eukaryotic organisms are homologous, it seems to us that many of the other regulatory characteristics of the molecules are also likely to be similar. We therefore predict that analogs of "regulatory subunits" will be present in eukaryotic transcriptive systems.

Highly purified (but not homogeneous) Polymerase I preparations have been obtained from calf thymus and rat liver. Preliminary analysis of the subunit composition suggests they have large subunits that are different from those of Polymerase II. The molecular structure of the two major classes of polymerase thus may be largely, if not entirely different.

Transcriptive Regulation

Regulation of transcription may occur by changes in the concentration, by modulation of the activity, or, by alteration of the specificity of one of the polym-

TABLE 6

Apparent structural homologies between prokaryotic and
eukaryotic DNA-dependent RNA polymerases

E. coli core polymerase		Eukaryotic Polymerase II	
		A*	B*
β'	155,000	β' 190,000	
		β"	170,000
β	145,000	β 150,000	150,000
α	40,000	α 40,000	40,000
		α' 25,000	25,000
	380,000	405,000	385,000
	$(\alpha_2\beta\beta')$	$(\alpha\alpha'\beta\beta')$	$(\alpha\alpha'\beta\beta")$

*The molecular weights are approximate.

erase molecules. All feasible mechanisms probably
have been subject to evolutionary trial, and it seems
likely to us that each of the regulatory modes mentioned will be employed in one biological system or
another. However, there is yet no instance in which
the mechanism of regulation of RNA synthesis in a eukaryotic system has been convincingly explained.

Certainly, substantial differences exist in the
polymerase levels in the nuclei of a number of tissues.
The polymerase concentration level in rat liver nuclei
is about ten-fold higher than the level in calf thymus.
From the data of Keller and Goor (1970), the nuclei of
the KB cell exhibit about five times more polymerase
activity than the liver. The levels found in cells
that do not synthesize RNA (e.g., the avian erythro-
cyte) are still lower than that of the calf thymus
(Smithyman and Rutter, unpublished observations).
Thus the polymerase levels in eukaryotic nuclei may
vary by at least two orders of magnitude. Even the
highest levels of enzymatic activity are lower than

those commonly found in prokaryotic organisms, especially when the activity is related to nuclear DNA
content. The level of polymerase in *E. coli*, for
example, is of the same order of magnitude as that
found in the eukaryotic nuclei, yet the DNA content of
the latter is approximately 1,000 times as great. It
seems possible that changes in polymerase level may
provide a general "coarse" control over transcription
in various cells. These differences in enzyme activity probably reflect changes in enzyme concentration,
but modulation of the catalytic activity of a constant
polymerase population is not ruled out.

A change in proportion of one of the polymerases
during a physiological transition implies independent
regulation and distinct biological roles of the enzymes. A selective increase in the activity of Polymerase I and, less generally, Polymerase II has been
observed in target tissues when stimulated with specific hormones or with a mitogenic stimulus (Blatti *et
al.*, 1970). We have previously shown there is an increase in Polymerase I activity in the liver after
treatment with glucocorticoids, or after partial hepa-
tectomy or in the uterus after estrogen treatment.
Figure 5 shows the changes in Polymerase I and II activity in uterine nuclei after estrogen treatment.
There is an increase in Polymerase I activity soon
after estrogen administration, while there is no significant change in Polymerase II activity. It is
conceivable in all these instances that the increased
ribosomal RNA synthesis is directly coupled with an
increased Polymerase I activity. Recently, Mainwaring,
Mangan and Peterken (1971) have demonstrated a similar
increase in Polymerase I activity in the prostate
gland after testosterone treatment. These experiments
again do not distinguish whether the increase in activity is due to the synthesis of new enzyme molecules,
or simply to the activation of those already present.
The uniform pattern of the response, however, suggests
a common mechanism for many physiological transitions,
involving an increase in Polymerase I activity and increased ribosomal RNA synthesis.

Another example of selective changes in polymerase
levels has been found during the development of the

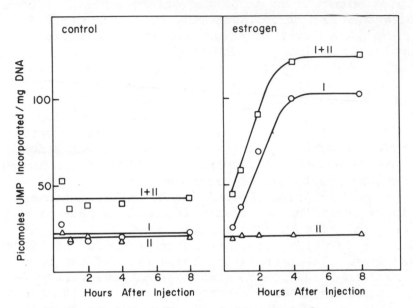

Figure 5. Estrogen stimulation of rat
uterine Polymerase I. Immature rats were
administered estrogen or control saline for
the indicated time prior to removal of the
uterus for preparation of nuclei. The isolated nuclei were assayed by standard
methods (Roeder and Rutter, 1969; Blatti *et
al.*, 1970) in the absence of α-amanitinn
(forms I and II) and in the presence of
α-amanitin (form I).

sea urchin embryo. We reported (Roeder and Rutter,
1970a) that, during developing of the mesenchyme
blastula to the prism stage, Polymerase I activity per
cell remained constant while Polymerase II and III activity declined. These changes in the relative proportions of these enzymes could be reasonably correlated with a progressively increased capacity for
ribosomal RNA synthesis. However, there is no direct
relationship in this system between the levels of
Polymerase I activity and ribosomal RNA synthesis.
Therefore the Polymerase I activity must be regulated.
The isolation and elucidation of the action of this
putative effector would be of great interest.

Paul Morris and I have recently found a dramatic modulation of the polymerase level which occurs on fertilization of the sea urchin egg (Table 7). In analogy with amphibian eggs, the sea urchin oocyte may

TABLE 7

Changes in RNA polymerase activity in developing sea urchin embryos

Hours	Stage	Total activity	Activity ratio $\left(\dfrac{I + III}{II}\right)$
00	Egg	100	25
1	1 cell	100	25
5	4 cells	100	20
7	16 cells	100	6
20	Blastula	200	2
46	Gastrula	340	1.5
93	Pluteus	810	0.5

*nMoles UMP incorporated/10^6 embryos.

synthesize substantial quantities of ribosomal RNA during development (Sponar, Sormona and Glisin, 1970). Amplification of the ribosomal genes similar to that demonstrated in *Xenopus* (Brown and Dawid, 1968) may occur. In the mature egg, however, little RNA synthesis occurs. We have found substantial quantities of I and/or III activity, but no detectable Polymerase II activity in the nuclei or cytoplasm of the mature egg. This observation indicates a regulation of Polymerase I and/or III in the mature oocyte. Shortly after fertilization, Polymerase II activity becomes detectable and continues to increase, until it becomes a relatively major component. All of this occurs during a period when the total RNA polymerase activity in the embryo remains essentially constant. This suggests turnover of one of the forms accompanied by synthesis of the other, or intra-conversion of the enzymes by exchanging subunits. Whatever the mechanism, there seems to be a dramatic change in the nature of the RNA polymerase activity coincident with the onset of embryological development. This is consistent with the hypothesis that RNA Polymerase II is required for transcription in early embryogenesis, and that initiation of this transcription requires the formation of Polymerase II.

The question of specific modulation of transcriptive specificity of the polymerases by interacting specifier subunits remains a tantalizing prospect, but one that has, as yet, received little direct experimental support. Stein and Hausen (1970) and later Seifart (1970) have reported that a factor present in thymus and liver homogenates increases Polymerase II activity on native but not denatured DNA templates. Similar activities now have been found in a number of other systems (F. Weinberg and W. Rutter, unpublished observations). Whether this factor also influences template specificity is unknown. Both negative (repressor-like) and positive (ψ-like) regulatory factors affecting ribosomal RNA synthesis in *Xenopus* oocytes and developing embryos have been reported by Toccini-Valentini and Crippa (1970). The isolation and elucidation of the role of these putative effectors would be of great interest.

The paucity of convincing evidence for regulatory factors does not necessarily indicate a scarcity of such molecules. There are major experimental difficulties in testing for changes in template specificity. The entire nuclear complement (DNA or chromatin) appears too complex a system for an analysis of subtle regulating of specific gene sets. The utilization of viral templates seems attractive, especially in view of the highly informative studies using the bacteriophage systems. Other alternative systems (ribosomal (nucleolar); 5S, tRNA, mitochondrial, chloroplast, etc.).

With satisfactory progress in our attempts to understand the structure of the polymerase molecules, it is propitious to focus on the details of regulation.

Is transcriptive specificity determined primarily by
specific regulatory (specifier) subunits, or by the
nature of the chromosomal template?

ACKNOWLEDGMENTS

This work was carried out with support from the National Institutes of Health (HD 04617) and the National Science Foundation (GB 14896). Stanley P. Blatti and Robert Weaver received postdoctoral fellowships from the National Institues of Health and Paul W. Morris from the American Cancer Society.

REFERENCES

Bekhor, I., Kung, G., and Bonner, J. (1969). *J. Mol. Biol. 39*, 351.

Blatti, S. P., Ingles, C. J., Lindell, T. J., Morris, P. W., Weaver, R. F., Weinberg, R., and Rutter, W. J. (1970). *Cold Spring Harbor Symp. Quant. Biol. 35*, 649.

Blobel, I. G., and Potter, V. R. (1966). *Science 154*, 1622.

Bonner, J., Dahmus, M. E., Fambrough, D., Huang, R. C., Marushige, K., and Tuan, D. Y. H. (1968). *Science 159*, 47.

Brown, D. D., and Dawid, I. B. (1968). *Science 160*, 272.

Chesterton, C. J., and Butterworth, P. H. W. (1971a). *Eur. J. Biochem. 19*, 232.

Chesterton, C. J., and Butterworth, P. H. W. (1971b). *FEBS Letters 15*, 181.

Huang, R.,C., and Huang, P. C. (1969). *J. Mol. Biol. 39*, 365.

Jacob, S. T., Sajdel, E. M., and Munro, H. N. (1970). *Nature 225*, 60.

Kedlinger, C. M., Gniazdowski, J., Mandel, L., Gissenger, F., and Chambon, P. (1970). *Bioch m. Biophys. Res. Commun. 38*, 165.

Kedlinger, C., Nuret, P., and Chambon, P. (1971). *FEBS Letters 15*, 169.

Keller, W., and Goor, R. (1970). *Cold Spring Harbor Symp. Quant. Biol. 35*, 671.

Leighton, T. J., Freese, P. K., and Doi, R. H. (1971). *Fed. Proc. 30*, 1069 (Abstract).

Lindell, T. J., Weinberg, F., Morris, P. W., Roeder, R. G., and Rutter, W. J. (1970). *Science 170*, 447.

Mainwaring, W. I. P., Mangan, F. R., and Peterken, B. M. (1971). *Biochem. J. 123*, 619.

Mandel, J. L., and Chambon, P. (1971). *FEBS Letters 15*, 175.

Muramatsu, M., Hodnett, J. L., and Busch, H. (1966). *J. Biol. Chem. 241*, 1544.

Paul, J., and Gilmour, R. S. (1968). *J. Mol. Biol. 34*, 305.

Roeder, R. G., Reeder, R. H., and Brown, D. D. (1970). *Cold Spring Harbor Symp. Quant. Biol. 35*, 727.

Roeder, R. G., and Rutter, W. J. (1969). *Nature 224*, 234.

Roeder, R. G., and Rutter, W. J. (1970a). *Biochemistry 9*, 2543.

Roeder, R. G., and Rutter, W. J. (1970b). *Proc. Nat. Acad. Sci. USA 65*, 675.

Seifart, K. H. (1970). *Cold Spring Harbor Symp. Quant. Biol. 35*, 719.

Seifert, W., Rabussay, D., and Zillig, W. (1971). *FEBS Letters 16*, 175.

Shapiro, A. L., Vinuela, E., and Maizel, J. V., Jr. (1967). *Biochem. Biophys. Res. Commun. 28*, 815.

Smith, K. D., Church, R. B., and McCarthy, B. J. (1969). *Biochemistry 8*, 4271.

Sponar, J., Sormona, Z., and Glisin, V. (1970). *FEBS Letters 11*, 254.

Stein, H., and Hausen, P. (1970). *Cold Spring Harbor Symp. Quant. Biol. 35*, 709.

Toccini-Valentini, G. P., and Crippa, M. (1970). *Cold Spring Harbor Symp. Quant. Biol. 35*, 737.

Weaver, R. F., Blatti, S. P., and Rutter, W. J. (1971). *Proc. Nat. Acad. Sci. USA 68*, 2994.

Widnell, C. C., and Tata, J. R. (1966). *Biochim. Biophys. Acta 123*, 478.

RNA Polymerases During Amphibian Development

Robert G. Roeder

Department of Embryology
Carnegie Institution of Washington
Baltimore

INTRODUCTION

Eukaryotic cells contain at least three forms of DNA-dependent RNA polymerase, designated I, II and III (Roeder, 1969; Roeder and Rutter, 1969). Two lines of evidence suggest that these enzymes possess template specificity *in vivo*. First, they have specific subcellular locations (Roeder and Rutter, 1970). Second, the synthesis of DNA-like RNA, but not that of ribosomal RNA, is inhibited by α-amanitin (Blatti *et al.*, 1970; Reeder and Roeder, 1971), a toxin which inactivates RNA polymerase II, but not RNA polymerases I and III (Roeder, Reeder and Brown, 1970; Lindell *et al.*, 1970; Gniazdowski *et al.*, 1970).

In order to investigate the role of the RNA polymerases in transcriptional control, we have examined these enzymes in oocytes and embryos of *Xenopus laevis*, the South African clawed toad. At different develop-

mental stages the major classes of RNA are synthesized
at vastly different rates (Brown and Littna, 1964a;
Brown and Littna, 1964b; Davidson, Allfrey and Mirsky,
1964). We wished to know whether the levels and relative proportions of the various enzymes could be correlated with the *in vivo* patterns of RNA synthesis.

METHODS

Unfertilized eggs, embryos, and tissue culture cells
of *X. laevis* were collected for enzyme analysis (Brown
and Littna, 1964a). Mature females were induced to
ovulate by hormone treatment, and 5-7 days later the
ovaries were removed for oocyte isolation. Large
oocytes (about 1 mm in diameter) were removed with a
hair loop and freed of follicle cells by manual dis-
section with watchmaker's forceps (Davidson, Allfrey
and Mirsky, 1964). While immature ovaries were taken
from young females about 8-10 weeks after metamorphosis
as a source of small oocytes (≤ 0.25 mm). The RNA
polymerase activity was solubilized from cell homogenates and analyzed by DEAE-Sephadex chromatography
(Roeder and Rutter, 1969; Roeder and Rutter, 1970;
Roeder, Reeder and Brown, 1970).

RESULTS

The RNA polymerase pattern in large oocytes is shown
in Figure 1. Activity was measured in the presence and
absence of α-amanitin, which specifically inhibits the
polymerase forms here designated as enzyme II. This
α-amanitin sensitive activity is plotted in the upper
panel of Figure 1 as the difference between total and
α-amanitin insensitive activity. The polymerase activity which is sensitive to α-amanitin is chromato-
graphically separable into at least two components
(designated II_A and II_B). Heterogeneity of enzyme II
has also been reported by others (Kedinger, Nuret and
Chambon, 1971). Relative to II_B, form II_A is not as
prominent in tissue culture cells or older embryos
(see below) as it is in oocytes and early embryos.

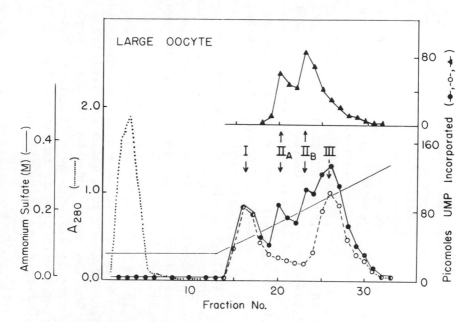

Figure 1.   DEAE-Sephadex chromatography of
RNA polymerase from *X. laevis* large oocytes.
RNA polymerase was solubilized from large
oocytes (about 1.0 mm in diameter), and a
sample containing 16 mg of protein was
chromatographed on an 0.8 x 12 cm column
as described in Methods.   Samples of 2.62
ml (Fractions 1-10) or 1.31 ml (Fractions
11-35) were collected, and aliquots were
assayed for activity under the conditions
described in Table 1.   Activity represents
picomoles UMP incorporated per 20 min/ml.
(●), Activity in the absence of α-amanitin;
(O), Activity in the presence of 2 µg/ml
α-amanitin.   The difference between total
activity and α-amanitin insensitive activity
is plotted in the upper graph (▲).

However, the most distinctive feature of this oocyte
enzyme profile compared to that of a normal somatic
cell is the presence of a very high proportion of RNA
polymerase III, the level of which approximates that
of enzyme I.   Since it is known that oocytes are very

TABLE 1

RNA polymerase activity at different stages of oogenesis and embryonic development

			Activity (Units/10⁴ cells)*		
Enzyme form	Small oocyte**	Large oocyte	Unfert. egg	Swimming tadpole (640,000 cells)	Tissue culture
I	22	8,700	10,800	.153 (89,200)†	0.179
III	24	11,200	9,600	.019 (12,000)	0.018
II	27	12,100	19,600	.160 (102,000)	0.151

*One unit of activity represents one picomole UMP incorporated into RNA per 20 min at 30° under the conditions described previously (Roeder, Reeder and Brown, 1970), except that ³H-UTP was present at a concentration of 0.05 mM. The data were calculated from DEAE-Sephadex chromatographic analyses (cf. Figs. 1-3) of solubilized enzyme preparations from the various tissues. Large oocytes (1.0 to 1.1 mm in diameter), unfertilized eggs, and embryos were hand counted. The average number of cells per embryo and tissue culture cell numbers were determined by measurement of the DNA content in cell homogenates, using a value of 6 μμgm DNA per diploid cell.

**Whole immature ovaries containing oocytes ≤ 0.25 mm in diameter were analyzed. The activity per oocyte was calculated assuming the average number of oocytes per ovary to be 20,000. This number is based upon the average amount of rDNA which was isolated from comparable ovaries in previous experiments and the amount of rDNA (25 μμgm) per oocyte (Brown and Dawid, 1968). The relative enzyme levels were not affected by follicle cell contamination since small oocytes treated with pronase to remove these cells (Mairy and Denis, 1971) showed about the same DEAE-Sephadex enzyme profile.

†Values in parentheses denote total activity per 10⁴ embryos.

166

active in rRNA synthesis (Brown and Littna, 1964b; Davidson, Allfrey and Mirsky, 1964) and that the rRNA genes are transcribed by an α-amanitin insensitive activity (Reeder and Roeder, 1971), the enzymatic species responsible for rRNA synthesis in these oocytes could be either form I or III.

The unfertilized egg RNA polymerase pattern (Fig. 2)

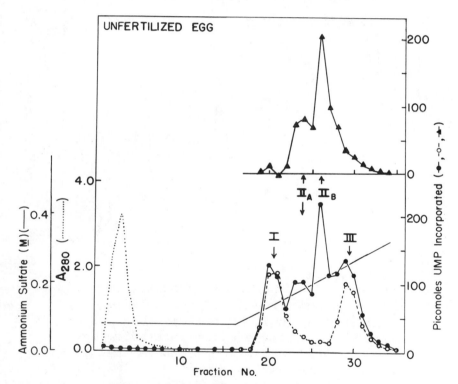

Figure 2. DEAE-Sephadex chromatography of RNA polymerase from X. *laevis* unfertilized eggs. RNA polymerase was solubilized from unfertilized eggs (dejellied), and a sample containing 12 mg protein was chromatographed as in Figure 1. Symbols as in Figure 1.

is similar to that of the large oocyte. Although enzyme profiles from all later developmental stages have not been analyzed, the α-amanitin sensitive and insen-

sitive RNA polymerase activities have been measured
from the two-cell stage through gastrulation (Roeder,
Reeder and Brown, 1970) and found to be constant.   Both
kinds of activity increase during subsequent development.   The RNA polymerase pattern for late swimming
tadpoles (640,000 cells) is shown in Figure 3.   It is

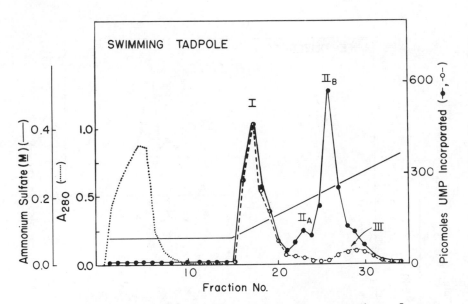

Figure 3.   DEAE-Sephadex chromatography of
RNA polymerase from *X. laevis* swimming
stage embryos (640,000 cells).   RNA polymerase was solubilized from late swimming
stage embryos, and a sample containing 13
mg protein was chromatographed as in
Figure 1.   Symbols as in Figure 1.

evident that all of the separable RNA polymerase components have not increased in equal proportion.   Enzymes I and II_B predominate just as they do in tissue
culture cells of *X. laevis*.

In Table 1 the activity of each form of RNA polymerase at the different stages has been expressed on a
per cell basis.   The enzyme levels present in small
non-pigmented, non-vitellogenic oocytes (≤ 0.25 mm in
diameter) and in actively dividing tissue culture cells

are shown for comparison. All forms of RNA polymerase
increase enormously during oogenesis so that the mature
oocyte and unfertilized egg have RNA polymerase levels
per cell about 4-5 orders of magnitude greater than
those found in tissue culture cells. During early development the total amount of enzyme activity per embryo remains constant for enzyme III while in the case
of enzymes I and II this value increases several fold.
However, the enormous increase in cell number which has
occurred by the swimming stage of the embryo greatly
reduces the concentrations per cell for all enzymatic
forms until each enzyme reaches about the level found
in tissue culture cells.

DISCUSSION

Maturing oocytes are specialized for rRNA synthesis
due to the selective amplification of their ribosomal
RNA genes (rDNA) (Brown and Dawid, 1968; Gall, 1968).
They also contain a much higher proportion of RNA polymerase III than is found in tissue culture cells.
This raises the possiblity that RNA polymerase III and
not RNA polymerase I is responsible for rRNA transcription or that these two enzyme forms may both be involved—perhaps in a precursor-product relationship.
The presumption that ribosomal RNA synthesis is mediated via enzyme I has been based upon the observations
that polymerase I is localized in isolated rat liver
nucleoli (Roeder and Rutter, 1970) and that rRNA synthesis in isolated nuclei is completely resistant to
α-amanitin (Blatti *et al.*, 1970; Reeder and Roeder,
1971). While these observations rule out the amanitin-
sensitive forms, RNA polymerase III might be involved
in rRNA synthesis. The high lability of RNA polymerase
III could account for its absence from isolated nucleoli, and α-amanitin cannot distinguish betweeen enzymes
I and III (Roeder, Reeder and Brown, 1970; Lindell *et
al.*, 1970). Studies of the *in vitro* transcription of
purified rDNA have thus far failed to indicate which
is the natural enzyme, as both RNA polymerase I (Roeder, Reeder and Brown, 1970) and RNA polymerase II (R.

Roeder and R. Roeder, unpublished observation) transcribe this DNA nonspecifically.

Oocytes contain extraordinarily high levels (per
cell) of all forms of RNA polymerase. With respect to
rRNA synthesis, a direct comparison of enzyme levels
in oocytes and somatic cells is misleading since each
oocyte has about 2,000 times more gene copies than a
single diploid somatic cell (Brown and Dawid, 1968).
However, even when the oocyte enzyme activity is expressed per rRNA cistron the level for each of the
enzymes is still much greater than that found in a
single somatic cell. Thus, it may be that the various
RNA polymerases are synthesized during oogenesis and
stored for use during early embryogenesis. Alternatively, if a minimal nuclear or cytoplasmic RNA polymerase concentration is necessary for transcription,
then the oocyte, by virtue of its enormous nuclear and
cytoplasmic volumes, would require a higher polymerase
content than would a normal-sized somatic cell.

Whether the oocyte enzymes function exclusively
during oogenesis or whether these same enzyme molecules
function during embryogenesis cannot yet be distinguished. However, some support for the latter possibility is provided by the present and previous (Roeder,
Reeder and Brown, 1970) studies of enzyme levels during embryogenesis. Most or all of the enzyme components present in the large oocyte appear to be retained
in the unfertilized egg, and although turnover has not
been ruled out, these same components are probably
maintained (at constant levels) through gastrulation,
even though the rates of synthesis of various kinds of
RNA (DNA-like, 4S and ribosomal) change drastically
during this developmental period (Brown and Littna,
1964a). The synthesis and accumulation of new enzyme
components may be necessary only when rapid cell division within the embryo reduces the enzyme levels per
cell to certain critical concentrations. If the enzymes inherited by the egg are in fact utilized during
subsequent development, then it appears that none of
the various RNA polymerase species is rate limiting
within the early embryo and that additional factors
(either enzyme associated or template associated) may
be necessary for transcriptional control *in vivo*.

SUMMARY

The amounts of the various forms of RNA polymerase (I, II and III) have been measured in oocytes, eggs, and embryos of *Xenopus laevis*. Although the relative proportions of these enzymes remain nearly constant during oogenesis, the absolute levels of all forms increase enormously so that a mature oocyte contains 4-5 orders of magnitude more activity than does an individual somatic cell. Oocytes and eggs contain unusually high levels of form III (25-35% of the total activity) in addition to the normal forms I and II which predominate in somatic cells. During embryogenesis the total amounts of forms I and II increase several fold while the amount of enzyme III per embryo remains constant. The rapid increase in cell number during early development reestablishes the same relative and absolute enzyme levels per cell in the swimming embryo as are found in adult somatic cells.

ACKNOWLEDGMENTS

I thank Drs. D. D. Brown and R. H. Reeder for their critical reading of this manuscript and Dr. Th. Wieland for a generous gift of α-amanitin. This work was carried out during the tenure of a postdoctoral fellowship from the American Cancer Society (DF-539).

REFERENCES

Blatti, S. P., Ingles, C. J., Lindell, T. J., Morris, P. W., Weaver, R. F., Weinberg, F., and Rutter, W. J. (1970). *Cold Spring Harbor Symp. Quant. Biol. 35*, 649.

Brown, D. D., and Dawid, I. B. (1968). *Science 160*, 272.

Brown, D. D., and Littna, E. (1964a). *J. Mol. Biol. 8*, 669.

Brown, D. D., and Littna, E. (1964b). *J. Mol. Biol. 8*, 688.

Davidson, E. H., Allfrey, V. G., and Mirsky, A. E.
    (1964). *Proc. Nat. Acad. Sci. USA 52*, 501.
Gall, J. G. (1968). *Proc. Nat. Acad. Sci. USA 60*, 553.
Gniazdowski, M., Mandel, J. L., Jr., Gissinger, F.,
    Kedinger, C., and Chambon, P. (1970). *Biochem. Biophys. Res. Commun. 38*, 1033.
Kedinger, C., Nuret, P., and Chambon, P. (1971). *FEBS
    Letters 15*, 169.
Lindell, T. J., Weinberg, F., Morris, P. W., Roeder,
    R. G., and Rutter, W. J. (1970). *Science 170* , 447.
Mairy, M., and Denis, H. (1971). *Devel. Biol. 24*, 143.
Reeder, R. H., and Roeder, R. G. (1971). *Carnegie Inst.
    Wash. Year Book 70*, p. 32.
Roeder, R. G. (1969). Doctoral Thesis, University of
    Washington.
Roeder, R. G., Reeder, R. H., and Brown, D. D. (1970).
    Cold Spring Harbor Symp. Quant. Biol. 35, 727.
Roeder, R. G., and Rutter, W. J. (1969). *Nature 224*,
    234.
Roeder, R. G., and Rutter, W. J. (1970). *Proc. Nat.
    Acad. Sci. USA 65*, 675.

The Control of Ribosomal RNA Synthesis

Andrew Travers

Medical Research Council
Laboratory of Molecular Biology
Cambridge, England

During the exponential growth of bacterial cells the synthesis of stable RNA species can account for nearly half of the instantaneous rate of RNA synthesis (Salser, Janin and Levinthal, 1968; Lazzarini and Winslow, 1970), yet the cistrons coding for these RNA species comprise less than 0.5% of the bacterial genome (Yankofsky and Spiegelman, 1962; Kennel, 1968). Thus under these growth conditions about half the actively transcribing RNA polymerase molecules are engaged in copying a highly restricted set of RNA species. Considering only the ribosomal RNA (rRNA) cistrons, of which there are about six per genome (Spadari and Ritossa, 1970) this high rate of synthesis implies that rRNA molecules are initiated at the rate of about one per second, a rate comparable with that for the initiation of phage mRNA species (Bremer, 1970).

The synthesis of rRNA at high growth rates approximates to the theoretical maximum. However, the param-

eter which is normally experimentally determined is
the *accumulation* of rRNA. This quantity in bacteria
is regulated in accordance with the physiological state
of the cell (this subject is reviewed by Neidhardt,
1964; Maaløe and Kjeldgaard, 1966; Edlin and Broda,
1968; Ryan and Borek, 1971). Clearly such regulation
could occur at the level of synthesis or of degradation
(or perhaps at both) of rRNA. At least two modes of
regulation of rRNA accumulation may be arbitrarily
distinguished. One such mode is the response correlated with a growth rate transition consequent upon a
change in carbon source. During such balanced growth
the rate of rRNA accumulation is positively correlated
with the growth rate of the cell (Schaecter, Maaløe
and Kjeldgaard, 1958; Neidhardt and Magasanik, 1960),
more rRNA accumulating per genome equivalent of DNA at
high growth rates than at low. Regulation of this
type may be regarded as 'fine tuning' (Gallant *et al.*,
1970).

A second mode of regulation of rRNA accumulation,
'stringent' control, is defined genetically by the
or RC locus (Stent and Brenner, 1961). rel^+ strains
of *Escherichia coli* when functionally deprived of an
amino acid, either by starvation of an amino acid aux-
otroph (Gale and Folkes, 1953; Pardee and Prestidge,
1956; Gros and Gros, 1958) or by restricting aminoacyl-
ation of a transfer RNA (tRNA) species even in the
presence of the required amino acid (Neidhardt, 1966)
reduce the rate of rRNA accumulation about ten-fold
(Lazzarini and Winslow, 1970). Mutants of the rel
gene (rel^-) exist which do not restrict rRNA accumulation during amino acid starvation and are said to possess 'relaxed' RNA control. In diploids rel^+ is dominant to the rel^- allele (Fiil, 1969), consistent with
the existence of a functional inhibitor of rRNA synthesis. Although rel^- strains are unable to control
the rate of accumulation under conditions of amino
acid starvation it should be emphasized that most rel^-
strains regulate rRNA accumulation normally during a
carbon source transition (Neidhardt, 1963).

To what extent are other species of RNA regulated
in a manner similar to rRNA? The accumulation of tRNA
and rRNA are restricted to a comparable extent during

the stringent response (Lazzarini and Winslow, 1970; Primakoff and Berg, 1970); Lazzarini and Dahlberg, 1970). During the stringent response the instantaneous rate of RNA synthesis may be reduced by a factor of three or four (Winslow and Lazzarini, 1969). Since the synthesis of stable RNA species accounts for less than half this instantaneous rate (Salser, Janin and Levinthal, 1968; Winslow and Lazzarini, 1969; Pato and von Meyenburg, 1970) this decrease is too large to be accounted for by control of only stable RNA species. This suggests that the production of mRNA should also be subject to stringent control. In support of this argument Gallant *et al.* (1970) have demonstrated that during amino acid starvation, the production of the messenger RNA coding for ornithine transcarbamylase is about four times higher in *rel⁻* cells than in *rel⁺* cells. However, Primakoff and Berg (1970) were unable to demonstrate any effect of the stringent response on the synthesis of $\phi80$ specific mRNA, although in the same cells the production of tRNA^{tyr} was under stringent control. Other studies on the expression of mRNA from the tryptophan (Edlin *et al.*, 1968; Stubbs and Hall, 1968; Lavalle and de Hauwer, 1968) and *lac* (Morris and Kjeldgaard, 1968) operons demonstrate that during the stringent response the accumulation of stable RNA species is preferentially reduced compared with the production of these mRNA species. Although this latter conclusion may be generally applicable the data presently available is too scanty to resolve the question of whether different mRNA species are subject to varying degrees of stringent control.

A very early and apparently invariant manifestation of *rel⁺* gene function during amino acid starvation is the rapid accumulation of an unusual nucleotide MS 1 (Cashel and Gallant, 1969; Cashel, 1969). The structure proposed for this compound is guanosine-5'-diphosphate-2'-(or 3')-diphosphate (Cashel and Kalbacher, 1970). The accumulation of MS 1 after amino acid starvation is very rapid in cells bearing *rel⁺* but not *rel⁻* alleles, the nucleotide attaining and usually exceeding the levels of GTP within two minutes (Cashel, 1969; Gallant *et al.*, 1970). Conversely upon amino acid resupplementation MS 1 rapidly disappears (Cashel, 1969).

In the case of carbon source deprivations and growth rate transitions rel^+ as well as rel^- strains restrict rRNA accumulation and both cell types also accumulate MS 1 (Lazzarini, Cashel and Gallant, 1971). Furthermore, even the basal level of MS 1 found in rel^+ and rel^- strains during balanced growth can be correlated with the RNA content and growth rate of the cells (Lazzarini, Cashel and Gallant, 1971). Such correlations would be obtained if MS 1 accumulation were either a cause or a consequence of restricted rRNA synthesis.

The accumulation of specific RNA species could be regulated either at the level of synthesis or of degradation. In the case of the most drastic reduction in the accumulation of the stable RNA species, i.e., during the stringent response, Lazzarini and Dahlberg (1971) have argued that this effect most probably is a consequence of the restriction of synthesis. The basis of their argument is that the rate of accumulation of RNA is that which would be predicted from the measured instantaneous rate of synthesis and further that under these conditions of growth, no turnover of newly synthesized rRNA could be detected. Nevertheless, degradation of rRNA and tRNA has been observed under certain conditions (Dubin and Elkart, 1965; Lazzarini and Santangelo, 1968; Lazzarini and Peterkofsky, 1965) and therefore it is still conceivable that degradative processes could play a role, albeit perhaps a minor one, in the control of the levels of the stable RNA species.

Control of RNA synthesis can be exerted principally during initiation or elongation of RNA molecules, and Winslow and Lazzarini (1969) have concluded that the rate of RNA chain elongation is lowered on average by a factor of 3-4 (and perhaps up to nine-fold) during the stringent response. However, the data of Primakoff and Berg (1970) showing that the total rate of $\phi 80$ mRNA synthesis is unaffected by the stringent response argues that if this reduction in the rate of elongation is general it is exactly compensated for by an increase in the rate of initiation of $\phi 80$ mRNA. An alternative explanation is that the rates of elongation of different RNA species are affected to varying extents by the

stringent response. The most compelling argument that
the principal action of the stringent response is at
the level of chain initiation is provided by Pettijohn,
Kossman and Stomato (1971) who show that the synthesis
of 16S RNA resumes within 40 sec after the restoration
of a required amino acid to a starved culture of rel^+
cells. The synthesis of 23S RNA then succeeds that of
16S RNA. Since the 16S sequence is proximal to the
rRNA promoter (Pato and von Meyenburg, 1970) it is
probably that the resumption of rRNA synthesis results
from an increased rate of initiation.

What is the molecular mechanism for the control of
rRNA synthesis? Dennis (1971) has presented evidence
that the control of rRNA synthesis is not dependent on
gene dosage and suggests that cytoplasmic elements may
therefore be involved in its regulation. This conclusion is, however, dependent on the genes for rRNA being
clustered.

One possibility which cannot be excluded is that the
observed pattern of control is a consequence of the
independent action of several regulatory mechanisms
directly affecting the activity of the rRNA promoters.
More attractive is the hypothesis that initiation of
rRNA synthesis is primarily controlled by a simple
regulatory protein. If it is assumed that control of
RNA synthesis during the stringent response and during
carbon source transitions is mediated by the same regulatory system, this system would have as its primary
target the control of stable RNA species. Although
the synthesis of the stable RNA species is preferentially curtailed during the stringent response, the
synthesis of certain mRNA species may also be reduced,
although to a lesser degree. Accordingly, it is possible that the regulatory system for stable RNA species
might also control the synthesis of certain mRNA species. In this case the regulatory system would lack
the specificity which characterizes such transcriptional regulatory protein as the *lac* represssor (Gilbert and Müller Hill, 1967) and the CAP protein (Pastan *et al.*, 1971). Such a lack of specificity would
be an inherent property of the RNA polymerase itself
or a transcriptional factor interacting directly with
the polymerase.

The rapidity of the response of rRNA synthesis to amino acid starvation or to its relief suggests that the control may be effected by a low molecular weight compound interacting directly with a component of the transcription system. A candidate for such an effector is MS 1 whose appearance and disappearance parallels alterations in the rate of RNA synthesis. In fact this compound has been shown to interact with at least two components of the *E. coli* transcription system, the DNA-dependent RNA polymerase (Cashel, 1970) and the ψ factor (Travers, Kamen and Cashel, 1970). Accordingly, we must ask whether the observed *in vitro* interactions are relevant to the control of transcription *in vivo*. Cashel (1970) has shown that MS 1 inhibits RNA synthesis by highly purified RNA polymerase *in vitro* on a variety of DNA templates by inhibiting both the rate of chain elongation and also by inhibiting chain initiation. Interestingly MS 1 affects initiation by selectively inhibiting a fraction of the RNA chains that contain a 5' terminal pppG residue. The effect of MS 1 on chain elongation could account, is part, for the observed decrease in the rate of chain elongation during the stringent response. However, the significance of the effect of interaction of MS 1 with RNA polymerase on chain initiation is less clear. Using a crude cell free system Zubay, Cheong and Gefter (1971) have demonstrated that MS 1 has no preferential effect on the DNA dependent synthesis of tRNA^{tyr}. This synthesis was carried out at a relatively high concentration of divalent cations (magnesium ion, 15 mM, calcium ion, 7 mM), conditions under which the ψ factor does not stimulate RNA synthesis (Travers, manuscript in preparation). Accordingly, the effect of MS 1 in this system may well reflect its action on RNA polymerase alone.

The second component of the transcription system with which MS 1 is known to interact is the psi factor. This factor has been characterized both in crude bacterial extracts and as a host specified component of the Qβ replicase (Travers, Kamen and Schleif, 1970). The psi factor prepared from Qβ replicase contains two polypeptide chains of molecular weight of 45,000 and 35,000 (Kamen, 1970). It is not known, however, whether

both or only one of these chains is required for psi
activity. In the *in vitro* system psi selectively
stimulates RNA synthesis with *E. coli* DNA as a template
but does not significantly affect transcription from
T7 or T4 DNA templates. MS 1 inhibits ψ stimulation
of RNA synthesis (Travers, Kamen and Cashel, 1970) but
this inhibition only occurs if MS 1 is added prior to
the formation of a ψ-dependent rifampicin-resistant
complex between *E. coli* DNA and RNA polymerase (Travers,
manuscript in preparation). This argues that MS 1 inhibits ψ activity at the level of initiation of RNA
polymerase. At the optimum conditions for ψ activity
RNA synthesis on *E. coli* DNA may be stimulated up to
five-fold. Of this RNA synthesis, rRNA has been estimated to comprise maximally about 15%. Thus, under
these conditions, at least 80% of the rRNA synthesis
observed *in vitro* is sensitive to MS 1. At present no
further analysis of the spectrum of RNA species transcribed from *E. coli* DNA *in vitro* has been performed
and consequently the exact specificity of the ψ factor
remains in doubt.

Thus the observed *in vitro* effects of the interaction of MS 1 with RNA polymerase and with ψ can explain
in large part the regulation of rRNA synthesis observed
in vivo. Such a correspondence does not, however, constitute rigorous proof that MS 1 is the responsible
low molecular weight effector or that psi and RNA
polymerase itself are regulatory proteins in this control. A final elucidation of this problem will require the isolation of a suitable mutant in an identifiable protein.

REFERENCES

Bremer, H. (1970). *Cold Spring Harbor Symp. Quant.
Biol. 35*, 109.
Cashel, M. (1969). *J. Biol. Chem. 244*, 3133.
Cashel, M. (1970). *Cold Spring Harbor Symp. Quant.
Biol. 35*, 407.
Cashel, M., and Gallant, J. (1969). *Nature 221*, 838.
Cashel, M., and Kalbacher, B. (1970). *J. Biol. Chem.
245*, 2309.

Dennis, P. P. (1971). *Nature 232*, 43.

Dubin, S. T., and Elkart, A. T. (1965). *Biochim. Biophys. Acta 103*, 355.

Edlin, G., and Broda, P. (1968). *Bact. Rev. 32*, 206.

Edlin, G., Stent, G. S., Baker, R. F., and Yankofsky, C. (1968). *J. Mol. Biol. 37*, 257.

Fiil, N. (1969). *J. Mol. Biol. 45*, 195.

Gale, E. F., and Folkes, J. P. (1953). *Biochem. J. 53*, 493.

Gallant, J., Erlich, H., Hall, B., and Laffler, T. (1970). *Cold Spring Harbor Symp. Quant. Biol. 35*, 397.

Gilbert, W., and Müller Hill, B. (1967). *Proc. Nat. Acad. Sci. USA 64*, 962.

Gros, F., and Gros, F. (1958). *Expt. Cell Res. 14*, 104.

Kamen, R. I. (1970). *Nature 228*, 527.

Kennel, D. (1968). *J. Mol. Biol. 34*, 85.

Lavalle, R., and de Hauwer, G. (1968). *J. Mol. Biol. 37*, 269.

Lazzarini, R. A., Cashel, M., and Gallant, J. (1971). *J. Biol. Chem. 246*, 4381.

Lazzarini, R. A., and Dahlberg, A. E. (1971). *J. Biol. Chem. 246*, 420.

Lazzarini, R. A., and Peterkofsky, A. (1965). *Proc. Nat. Acad. Sci. USA 53*, 549.

Lazzarini, R. A., and Santangelo, E. (1968). *J. Bacteriol. 95*, 1212.

Lazzarini, R. A., and Winslow, R. M. (1970). *Cold Spring Harbor Symp. Quant. Biol. 35*, 383.

Maaløe, O., and Kjeldgaard, N. O. (1966). *Control of Macromolecular Synthesis* (New York: W. A. Benjamin, Inc.).

Morris, D. M., and Kjeldgaard, N. O. (1968). *J. Mol. Biol. 31*, 145.

Neidhardt, F. C. (1963). *Biochim. Biophys. Acta 68*, 365.

Neidhardt, F. C. (1964). *Prog. Nucleic Acid Res. 3*, 145.

Neidhardt, F. C. (1966). *Bact. Rev. 30*, 701.

Neidhardt, F. C., and Magasanik, B. (1960). *Biochim. Biophys. Acta 42*, 99.

Pardee, A., and Prestidge, L. (1950). *J. Bacteriol. 71*, 677.

Pastan, I., de Crombrugghe, B., Chen, B., Anderson, W., Parks, J., Nissley, P., Straub, M., and Perlman, R. L. (1971). *Proc. Third Ann. Biochem.— PCR1 Winter Symp.* (Florida: University of Miami Press).

Pato, M. L., and von Meyenburg, K. (1970). *Cold Spring Harbor Symp. Quant. Biol. 35*, 497.

Pettijohn, D. E., Kossman, C. R., and Stomato, T. D. (1971). *Fed. Proc. 30*, 1161.

Primakoff, P., and Berg, P. (1970). *Cold Spring Harbor Symp. Quant. Biol. 35*, 391.

Ryan, A., and Borek, E. (1971). *Prog. Nucleic Acid Res. 11*,

Salser, W., Janin, J., and Levinthal, C. (1968). *J. Mol. Biol. 31*, 237.

Schaecter, M., Maaløe, O., and Kjeldgaard, N. O. (1958). *J. Gen. Microbiol. 19* , 592.

Spadari, S., and Ritossa, F. (1970). *J. Mol. Biol. 53*, 357.

Stubbs, J. D., and Hall, B. D. (1968). *J. Mol. Biol. 37*, 303.

Stent, G. S., and Brenner, S. (1961). *Proc. Nat. Acad. Sci. USA 47*, 2005.

Travers, A. A., Kamen, R. I., and Cashel, M. (1970). *Cold Spring Harbor Symp. Quant. Biol. 35*, 415.

Travers, A. A., Kamen, R. I., and Schleif, R. F. (1970). *Nature 228*, 748.

Winslow, R. M., and Lazzarini, R. A. (1969). *J. Biol. Chem. 244*, 3387.

Yankofsky, S. A., and Spiegelman, S. (1962). *Proc. Nat. Acad. Sci. USA 48*, 1465.

Zubay, G., Cheong, L., and Gefter, M. L. (1971). *Proc. Nat. Acad. Sci. USA 68*, 2195.

Visualization of
Genetic Transcription

O. L. Miller, Jr., and Barbara A. Hamkalo

Biology Division
Oak Ridge National Laboratory
Oak Ridge, Tennessee

INTRODUCTION

In this paper we describe electron microscopic observations of structural aspects of gene action in two
vastly different cell types: first in a eukaryotic
cell, the amphibian oocyte, and secondly in prokaryotic bacterial cells. In eukaryotes (with the possible
exceptions of their mitochondria or chloroplasts), synthesis of the various RNAs and translation of messenger
RNAs into protein typically are compartmentalized within the nucleus and cytoplasm, respectively. In bacteria, transcription and translation occur simultaneously and are intimately coupled. Previous reports of
these investigations can be found in Miller and Beatty
(1969a, b, c), Miller, Hamkalo and Thomas (1970), and
Miller *et al*. (1970).

AMPHIBIAN OOCYTES

Amphibian oocytes are highly favorable cells for studies of gene action. During early oogenesis, the genes coding for ribosomal RNA (rRNA), normally localized at the nucleolus organizer, are amplified in a process that results in hundreds of extrachromosomal nucleoli per nucleus which are actively synthesizing ribosomal RNA precursor (rpRNA) molecules. The oocyte chromosomes are in a highly extended, highly metabolic state, the so-called "lampbrush chromosome" stage. In addition, amphibian oocyte nuclei are so large (ranging up to 1 mm in diameter in some species) that one can isolate the nuclear contents by simple manual methods and rapidly disperse nucleoli and chromosomes for electron microscopy. (General procedures for handling oocyte nuclei for light and electron microscopy are given in Callan and Lloyd, 1960; Gall, 1966a; Miller, 1965; and Miller and Beatty, 1969b; and current techniques are summarized in the legend for Fig. 2.)

Nucleolar Genes

The extrachromosomal nucleoli of amphibian oocytes are located close to the nuclear envelope and show two morphological components, a fibrous core and a surrounding granular cortex (Fig. 1). When these nucleoli are rapidly isolated and placed in distilled water (adjusted to pH 7–9), the cortices disperse, allowing the compact cores to expand. Unwound cores prepared for electron microscopy are seen to consist of a thin axial fiber in the form of a circle with discrete, repeating matrix segments (Fig. 2). The matrix segments consist of 80–100 slender fibrils arranged in a gradient of shorter to longer lengths. The fibril gradients of all matrices in a single core show the same polarity. The number of matrix units per axis reflects the size of the nucleolus from which the core was derived and ranges from an observed low of five units to an estimated 1,000 or more.

Unless stretched during isolation, the matrix units, which are separated from neighboring units by matrix-

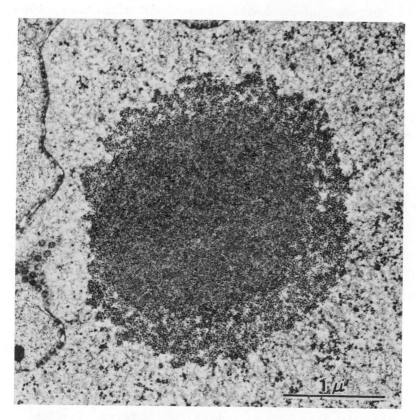

Figure 1. Thin section of an extrachromosomal nucleolus in an oocyte on the spotted newt, *Triturus viridescens*.

free segments, are 2.2—2.4 μ long. Matrix-free (spacer) segments usually are about one-third as long as a
matrix unit but may range in length to ten or more
times that of a matrix unit. Enzymatic digestion and
electron staining procedures demonstrate that the con-
tinuity of the core axis within both matrix units and
spacer segments is maintained by DNA, and that the matrix fibrils consist of ribonucleoprotein (RNP).
     Electron microscopic autoradiography shows that RNA
synthesis is limited to the matrix units, and several
lines of evidence strongly suggest that this RNA consists of rpRNA molecules. Most, if not all, of the
DNA amplified in the early oocyte is contained in the

extrachromosomal nucleoli (Evans and Birnstiel, 1968);
saturation values for rRNA hybridization with this DNA
are equivalent to those for nucleolar genes from somatic cells (Gall, 1969); and combined biochemical (Gall,

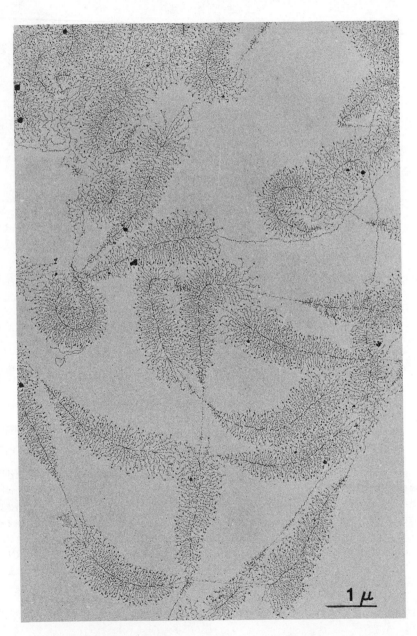

1 μ

1966b) and autoradiographic (Lane, 1967; Macgregor, 1967) data indicate that the synthesis of 40S amphibian rpRNA molecules in oocytes is localized in the core regions of the extrachromosomal nucleoli. Consequently, we conclude that the DNA segment forming the backbone of each matrix unit is a gene coding for rpRNA, and that the gradient of matrix fibrils contain rpRNA molecules in progressive stages of synthesis. If matrix units are slightly stretched during preparation, RNA polymerase molecules can be visualized as granules on the gene at the attachment site of each RNP fibril (Fig. 3).

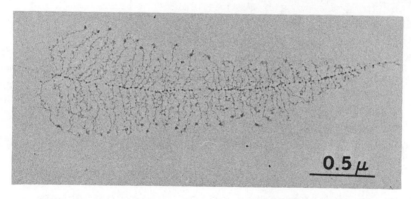

Figure 3. Portion of a stretched nucleolar gene showing individual RNA polymerase molecules on the gene at the base of each nascent RNP fibril. *T. viridescens.*

Figure 2. (Opposite page.) Portion of a dispersed nucleolar core showing repeating matrix units composed of gradients of thin fibrils. *T. viridescens.* Nuclear contents were dispersed in water adjusted to pH 9, then centrifuged onto carbon-coated grids (3-7 min, ~2500 x *g*) through 0.1 M sucrose plus 10% formalin (adjusted to pH 9). The grids were rinsed in 0.4% Kodak Photoflo, dried, stained for 1 min with 1% phosphotungstic acid in 70% ethanol (unadjusted pH ~2), and dried. In some cases, preparations are double-stained for 1 min with 1% uranyl acetate in 70% ethanol following the phosphotungstic acid staining.

The molecular weight of the amphibian rpRNA molecule is about 2.6×10^6 daltons (Loening, Jones and Birnstiel, 1969; Perry *et al.*, 1970), and about 2.6 μ of B-conformation DNA would be required to encode this RNA. Unstretched genes are about 0.2—0.4 μ shorter than this, suggesting that the simultaneous activity of the closely packed RNA polymerases may foreshorten the gene from its B-conformation length.

When stretched the full length of its polynucleotide backbone, an rpRNA molecule should be about 5.2 μ long. The fibrils at the termination ends of the rpRNA genes (which presumably contain essentially completed rpRNAs), however, are only about 0.5 μ long, or one-tenth the full length of the rpRNA molecule. This reduction presumably results from coiling of the rpRNA molecule within protein that becomes associated with the nascent RNA.

Electron-dense granules abruptly appear on the ends of the RNP fibrils a short length from the initiation site of each gene. The role of these granules is unknown, as is the function of the spacer segments intercalated between the repetitious rpRNA genes.

Lampbrush Chromosomes

Lampbrush chromosomes of amphibian oocytes are in an extended diplotene stage of meiosis. Figure 4 is a phase-contrast photograph of a portion of a chromosome pair isolated from an oocyte of the spotted newt, *Triturus viridescens*. The main axis of each homologous chromosome contains two chromatids and consists of a series of Feulgen-positive granules, most of which exhibit pairs of Feulgen-negative lateral loops. Enzymatic digestion and staining procedures show that a lateral loop consists of an axis of DNA coated with an RNP matrix. The matrix forms a single thin-to-thick gradient along the entire loop axis. Autoradiographic studies demonstrate that RNA synthesis occurs on essentially every loop during oogenesis, and that newly synthesized protein becomes associated with the nascent RNA (Gall and Callan, 1962).

Figure 5 is an electron micrograph of a portion of a lateral loop close to the initiation site for RNA

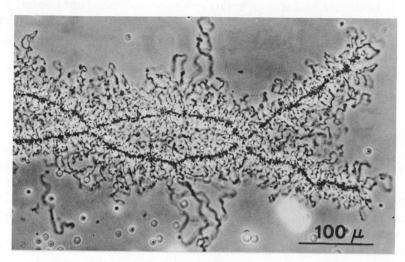

Figure 4.   Phase-contrast photograph of portion of a pair of lampbrush chromosomes isolated from an oocyte of *T. viridescens*.
(Courtesy of Dr. J. G. Gall, Yale University.)

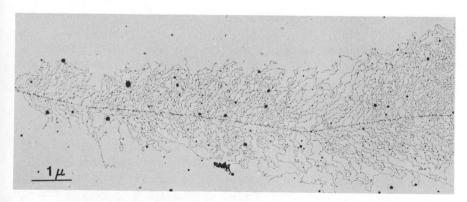

Figure 5.   Electron micrograph of a portion
of a lampbrush chromosome loop near the initiation site for RNA synthesis. *T. viridescens*.

synthesis.   As with the nucleolar genes described previously, the RNP matrix consists of individual slender

fibrils attached to RNA polymerase molecules closely spaced on a deoxyribonuclease-sensitive axis. In this case, however, an active locus is many times longer than a nucleolar gene, and the RNP fibrils also are commensurately longer. Again, the nascent RNA molecules appear to be considerably foreshortened from their full length by associated protein.

The lateral loops of lampbrush chromosomes of *T. viridescens* average about 50 μ in length (Gall, 1956), and a few loops range up to 300 μ. To date, we have not been able to observe critically single RNP fibrils at thick ends of lateral loops, but we have measured RNP fibrils over 25 μ long at intermediate points along loop axes. Since the nascent RNA is foreshortened at least to one-fifth its actual length by associated protein, such fibrils presumably contain RNA molecules well over 100 μ long (or over 50 x 10^6 daltons molecular weight). Thus, it appears that synthesis of extremely long RNA molecules occurs on each of the thousands of lateral loops of lampbrush chromosomes of amphibian oocytes. However, the genetic role, if any, of such molecules remains to be elucidated.

BACTERIAL GENES

The structural aspects of active bacterial genes have been studied using lysozyme-treated, osmotically shocked cells prepared for electron microscopy by the techniques described above (legend, Fig. 2) for the visualization of amphibian oocyte genetic activity. Figure 6 shows a low-magnification micrograph of a burst cell. Portions of the bacterial chromosome and associated structures are extruded from the cell by osmotic shock. In such preparations, it is possible to identify active structural genes—i.e., genes coding for messenger RNA (mRNA) molecules that subsequently direct the amino acid sequences of proteins, and genes actively synthesizing the RNAs that are structural parts of ribosomes (rRNAs).

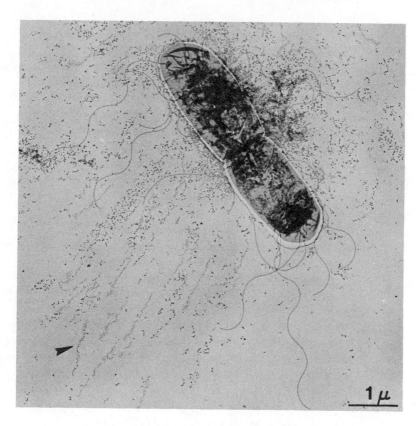

Figure 6. Electron micrograph of an osmotically shocked *Salmonella typhimurium* cell,
showing extruded genome with attached polyribosomes and active ribosomal RNA genes
(arrow).

Structural Genes

In microorganisms, the synthesis of mRNA is closely
coupled in time and space with the translation of messengers by polyribosomes (Stent, 1964); therefore, active structural genes are segments of the chromosome
with attached polyribosomes. Figure 7 illustrates
such a region. Ribosomes are attached at irregular
intervals to the genome through mRNAs, singly and as
polyribosomes, forming a short-to-long polysome gradient. The length of this active region (3 μ) is between

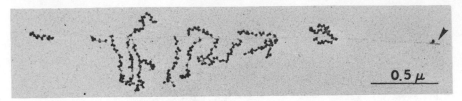

0.5 μ

Figure 7. Genetically active segment of an
E. coli chromosome with attached polyribosomes. Arrow indicates an RNA polymerase
on the presumptive initiation site of the
operon.

those estimated for the five-gene tryptophan operon
(2.3 μ; Imamoto and Yanofsky, 1967) and for the nine-
gene histidine operon (3.75 μ; Benzinger and Hartman,
1962). Therefore, we conclude that Figure 7 illustrates an active, though not specifically identified
Escherichia coli operon.

Assuming that all RNA polymerases transcribe at
about the same rate, the irregular spacing between the
polyribosomes in Figure 7 suggests that initiation of
transcription is aperiodic. Occasionally, a few shorter polyribosomes are seen distal to the longest polysome of a gradient. These shorter polysomes could
reflect mRNA degradation. There is biochemical evidence that degradation proceeds from the 5'-PO4 to the
3'-OH end of the messenger (Kuwano *et al.*, 1969; Morikawa and Imamoto, 1969; Morse *et al.*, 1969). Since no
free polyribosomes are seen in our preparations, it is
likely that the 5'-to-3' degradation occurs while
mRNAs are attached to the genome.

The first ribosome of a polyribosome is closely
associated with the genome, and granules (~80 Å in diameter) are seen on the DNA at each such association
site (Figs. 7 and 8). These granules are almost certainly RNA polymerase molecules that were actively
transcribing the bacterial chromosome at the time of
isolation. In negatively stained preparations (Fig.
8), these granules are quite evident.

Although it is not difficult to locate genetically
active segments of the bacterial genome, the active DNA

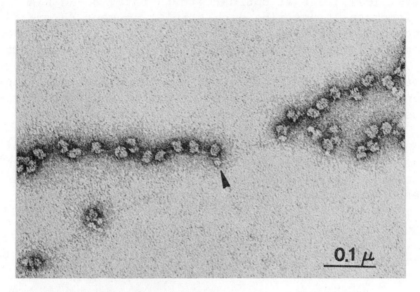

Figure 8.   A negatively stained segment of
E. coli chromosome showing polyribosomes
attached to chromosome by RNA polymerase
molecules (arrow).

is only a small portion of the DNA extruded from a
burst cell. This observation agrees with the DNA–RNA
hybridization data of Kennel (1968), which indicate
that only a small percentage of the *E. coli* chromosome
is active in transcription at any one instant. We
have, however, observed small granules about the size
of the RNA polymerases of active loci distributed at
random along inactive chromosome segments, suggesting
that much of the nontranscribing RNA polymerase in a
cell is bound to the genome, awaiting the proper signal and factors required for initiation.

Ribosomal RNA Genes

In the *E. coli* chromosome, there are about six
genes coding for each of the three species of rRNA
(Purdom, Bishop and Birnstiel, 1970). Biochemical and
genetic data indicate that these genes are arranged on
six chromosomal segments, which contain one 16S, one
23S, and one 5S cistron, in that order (Doolittle and

Pace, 1971); and at optimal growth rate, 80-90 RNA
polymerases transcribe each triplet segment (Bremer
and Yuan, 1968; Manor, Goodman and Stent, 1969). The
combined molecular weights of the three RNAs is about
1.7×10^6 daltons; therefore, about 1.7 μ of B-conformation duplex DNA is required to encode each rRNA segment. Also, ribosomal proteins rather than mature ribosomes associate with nascent rRNA chains (Mangiarotti
et al., 1968).

These data can be used to predict the structure of
the active rRNA segments: 1.7 μ of DNA covered by many
closely spaced RNA polymerases, with an RNP fibril attached to each polymerase. Figure 9 shows an active

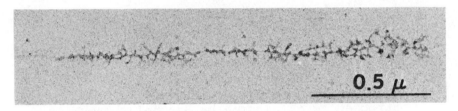

Figure 9. An rRNA region of the *E. coli*
genome, showing polymerase activity on the
16S rRNA (shorter gradient of fibrils) and
23S rRNA cistrons.

region of the genome, with fibril-containing matrices
rather than polyribosomes attached. Each matrix
(about 1.3 μ long) is composed of 60-70 closely spaced
RNP fibrils, which form two short-to-long fibril gradients. The absence of such regions in preparations
of a temperature-sensitive mutant of *E. coli* grown at
42°, when no RNA synthesis is detected biochemically
(Atherly, unpublished data), supports the conclusion
that these regions are active rRNA genes.

Since the rRNA segments measure only 1.3 μ, rather
than the expected 1.7 μ, it is possible that DNA which
is being transcribed by many closely spaced RNA polymerases is foreshortened from the B-conformation length,
as we have suggested for the nucleolar genes of amphibian oocytes.

Occasionally, some rRNA regions can be observed
with a few polyerases bound distal to the second RNP

fibril gradient. Since the 5S rRNA gene is so small
that it essentially would be covered by two or three
polymerases, these granules might be polymerases that
were active in 5S gene transcription at the time of
isolation.

The relative lengths of the two fibril gradients
agree with the fact that the 16S rRNA has a molecular
weight one-half that of the 23S rRNA. The existence
of two fibril gradients suggests a separate initiation
site for the transcription of each of these closely
linked cistrons. If this is true, then with time
after addition of a drug, such as rifampin, which
selectively inhibits the initiation of transcription
(Lill *et al.*, 1969), both cistrons should simultaneously lose fibrils from the putative initiation ends. Alternatively, if there is a single initiation site at
the proximal end of the 16S gene, then the 16S gradient should become shorter and disappear with time after
drug addition, while the 23S gradient should remain intact until 16S transcription has ceased.

Figure 10 shows that 80 sec after rifampin is added,

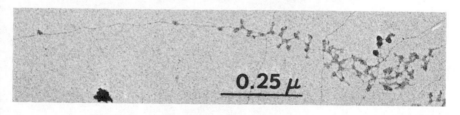

0.25 μ

Figure 10. An rRNA locus 80 sec after rifampin treatment. The gradient is essentially cleared of polymerases. *E. coli.*

a nearly complete 23S gradient remains, although there
is no evidence of 16S transcription. Such observations
argue for a single initiation site for the transcription of these two cistrons and suggest that the 16S
rRNA molecule is released from the polymerase, which
then continues along the DNA to transcribe the 23S
gene. This represents direct support for similar conclusions drawn from the biochemical experiments of
Pato and von Meyenburg (1970) and Kossman, Stomato and
Pettijohn (1971).

Although a closely adjacent arrangement of the six rRNA segments of *E. coli* has been suggested by Yu, Vermeulin and Atwood (1970), we find regions of the genome up to and greater than 10 µ long with a single rRNA region bracketed by structural gene activity (Fig. 11).

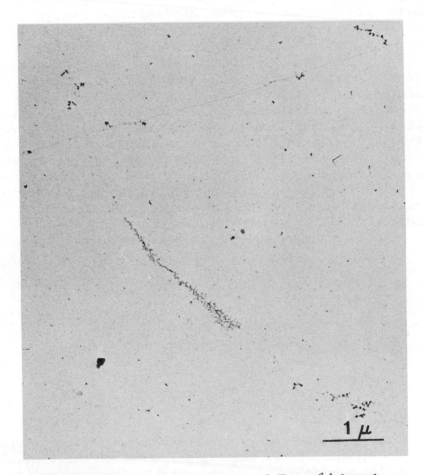

Figure 11.   An rRNA locus of *E. coli* bracketed by polyribosomes attached to active structural gene regions.

This observation is in agreement, rather, with a more scattered arrangement of the genes, as suggested by Purdom, Bishop and Birnstiel (1970) and Birnbaum and Kaplan (1971).

CONCLUSIONS

Using relatively simple preparative techniques, we have been able to visualize some ultrastructural aspects of genetic activity in both eukaryotic and prokaryotic cell types. In both cases, the genetic apparatus can be isolated with relatively simple manipulations. Similar studies with eukaryotic cells that contain relatively small nuclei will require novel isolation techniques. However, progress in this direction should make it possible to study molecular cytogenetics at the ultrastructural level.

ACKNOWLEDGMENT

This research was sponsored by the U.S. Atomic Energy Commission under contract with the Union Carbide Corporation.

REFERENCES

Benzinger, R., and Hartman, P. E. (1962). *Virology 18*, 614.

Birnbaum, L. S., and Kaplan, S. (1971). *Proc. Nat. Acad. Sci. USA 68*, 925.

Bremer, H., and Yuan, D. (1968). *J. Mol. Biol. 38*, 163.

Callan, H. G., and Lloyd, L. (1960). *Phil. Trans. Roy. Soc. London Ser. B. Biol. Sci. 243*, 135.

Doolittle, W. F., and Pace, N. R. (1971). *Proc. Nat. Acad. Sci. USA 68*, 1786.

Evans, D., and Birnstiel, M. L. (1968). *Biochim. Biophys. Acta 166*, 274.

Gall, J. G. (1956). *Brookhaven Symp. Biol. 8*, 17.

Gall, J. G. (1966a). In *Methods in Cell Physiology*, D. M. Prescott, ed., *Vol. II* (New York: Academic Press), p. 37.

Gall, J. G. (1966b). *Nat. Cancer Inst. Monogr. 23*, 475.

Gall, J. G. (1969). *Genetics (Suppl.) 61*, 121.

Gall, J. G., and Callan, H. G. (1962). *Proc. Nat. Acad. Sci. USA 48*, 562.

Imamoto, F., and Yanofsky, C. (1967). *J. Mol. Biol. 28*, 1.

Kennel, D. (1968). *J. Mol. Biol. 34*, 85.

Kossman, C. R., Stomato, T. D., and Pettijohn, D. E. (1971). *Nature New Biology 234*, 102.

Kuwano, M., Kwan, C. N., Apirion, D., and Schlessinger, D. (1969). *Lepetit Colloq. Biol. Med. 1*, 222.

Lane, N. J. (1967). *J. Cell Biol. 35*, 421.

Lill, H., Lill, U., Sippel, A., and Hartmann, G. (1969). *Lepetit Colloq. Biol. Med. 1*, 55.

Loening, U. E., Jones, K. W., and Birnstiel, M. L. (1969). *J. Mol. Biol. 45*, 353.

Macgregor, J. C. (1967). *J. Cell Sci. 2*, 145.

Mangiarotti, G., Apirion, D., Schlessinger, D., and Silengo, L. (1968). *Biochemistry 7*, 456.

Manor, H., Goodman, D., and Stent, G. S. (1969). *J. Mol. Biol. 39*, 1.

Miller, O. L., Jr. (1965). *Nat. Cancer Inst. Monogr. 18*, 79.

Miller, O. L., Jr., and Beatty, B. R. (1969a). *Science 164*, 955.

Miller, O. L., Jr., and Beatty, B. R. (1969b). *Genetics (Suppl.) 61*, 133.

Miller, O. L., Jr., and Beatty, B. R. (1969c). *J. C Cell. Physiol. 74 (Suppl. 1)*, 225.

Miller, O. L., Jr., Hamkalo, B. A., and Thomas, C. A., Jr. (1970). *Science 169*, 392.

Miller, O. L., Jr., Beatty, B. R., Hamkalo, B. A., and Thomas, C. A., Jr. (1970). *Cold Spring Harbor Symp. Quant. Biol. 35*, 505.

Morikawa, N., and Imamoto, F. (1969). *Nature 223*, 37.

Morse, D., Mosteller, R., Baker, R., and Yanofsky, C. (1969). *Nature 223*, 40.

Pato, M., and von Meyenburg, K. (1970). *Cold Spring Harbor Symp. Quant. Biol. 35*, 497.

Perry, R. P., Cheng, T., Freed, J. J., Greenberg, J. R. R., Kelly, D. E., and Tartof, K. D. (1970). *Proc. Nat. Acad. Sci. USA 65*, 609.

Purdom, I., Bishop, J. O., and Birnstiel, M. L. (1970). *Nature 227*, 239.

Stent, G. S. (1964). *Science 144*, 816.

Yu, M. T., Vermeulen, C. W., and Atwood, K. C. (1970).
   Proc. Nat. Acad. Sci. USA 67, 26.

The Formation of Messenger RNA
in HeLa Cells by Post-Transcriptional
Modification of Nuclear RNA

J. E. Darnell

*Career Research Scientist of
the City of New York*

R. Wall, M. Adesnik

Damon Runyon Memorial Fund

L. Philipson

*The Wallenberg Laboratory
University of Uppsala
Uppsala, Sweden*

INTRODUCTION

The expression of genes in bacterial cells is regulated largely, if not entirely, by controlled transcription of specific segments of DNA (cistrons or operons) followed by the immediate or nearly immediate translation of the resulting mRNA (Miller, p. 181-197). Regulatory proteins which either limit transcription (negative control) or in some way promote transcription (positive control) have now been discovered (for details see Travers, p. 171-179). The imaginative prediction of transcriptive control by Jacob and Monod (1961) followed by its experimental verification in bacteria, has provided at the same time a sensible framework and a shackle for discussions of regulation in mammalian cells. The temptation has not been resisted in many quarters to apply the undiluted Jacob-Monod model to mammalian cells or to eukaryotic cells

in general. This, in spite of suggestions many years
ago (Harris *et al.*, 1963) followed by positive experimental evidence, that there existed a class of nuclear
RNA (HnRNA, heterogeneous nuclear RNA) which was
rapidly synthesized (Harris, 1963; Scherrer, Latham
and Darnell, 1963; Scherrer and Marcaud, 1965; Houssais
and Attardi, 1966; Soeiro, Birnboim and Darnell, 1966)
and, in the majority, degraded in the nucleus without
participating in protein synthesis (Soeiro *et al.*,
1968). If a selected minority of this HnRNA could be
shown to serve as precursor to mRNA in eukaryotic
cells then the possibility of post-transcriptional
regulation of at least some genes must be entertained
(Darnell, 1968). Such a scheme might, of course,
exist in addition to transcriptive control.

The purpose of this paper will be to summon evidence that a fraction of the metabolically active HnRNA
does provide a reservoir from which mRNA is derived in
cultured mammalian cells and to suggest a mechanism
which might regulate such RNA processing.

There is a great deal of work which is consistent
with or which supports the above mentioned pathway of
mRNA formation. First, the initial RNA product in
ribosome biosynthesis by mammalian cells is a high
molecular weight precursor molecule which is specifically cleaved to give rise to the ribosomal RNA found
in the cytoplasmic ribosomes (Darnell, 1968). While
this finding has no necessary relationship to mRNA
formation it clearly demonstrates the capacity of
mammalian cell nuclei to perform specific steps of
post-transcriptional modification of RNA.

There are also two separate lines of recent experi-
mentation which indicate the processing of HnRNA into
mRNA.

Cells transformed by small DNA viruses produce high
molecular weight HnRNA molecules containing virus-
specific sequences while much smaller virus-specific
mRNA molecules are found functioning in cytoplasmic
polyribosomes (Benjamin, 1966; Lindberg and Darnell,
1970; Tonegawa, Walter and Dulbecco, 1970). Recently
it has been found that the HnRNA containing virus-specific regions also contains cellular sequences (Wall
and Darnell, 1971). Such molecules may ultimately

serve as excellent tools for the study of nuclear RNA processing. A second type of experiment in non-transformed cells indicates that a common sequence, albeit an unusual one, is shared by HnRNA and mRNA (Edmonds, Vaughan and Nakazato, 1971; Lee, Mendecki and Brawerman, 1971; Darnell, Wall and Tushinski, 1971). The majority of this paper will be devoted to the rapid advances this latter finding has spurred.

The peculiar sequence in question, a polyriboadenylic acid [poly(A)], of about 200 residues, was discovered a number of years ago (Hadjivassiliou and Brawerman, 1966) as was a nuclear enzyme capable of producing poly(A) (Edmonds and Abrams, 1960). These earlier findings were not properly appreciated until the past year, however, probably for two major reasons. The first reason is quite understandable in retrospect —the poly(A) was not shown to exist as a covalent part of either the HnRNA or of mRNA. In fact, HnRNA had not been clearly characterized at the time of the original work on poly(A). However, the second reason for ignoring the existence of poly(A) does no credit to the community of animal cell biologists and implies that those of us working on animal cells should not always be swayed by results obtained with bacteria. Poly(A) synthesizing enzymes were found in bacteria shortly after they were discovered in mammalian cells (August, Ortiz and Hurwitz, 1962; Gottesman, Canellakis and Canellakis, 1962). Because no physiologic function was ever discovered for the bacterial enzymes and perhaps especially because such activity disappeared after T-even bacteriophage infection (Ortiz *et al.*, 1965) poly(A) synthesis was relegated by the bacterial geneticists and biochemists to the curiosity shelf and largely forgotten. The effect of this turn of events was to essentially inhibit attempts to understand what poly(A) and poly(A) polymerase might be doing in *mammalian* cells.

The rejuvenation of interest in this problem began with several recent reports of poly(A) in various mammalian RNA fractions. Kates and his co-workers found that poly(A) was synthesized by enzymes contained in the "core" of *vaccinia* virions and that poly(A) appeared to comprise a substantial part, about 10%, of

the vaccinia mRNA (Kates and Beeson, 1970; Kates, 1970).
Edmonds and Caramela (1969) found a discrete segment of
poly(A) in the nuclei of tumor cells and Lim and Canellakis (1970) reported a very adenylate-rich ribonuclease-resistant fraction in hemoglobin mRNA. Efforts
then were made in several laboratories to determine
whether the poly(A) was a part of both HnRNA and mRNA
and to develop methods for the rapid and accurate
measurement of poly(A) so that it might be more easily
studied (Edmonds, Vaughan and Nakazato, 1971; Lee,
Mendecki and Brawerman, 1971; Darnell, Wall and Tushinski, 1971; Darnell et al., 1971; Sheldon, Jurale
and Kates, 1971). A summary of some of the salient
findings of these experiments will be given below,
followed by a theory of how poly(A) might be involved
in the regulation of genetic expression. The experiments presented come from our own laboratory but quite
a number of similar experiments have also been performed in the laboratories of Drs. M. Edmonds, G.
Brawerman and J. Kates.

Existence and Measurement of Poly(A)

A suggestion that RNA from growing HeLa cells contained poly(A) sequences was obtained by studying RNA
from cells labeled either with adenosine or uridine.
Samples of mRNA isolated from polyribosomes after
labeling with either of the two nucleosides were
tested for resistance to pancreatic ribonuclease (Fig.
1). Over 98% of the uridine-derived radioactivity was
converted to acid soluble material while 15 to 40% of
the adenosine labeled mRNA was resistant to ribonuclease treatment in moderate ionic strength (~0.1 M or
over) but sensitive to alkaline hydrolysis or to nuclease in low ionic strength (0.015 M or below) (Beers,
1960). It has been shown that this ribonuclease-resistant portion of the mRNA consisted mainly of a segment of poly(A) approximately 150-200 nucleotides long
(Darnell, Wall and Tushinski, 1971; Edmonds, Vaughan
and Nakazato, 1971) and that a segment of this size
was responsible for the ribonuclease in all size
classes of mRNA. It was further demonstrated that the
poly(A) was part of the messenger ribonucleotide chain

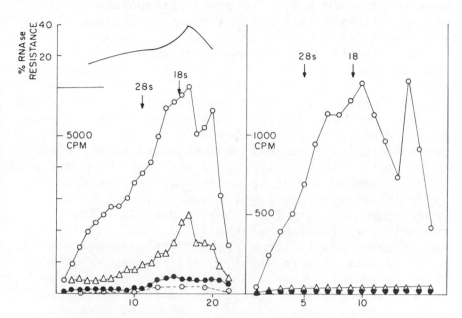

Figure 1. Ribonuclease sensitivity of
polysomal RNA labeled with adenosine or
uridine. HeLa cells treated with a low
concentration of actinomycin (0.05 γ/ml)
to prevent ribosomal RNA synthesis were
labeled for 3 hr either with ^{3}H adenosine
(left panel) or ^{3}H uridine (right panel).
Polyribosomes were isolated by zonal sedimentation and RNA released by sodium dodecyl sulfate before sedimentation through
a sucrose gradient. Samples of the RNA
from each gradient were assayed for acid
precipitable radioactivity after (1) no
treatment,(O), (2) pancreatic RNase digestion, 2 γ/ml, 30 min at 37° in .3 M NaCl,
0.3 M Na citrate (●), (3) same RNase treatment in 0.015 M NaCl, 0.0015 M Na citrate
(△), (4) 0.3 N KOH digestion for 18 hr at
37° (O). The insert to the left panel
plots the percent of RNase resistance
throughout the gradient. (For further detail see Darnell, Wall and Tushinski, 1971.)

because it could not be released by exposure to dimethyl sulfoxide, a treatment known to disrupt hydro-
gen bonded structures (Strauss *et al*., 1968).

The relative ribonuclease resistance of poly(A) in
moderate salt concentrations has formed the basis for
most of the methods of isolation of the poly(A) sequences. Figure 2 gives a demonstration of the method
we have used most commonly in our laboratory. ^{32}P-la-
beled HnRNA or mRNA was digested with ribonuclease,
precipitated with ethanol and subjected to gel electrophoresis. In both the HnRNA and mRNA there was a
ribonuclease resistant sequence which migrated as if
it were approximately 150-200 nucleotides long which
was composed of more than 85-90% adenylic acid. A
much higher percentage of the total radioactivity was
recovered as poly(A) from the mRNA than from the HnRNA.
Base analyses of more highly purified samples of this
RNase resistant sequence have shown the base composition to approach 100% adenylic acid (Hadjivassiliou
and Brawerman, 1966; Edmonds, Vaughan and Nakazato,
1971). This gel electrophoresis assay provided a
quick and easy quantitation of the specific 150-200
nucleotide poly(A) sequence without further purification. We have also taken advantage of the fact that
the poly(A) in a ribonuclease digest can be specifically bound to millipore filters under appropriate
ionic conditions then eluted for gel analysis (Lee,
Mendecki and Brawerman, 1971). This step rids samples
of low molecular weight ribonuclease digestion products or DNA fragments which contaminate RNA samples.
At the conclusion of these preliminary experiments,
it was clear that both HnRNA and various sized polysomal mRNA molecules contain as part of the polynucleotide chain a unit of polyadenylic acid approximately 150-200 nucleotides long.

Flow of Poly(A) from the Nucleus to
the Polyribosomes

The next question attacked was whether the poly(A)
originated in the nucleus or was added to RNA molecules in the cytoplasm. Cells were labeled with ^{3}H
adenosine for various periods of time and HnRNA and

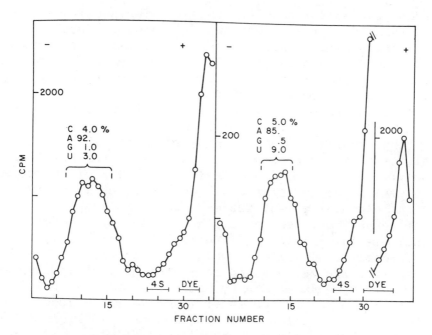

Figure 2. Polyacrylamide gel electrophoresis of RNase resistant fraction of HnRNA
and mRNA. RNA from HeLa cells labeled in
PO_4-free medium for 3 hr with ^{32}P was isolated from detergent cleaned nuclei or from
EDTA releasable fractions of polyribosomes
(Soeiro and Darnell, 1970). The purified
HnRNA and mRNA were digested for 30 min at
37° with 20 γ/ml DNase in 0.1 M NaCl, 0.001
M $MgCl_2$ and 0.01 M Tris followed by addition of 0.01 M EDTA and 2 γ/ml of pancreatic RNase and 5 units/ml of T1 RNase. The
digest was precipitated with ethanol along
with yeast RNA as a carrier, redissolved
and subjected to electrophoresis on 10%
polyacrylamide gels (Darnell *et al.*, 1971).
The gels were sliced and radioactivity
counted as Cerenkov radiation. RNA was extracted from the indicated gel slices, precipitated, hydrolyzed with alkali and base
composition determined.

mRNA assayed for poly(A) content. The general result
of these experiments was that after short label times
there was more total radioactive poly(A) in the HnRNA
than in the cytoplasmic RNA, whereas by label times in
excess of 150 min the balance was in the opposite direction. These experiments are consistent with the
origin of the poly(A) in the nucleus and its subsequent movement to the cytoplasm.

The experiments employing very brief label times
give strong indication that there is no pool of free
poly(A) molecules in the cell. About 91% of the total
poly(A) in cells after a 45 sec label was found in the
high molecular weight nuclear RNA. RNA samples in
this experiment were precipitated with LiCl before
analysis for poly(A). Virtually all of the poly(A)
was recovered in lithium chloride-precipitable RNA
both from the nucleus and the cytoplasm. Poly(A) prepared from cellular RNA by RNase digestion is only
partially precipitated under the same conditions,
thereby demonstrating its association with high molecular weights in the native cellular RNA.

It appears, therefore, that the initial point of
synthesis of poly(A) is in HnRNA. Whether poly(A) was
present in all size nuclear molecules or only in a
restricted size class was also determined. The extracted HnRNA was subjected to sedimentation analysis
and samples greater than 50S (from 30-50S), and less
than 30S were individually examined for the relative
proportion of poly(A) with the finding that even after
such very brief labels there was an equal percentage
of poly(A) in all size classes (Jelinek, Philipson,
Wall and Darnell, in preparation). This would indicate that per molecule of HnRNA, more poly(A) is found
in large molecules. This preference could be a result
of the fact that a large number of the shorter chains
were not completed during the very brief label or alternatively there might be some HnRNA chains in the
shorter size range in which poly(A) is not ever
present, or that some of the shorter chains represent
either true or experimentally induced broken molecules
in which no poly(A) exists. Since the 45 sec label
time is within the synthesis time of the HnRNA molecules [e.d., 45S rpreRNA, 14,000 nucleotides, takes

about 2 min to be made (Greenberg and Penman, 1966),
and polio RNA, 7,000 nucleotides about 1 min (Darnell
et al., 1967)] it seems likely that most of the short
molecules might represent nascent chains the synthesis
of which has not yet reached the 3' terminus.

Molecules greater than 50S which have an average
chain length of perhaps 20,000 nucleotides would be
expected to contain about 5,000 total adenylic acid
residues on the basis of a base composition of 28%
adenylic acid (Darnell, 1968). If each HnRNA contained one unit of poly(A) about 180 nucleotides long
[the length reported by two groups for the poly(A) sequence (Kates, 1970; and Brawerman, personal communication)] then the percent of adenosine label recovered
as poly(A) from HnRNA should be 3.1%. This assumes
very little or no label in guanylic acid which has
been experimentally verified (Darnell *et al.*, 1971).
In our experiments recovery of adenosine label in
poly(A) has ranged from 1.6 to 2.6% indicating that at
least half the largest HnRNA chains do contain a
poly(A) sequence just after synthesis (Darnell, Wall
and Tushinski, 1971). In summary, it appears that the
HnRNA may serve as a receptor for the addition of the
poly(A) and that molecules of various sizes contain
the same unit of poly(A).

How Does Poly(A) Become Attached to
Nuclear RNA?

Figure 3 gives two possible modes of addition of
poly(A) to the HnRNA. Either scheme might result in
poly(A) being only at the 3' terminus of the RNA
molecule, which location for the poly(A) is supported
by the results of Kates (1970) and Brawerman (private
communication). These workers have shown that when
adenosine labeled poly(A) is subjected to alkaline
hydrolysis one adenosine residue is recovered for
every 180 AMP residues. Thus, the poly(A) segment has
a free hydroxyl on the 3' end and must be at that
terminus of the RNA molecule.

In order to distinguish between the transcription
of poly(A) by RNA polymerase from a section of the DNA
made up of polydeoxy(T)—polydeoxy(A) as opposed to the

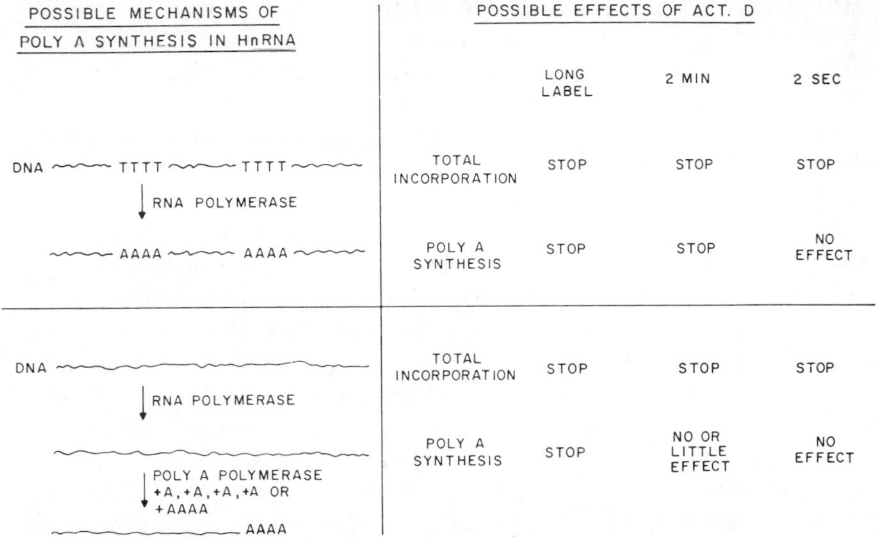

POSSIBLE MECHANISMS OF POLY A SYNTHESIS IN HnRNA | POSSIBLE EFFECTS OF ACT. D

	LONG LABEL	2 MIN	2 SEC
DNA ～～ TTTT ～～ TTTT ～～ ↓ RNA POLYMERASE ～～ AAAA ～～ AAAA ～～			
TOTAL INCORPORATION	STOP	STOP	STOP
POLY A SYNTHESIS	STOP	STOP	NO EFFECT
DNA ～～～～ ↓ RNA POLYMERASE ～～～～ ↓ POLY A POLYMERASE +A,+A,+A,+A OR +AAAA ～～～ AAAA			
TOTAL INCORPORATION	STOP	STOP	STOP
POLY A SYNTHESIS	STOP	NO OR LITTLE EFFECT	NO EFFECT

Figure 3. Two possible modes of the synthesis of poly(A) portion of HnRNA and predictions about the effect of actinomycin on each.

terminal addition of the poly(A) by another mechanism, we employed the drug actinomycin D. This drug stops all RNA-dependent DNA synthesis in cells by binding to deoxyguanylate residues scattered throughout the DNA (Reich and Goldberg, 1964; Sobell *et al.*, 1971). Actinomycin D inhibition for long periods of time either before or during exposure to labeled nucleosides would reduce total incorporation as well as any labeling of poly(A), because either the RNA polymerase would be blocked or the potential receptors for poly(A) namely, recently finished HnRNA molecules, would be depleted by a long actinomycin treatment. Likewise, the use of actinomycin for extremely short periods of time might lead to an erroneous conclusion since actinomycin binds only to deoxyguanylate residues and if an RNA polymerase were transcribing a region of polydeoxy(T)—polydeoxy(A) then the movement of the enzyme would not be affected by actinomycin. The appropriate experimental design which takes these difficulties

into account involves exposure of cells to actinomycin
treatment for a very short time followed by a brief
period of label (Darnell *et al.*, 1971). The ideal
situation would be to perform the entire experiment
for a time equal to the synthesis time of the RNA
molecules in question, i.e. for a time which allows
RNA polymerase to move over 10-20,000 base pairs.
Since the heterogeneous nuclear RNA is composed of
molecules predominantly larger than 7,000 nucleotides
[synthesis time of 1 min, and perhaps even longer than
that for 45S ribosomal precursor RNA (14,000 nucleotides—approximately 2 min synthesis time)], cells
were treated with actinomycin for 1 min followed by a
2 min label period with ^{3}H adenosine. The HnRNA was
extracted and assayed for total radioactivity and incorporation into poly(A). The results given in Table
1 show that there is a much greater depression of
total RNA synthesis than there is of poly(A) synthesis.

TABLE 1

Experimental conditions			HnRNA		Poly(A) from HnRNA	
-1.5 min	0	2 min	Total CPM	% Reduction by actinomycin	CPM	% Reduction by actinomycin
—	^{3}H adenosine	stop	3.2×10^5		2700	
Actinomycin D 7.5 μg/ml	^{3}H adenosine	stop	3.6×10^4	89%	1350	50%

This is in accord with the second model of the addition of poly(A)—i.e., terminal (3') addition to finished HnRNA molecules by a second enzymatic step. The
enzyme discovered by Edmonds and Abrams (1960) could
perform such a modification. These experiments with
actinomycin (Darnell *et al.*, 1971), though not deci-
sive, are strong evidence against the necessity for
progressive movement of RNA polymerase along the chain
of DNA in order for poly(A) synthesis to occur.

Longer periods of treatment with actinomycin do reduce
the synthesis of poly(A) which is in accord with
either model.

An additional reason for believing that the poly(A)
is added post-transcriptionally comes from experiments
with cells infected by adenovirus (Philipson *et al.*,
1971). This DNA virus multiples in the cell nucleus
where large virus-specific RNA molecules are transcribed and converted to smaller mRNA which functions
in the cytoplasm (Green *et al.*, 1970; Parsons and
Green, 1971). Both nuclear and cytoplasmic virus-
specific RNA contains poly(A) sequences which are very
similar if not identical to poly(A) sequences in cellular RNA molecules (Philipson *et al.*, 1971). Cellular DNA has a very low capacity to hybridize isolated
poly(A) and viral DNA will not bind poly(A) at all.
Thus, it appears likely that the poly(A) sequences do
not arise as a result of transcription from DNA but
must be added post-transcriptionally (Philipson *et al.*,
1971).

Cordycepin (3'-deoxyadenosine) has also been very
useful in elucidating the mechanism of poly(A) synthesis in addition to suggesting some role for poly(A)
in the successful conversion of HnRNA to mRNA. This
drug is known to cause premature termination of RNA
chains *in vitro* (Klenow and Frederiksen, 1964; Shigeura
and Boxer, 1964). Work in Penman's laboratory has
shown that *in vivo* cordycepin stops ribosomal precursor
RNA synthesis almost completely and immediately, but
does not appear to affect HnRNA formation greatly, if
at all (Siev, Weinberg and Penman, 1969; Penman, Rosbash and Penman, 1970). When it was found that the
mRNA labeling was severely limited by cordycepin treatment, whereas HnRNA formation was not, it was concluded that HnRNA has no relationship to mRNA (Penman,
Rosbash and Penman, 1971). We have examined the affect
of cordycepin on poly(A) synthesis and are forced to a
different conclusion. Cordycepin markedly inhibits the
formation of poly(A) and therefore in cordycepin-
treated cells, labeled poly(A) never appears in mRNA,
at least not as segments 200 nucleotides long (Fig. 4).
This inhibition of poly(A) synthesis is associated
with the drastic reduction in the total incorporation

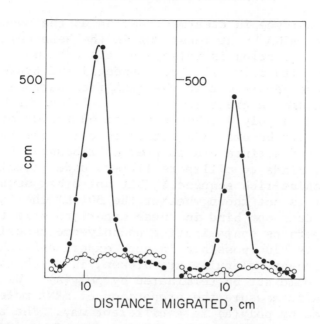

Figure 4.   Electrophoresis of RNase resistant portions of HnRNA and mRNA from normal
and cordycepin treated cells.   HeLa cells
were labeled for 30 min with ^{3}H adenosine
in the presence or absence of 50 γ/ml of
cordycepin (3' deoxyadenosine).   HnRNA and
mRNA from each culture was prepared and
poly(A) assayed by gel electrophoresis as
described in Figure 2.   Left panel, mRNA;
right panel, HnRNA; (●), control; (O), cordycepin treated.

of label into mRNA.   The demonstration of the effect
of cordycepin on cellular RNA is paralleled by results
with adenovirus-specific RNA (Philipson *et al.*, 1971).
The production of virus-specific mRNA which has also
been shown to be terminated by poly(A), is also de-
pressed in the presence of cordycepin.   These experiments indicate that a close association exists between
proper poly(A) addition and subsequent appearance of
mRNA in polysomes.   They do not necessarily indicate a
direct role for poly(A) in processing but it does seem
clear that proper mRNA processing and transport and
poly(A) addition are in some way associated.

Since cordycepin treatment decreases the appearance of labeled mRNA by at least 80% in the HeLa cell cytoplasm, the question is raised—do all mRNA molecules terminate with poly(A)? Lee, Mendecki and Brawerman (1971) have reported that the mRNA from sarcoma 180 cells contains a great number of molecules which are terminated in poly(A), because from 60 to 70% of the labeled polyribosomal RNA formed in cells treated by low doses of actinomycin to prevent ribosomal RNA labeling, binds to millipore filters under conditions where adenine-rich sequences, but not other sequences, bind. It is not known whether the 30% of the labeled RNA that does not bind in these experiments is truly mRNA or perhaps contaminating nonpolysome associated RNA. It is also possible in such experiments that some messenger molecules are broken, in which case one portion would not be terminated by poly(A). We have attempted to measure the proportion of mRNA molecules terminated by poly(A) in a different way. The average size of the mRNA in HeLa cells is approximately 2000 nucleotides (Darnell, 1968). If the base composition of the 2000 nucleotides, excluding the poly(A), were balanced (i.e., C=A=G=U), one would expect approximately 500 internal adenylate residues and approximately 180-200 terminal adenylate residues: about 28% of the total adenosine label as poly(A). Cells were labeled with ^{3}H adenosine for 30 min and mRNA was isolated from purified polyribosomes by EDTA release. The released phenol-extracted RNA was digested and the total digestion products examined on a 20% polyacrylamide gel which separates all digestion products from the large poly(A) segment. The results show that 25% of the radioactivity in the mRNA can, in fact, be recovered in the 9S peak of poly(A) (Fig. 5). Moreover, it is clear from these results that there are no intermediate sequences larger than 8-10 nucleotides which are resistant to pancreatic and T1 RNase. Thus, it does appear that most of the rapidly labeled mRNA molecules in the HeLa cell cytoplasm are in fact terminated with the poly(A) which is 150-200 nucleotides in length. That this is not absolutely the case has been shown by isolating histone mRNA labeled with

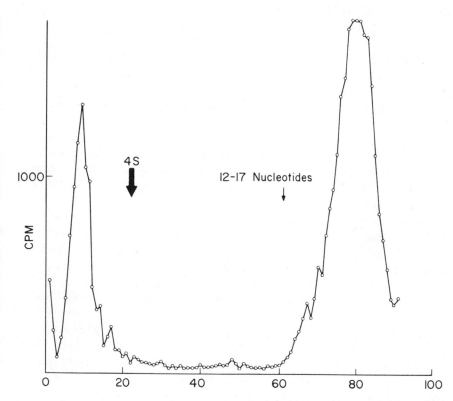

Figure 5. RNase digestion products of mRNA.
Adenosine labeled mRNA was digested with
pancreatic and T1 RNase (see Fig. 2) and
directly subjected to electrophoresis on
20% polyacrylamide gels. Markers present
either in the same or parallel gels were
[14]C HeLa cell tRNA, adenosine-rich ^{32}P or
oligonucleotide from reovirus, 12-17 nucleotides (courtesy of Dr. A. Shatkin).

adenosine. This particular mRNA species contains no
poly(A) at all, either large of small pieces. The
finding with histone mRNA will be reported in detail
elsewhere.

How is Poly(A) Related to Signals
that Exist in the Genome?

For the past several years many laboratories have
been studying DNA sequences in eukaryotic cells which
differ from the majority of the DNA by rapid reassociation after denaturation (Britten and Kohne, 1968;
McCarthy, 1968; Walker, 1968; Britten and Davidson,
p. 5-27). The interpretation of this work is that
such rapidly reannealing sequences are repeated copies
of similar or related sequences. In HeLa cells, some
of these "reiterated" regions are transcribed into RNA
which appears both in the HnRNA and in the mRNA
(Pagoulatos and Darnell, 1970; Soeiro and Darnell,
1970). However, there are some sequences in HnRNA
which hybridize more rapidly than any sequences which
exist in mRNA (see Fig. 6, and Darnell and Balint,
1970) indicating that some of the nuclear fraction
never appears in the cytoplasm. During the examination of the base composition of such rapidly hybridizing RNA, we were led into a study of the existence of
poly(A) in HeLa cell RNA fractions (Darnell, Wall and
Tushinski, 1971). Table 2 shows the base composition
of HnRNA and of hybrids between HnRNA and HeLa cell
DNA. Both have similar base compositions which resemble the overall base composition of DNA. Such is
not the case for hybrids from the cytoplasmic mRNA.
Here the hybridized RNA was found to contain a very
high content of adenylic acid. It should be pointed
out that the poly(A) itself under these conditions
(65°, 2 x SSC) will not hybridize to the DNA and
therefore it appears that there is a rapidly hybridizing sequence within the messenger which has some intimate relationship with the poly(A). Another type
of experiment which verifies this conclusion has been
recently performed. The rate of hybridization of
adenosine-labeled RNA is 2.5-5 times faster than that
of uridine-labeled mRNA (Fig. 7). This more rapid
rate of hybridization of the adenosine-labeled material is abolished if the hybrids are scored after treatment with T2 ribonuclease which will attack poly(A) at
high salt concentrations. Apparently, there are
labeled sequences of unpaired adenine-containing ma-

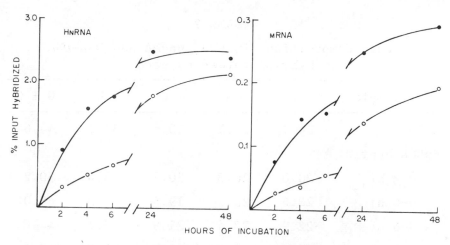

Figure 6.   Hybridization of HnRNA and mRNA.
[3]H uridine labeled mRNA and HnRNA were
prepared from HeLa cells and hybridized
either at 65° in 0.3 M NaCl (0.03 M Na
citrate, 2 x SSC) or at 45° in the same
buffer containing 30% formamide to 20 γ of
HeLa cell DNA, immobilized on a nitrocellulose filter.   Left panel, HnRNA; right panel
panel, mRNA; (●), 2 x SSC at 65°; (O),
2 x SSC with formamide 30%. (See Darnell
and Balint, 1970, for details.)

terial in the hybrids when the mRNA has been labeled
with adenosine, which is not the case when the RNA is
labeled with uridine.   Thus sequences exist in mRNA
which hybridize more rapidly than the total RNA within
the messenger, but somewhat more slowly than the
majority of the HnRNA.   These results with mRNA explain the finding that rapidly hybridizing HnRNA sequences do not result in hybrids with a high adenylate
content since fewer rapidly hybridizing nuclear
regions are neighbors to poly(A).
perhaps calls for the addition of poly(A).   HnRNA containing such sequences appears to be transcribed as
various sized molecules based on the finding that even

TABLE 2

Base composition of mRNA, HnRNA and RNA-DNA
hybrids formed by each

RNA sample	C	A	G	U	G + C
HnRNA	24.1	24.1	20.5	31.1	44.6
HnRNA hybrid					
0-3 hr	30.0	26.3	20.2	23.4	50.2
0-6 hr	28.6	25.1	19.4	26.9	48.0
0-6 hr	22.9	27.4	21.9	27.1	44.8
6-24 hr	25.3	28.3	21.9	25.3	47.2
mRNA	24.4	30.1	21.8	23.8	46.2
mRNA hybrid					
0-6 hr	18.7	44.0	17.9	19.4	36.6
0-6 hr	17.8	51.3	16.8	14.1	34.2

^{32}P labeled HnRNA or mRNA was hybridized to a
series of DNA filters containing 40 μg/filter for 3 or
6 hr. The HnRNA remaining in the supernatant was further hybridized from 6 to 24 hr. RNase-resistant hybrids (0.3-1% of input; total cpm in various samples
from 5000 to 50,000) were recovered and base composition was determined after alkaline hydrolysis.

Summary and Model of mRNA Manufacture
in Mammalian Cells

The above results can all be accomodated within the
model given in Figure 8. According to this model,
some genes within the mammalian cells, apparently most
of those which are expressed in cytoplasmic mRNA in
HeLa cells, may be terminated in a repeated sequence
(the rapidly hybridizing portion of mRNA) which serves
to ensure the correct termination of the RNA and/or

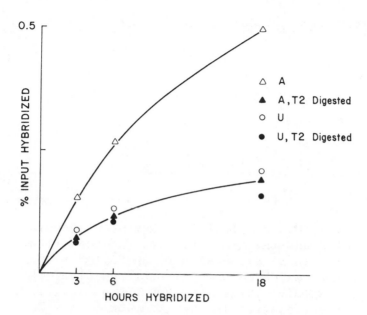

Figure 7. Hybridization of mRNA. Adenosine (A) or uridine (U) labeled mRNA was hybridized at 65° (see Fig. 6) and hybrid measured after the usual pancreatic or T1 ribonuclease digestions (Darnell and Balint, 1970). Other filters were subsequently also treated with T2 ribonuclease, 0.1 unit/ml at 37° in 0.9 M NaCl, 0.01 M EDTA, 0.1 M acetate buffer, pH 4.5.

after a 45 sec label period, poly(A) is found in all sizes of HnRNA. A specific *termination control* would appear a reasonable way for the RNA polymerase to stop immediately after transcription of a gene which is called for at that particular moment in the cytoplasm of the cell. Such a regulatory mechanism requires an analogous type of nucleic acid-protein specificity as is known to exist for bacterial repressor proteins and their specific loci on DNA. The difference in the proposed scheme is that recognition would exist at a termination signal for each gene to be translated.

Upon completion of transcription of the 3' end of such a gene the recognition would be made that poly(A)

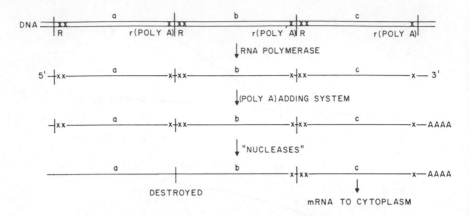

Figure 8. Model of DNA sequences encoding
HnRNA and the transcription and post-transcriptional processing of HnRNA. R = DNA
sequences which when transcribed give rise
to rapidly hybridizing HnRNA sequences that
are not present in the cytoplasm. r = DNA
sequences which when transcribed give rise
to rapidly hybridizing sequences that are
present in cytoplasmic mRNA.

must be added. This might be accomplished by either
modifying the RNA polymerase so that it becomes a
poly(A) polymerase, or by discharging the RNA polymerase and engaging a poly(A) polymerase. After completion of the poly(A) synthesis, specific nucleolytic
cleavages would be required resulting in the messenger
being preserved and the remainder of the HnRNA including certain rapidly hybridizing sequences would be
destroyed. The mRNA then would find its way into the
cytoplasm through presumably specific channels either
involving binding factors, ribosomes or virtually any
conceivable other model. The advantage to the cell of
this scheme would be that by proper termination at any
point along the large HnRNA, the mRNA could be produced for any included gene which is demanded in the
cytoplasm. Thus, transcription of the particular
block of genes involved in this type of regulation
would not have to be individually regulated in order
to allow the cell the possibility of using any one of

such a group of genes at any time. This is not to say,
of course, that transcriptional control might not also
be possible in such regions with termination and/or
processing control added secondarily.

The final figure (Fig. 9) serves to illustrate that

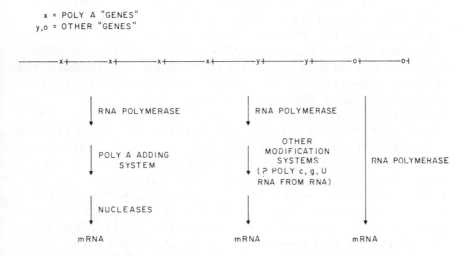

Figure 9. General model of HeLa cell DNA
showing genes from which HnRNA might arise
that required poly(A) addition for post-
transcriptional regulation as well as two
other types of genes where other mechanisms
might obtain. Y indicates other processing
mechanisms; "0" indicates bacterial type
(? operon) regulation.

whereas some mRNA is very likely processed by a system
using some of the elements of the model presented
above, it is clear one must keep an open mind about
proposing a model that purports to account for all
possibilities of mRNA biogenesis in mammalian cells.
For example, some discussion of RNA-dependent RNA
synthesis has been generated by the finding that HnRNA
may hybridize to a portion of mRNA as detected by conversion of labeled mRNA from ribonuclease sensitivity
to insensitivity. This might mean that some HnRNA is
copied into mRNA. Such RNA-dependent mechanisms

clearly exist for two groups of viruses (See Baltimore, p. 245-257).

In addition signals other than those involved in poly(A) addition might be used to serve the function of specifying termination and proper cleavage. Still another possibility as indicated in the third part of Figure 9 is bacterial type regulation, i.e., strictly transcriptional regulation. Such genes when transcribed would be immediately translated into protein. A candidate for this type of control is histone mRNA, which as has been stated earlier, does not contain the poly(A) sequence and has been found to migrate to the cytoplasm almost immediately upon transcription. This latter fact was first found by Schochetman and Perry (personal communcation) and was also observed in this laboratory (Adesnik and Darnell, in preparation).

It is certainly wise at this point to keep a completely open mind and say that virtually any control mechanism which has evolved in any eukaryotic cell might be expected to be available to the most complex nucleus of all, that of the mammalian cell. The excitement at the present time is generated by the apparent certainty that at least one mechanism of mRNA biogenesis in eukaryotes has at last been described and that this mechanism—poly(A) addition to large HnRNA followed by specific cleavage—is very different from mRNA biogenesis in bacteria.

ACKNOWLEDGMENTS

This work was supported by grants from the National Institutes of Health, the National Science Foundation and the American Cancer Society. J. E. Darnell is a Career Research Scientist of the City of New York. R. Wall and M. Adesnik are Damon Runyon Memorial Fund for Cancer Research Fellows, and L. Philipson has an Eleanor Roosevelt International Cancer Fellowship awarded by the International Union Against Cancer.

REFERENCES

August, J. T., Ortiz, P., and Hurwitz, J. (1962). *J. Biol. Chem. 237*, 3786.

Beers, R. F. (1970). *J. Biol. Chem. 235*, 2393.

Benjamin, T. (1966). *J. Mol. Biol. 16*, 359.

Britten, R. J., and Kohne, D. (1968).*Science 161*, 529.

Darnell, J. E. (1968). *Bact. Revs. 32*, 262.

Darnell, J. E., and Balint, R. (1970). *J. Cell. Phys. 76*, 349.

Darnell, J. E., Girard, M., Baltimore, D., Summers, D. F., and Maizel, J. V. (1967). In *The Molecular Biology of Viruses*, J. Colter, ed. (New York: Academic Press), p. 375.

Darnell, J. E., Philipson, L., Wall, R., and Adesnik, M. (1971). *Science* (in press).

Darnell, J. E., Wall, R., Tushinski, R. J. (1971). *Proc. Nat. Acad. Sci. USA 68*, 1321.

Edmonds, M., and Abrams, R. (1960). *J. Biol. Chem. 235*, 1142.

Edmonds, M., and Caramela, M. G. (1969). *J. Biol. Chem. 244*, 1314.

Edmonds, M., Vaughan, M. H., and Nakazato, H. (1971). *Proc. Nat. Acad. Sci. USA 68*, 1336.

Gottesman, M. E., Canellakis, Z. N., and Canellakis, E. S. (1962). *Biochim. Biophys. Acta 61*, 34.

Green, M., Parsons, J. J., Pina, M., Fujinaga, K., Caffier, H., and Landgraf-Leurs, I. (1970). *Cold Spring Harbor Symp. Quant. Biol. 35*, 803.

Greenberg, H., and Penman, S. (1966). *J. Mol. Biol. 21*, 527.

Hadjivassilious, A., and Brawerman, G. (1966). *J. Mol. Biol. 20*, 1.

Harris, H. (1963). *Nature 198*, 184.

Harris, H., Fisher, H. W., Rodgers, A., Spencer, T., and Watts, J. W. (1963). *Proc. Royal Soc. Ser. B 157*, 177.

Houssais, J. F., and Attardi, G. (1966). *Proc. Nat. Acad. Sci. USA 56*, 616.

Jacob, F., and Monod, J. (1961). *J. Mol. Biol. 3*, 318.

Kates, J. (1970). *Cold Spring Harbor Symp. Quant. Biol. 35*, 743.

Kates, J., and Beeson, J. (1970). *J. Mol. Biol. 50*, 19.

Klenow, H., and Frederiksen, S. (1964). *Biochim. Biophys. Acta 87*, 495.

Lee, S. Y., Mendecki, J., and Brawerman, G. (1971). *Proc. Nat. Acad. Sci. USA 68*, 1331.

Lim, L., and Canellakis, E. S. (1970). *Nature 227*, 710.

Lindberg, U., and Darnell, J. E. (1970). *Proc. Nat. Acad. Sci. USA 65*, 1089.

McCarthy, B. J. (1968). *Biochem. Genetics 2*, 37.

Miller, O. (1972). This volume, p. 181.

Ortiz, P., August, J. T., Watanabe, M., Kaye, A. M., and Hurwitz, J. (1965). *J. Biol. Chem. 240*, 432.

Pagoulatos, G. N., and Darnell, J. E. (1970). *J. Mol. Biol. 54*, 517.

Parsons, J. T., and Green, M. (1971). *Virology 45*, 154.

Penman, S., Rosbash, M., and Penman, M. (1970). *Proc. Nat. Acad. Sci. USA 67*, 1878.

Philipson, L., Wall, R., Glickman, R., and Darnell, J. E. (1971). *Proc. Nat. Acad. Sci. USA* (in press).

Reich, E., and Goldberg, I. H. (1964). *Prog. in Nucl. Acid Res. 3*, 183.

Scherrer, K., and Marcaud, L. (1965). *Bull. Soc. Chim. Biol. 47*, 1697.

Scherrer, K., Latham, H., and Darnell, J. E. (1963). *Proc. Nat. Acad. Sci. USA 49*, 240.

Sheldon, R., Jurale, C., and Kates, J. (1971). *Proc. Nat. Acad. Sci. USA* (in press).

Shigeura, H. T., and Boxer, G. A. (1964). *Biochem. Biophys. Res. Commun. 17*, 758.

Siev, M., Weinberg, R., and Penman, S. (1969). *J. Cell Biol. 41*, 510.

Sobell, H. M., Jain, S. C., Sakore, T. D., and Nordman, C. E. (1971). *Nature New Biology 231*, 200.

Soeiro, R., and Darnell, J. E. (1970). *J. Cell Biol. 44*, 467.

Soeiro, R., Birnboim, H. C., and Darnell, J.E. (1966). *J. Mol. Biol. 19*, 362.

Soeiro, R., Vaughan, M. H., Warner, J. R., and Darnell, J. E. (1968). *J. Cell Biol. 39*, 112.

Strauss, J. H., Kelly, R. B., and Sinsheimer, R. L. (1968). *Biopolymers 6*, 793.

Tonegawa, S., Walter, G., and Dulbecco, R. (1970). In *The Biology of Oncogenic Viruses*, L. Silvestri, ed. (Amsterdam: North-Holland Publ. Co.), p. 65.

Travers, A. (1972). This volume, p. 171.

Walker, P. M. B. (1968). *Nature 219*, 228.

Wall, R., and Darnell, J. E. (1971). *Nature New Biology 232*, 73.

Polyadenylic Acid Sequences in the RNA of HeLa Cells and the Virion RNA of Polio and Eastern Equine Encephalitis Viruses

Mary Edmonds, Maurice H. Vaughn, Hiroshi Nakazato

Department of Biochemistry
Faculty of Arts and Sciences
University of Pittsburgh

John A. Armstrong

Department of Epidemiology and Microbiology
Graduate School of Public Health
University of Pittsburgh

Bruce A. Phillips

Department of Microbiology
School of Medicine
University of Pittsburgh

Polyadenylic acid (Poly A) sequences of 150 to 200 nucleotides have recently been shown to be covalently bound to heterogeneous nuclear DNA (HnRNA) and messenger RNA (mRNA) of HeLa cells (Edmonds, Vaughan and Nakazato, 1971; Darnell, Wall and Tushinski, 1971) and mouse sarcoma cells (Lee, Mendecki and Brawerman, 1971). Among the functions suggested for sequences of this homogeneity and length are to serve as recognition sites either for the release of mRNA from the large HnRNAs of the nucleus, or for the transport of mRNA out of the nucleus, or for the translation of mRNA or for some combination of these functions.

The experimental evidence for the existence, location, and size of these Poly A sequences will be reviewed before new measurements of the relative size of these sequences in HeLa nuclei and cytoplasm are presented. In addition, two animal RNA viruses have been examined for Poly A sequences (Armstrong *et al.*, 1972)

227

in an attempt to narrow the range of the possible functions for Poly A sequences noted above. Certain of
these functions would necessarily be restricted particularly in the case of the two selected, Polio and Eastern Equine Encephalitis viruses (EEE), since both
replicate only in the cytoplasm of the host and both
are known to be infectious as purified virion RNA.
This latter fact allows us to consider these virion
RNAs as analogues of host cell mRNA since they depend
on the translational components of the cell for the
production of new virus. The detection of a Poly A
sequence in the virion RNA of these two viruses which
are otherwise dissimilar suggests that Poly A sequences
play a role in the translation process.

POLY A SEQUENCES IN HELA CELLS

The Poly A content of ^{32}P-labeled RNAs was measured
by a technique dependent on the hybridization to poly-
thymidylate cellulose of Poly A sequences released
from RNA by ribonuclease digestion (Edmonds, 1971). A
portion of the Poly A eluted from the cellulose is
used to measure the amount of the ^{32}P-labeled RNA recovered as Poly A, while the remainder recovered by
ethanol precipitation is used for electrophoretic or
compositional analysis (Edmonds, Vaughan and Nakazato,
1971).

A broad distribution of Poly A sequences was found
within two classes of RNA of HeLa cells which are both
characteristically heterogeneous, i.e., the rapidly
labeled, polydisperse HnRNA of the nucleus and the
polyribosome associated mRNA. The Poly A sequences
recovered from each class of RNA were rather homogeneous and similar in length.

Poly A Sequences in the Nucleus

The absence of Poly A sequences from ribosomal RNA
and its precursors was established when none could be
detected in the RNA of nucleoli isolated from the HeLa
nucleus. This simplified interpretation of the distribution of Poly A sequences is shown in Figure 1.

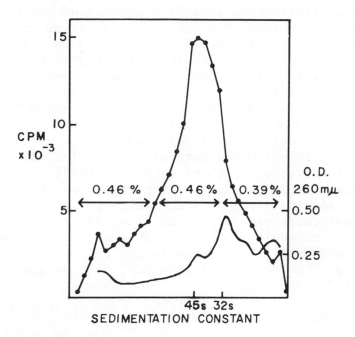

Figure 1. Sedimentation analysis and Poly
A content of total nuclear RNA after a 60
min label with $^{32}P-PO_4$. Details of labeling, and cell fractionation procedures can
be found in Edmonds, Vaughan and Nakazato
(1971). RNA extraction with hot phenol-SDS
is that described by Warner *et al.* (1971).
The RNA was analyzed by sedimentation on a
15-30% sucrose gradient in NETS (0.2% SDS,
0.1 M NaCl, 0.01 M EDTA, 0.01 M Tris, pH
7.4) for 14.5 hr at 19,000 rpm, 18°, in the
SW40 rotor of the Beckman ultracentrifuge.
Fractions were divided into portions for TCA
precipitation and Poly A analysis as described by Edmonds, 1971.

Allowing for the large 32 to 45S labeled peak seen in
this sedimentation profile of total nuclear RNA, it
appeared that the distribution of Poly A sequences
paralleled that of the HnRNA (not shown) which is
characteristically polydisperse with a DNA-like base
composition and a rapid rate of turnover.

Since electrophoretic analysis revealed the Poly A
sequences recovered from all size classes of HnRNA to
be similar in length, the existence of a heterogeneous
population of Poly A molecules including very large
ones appeared unlikely, and suggested rather that HnRNA
molecules contained covalently linked Poly A sequences
which were released by the ribonuclease treatment which
preceded hybridization to Poly T cellulose.

This conclusion was established by a competition
experiment in which the largest ^{32}P-labeled HnRNA molecules (>45S) were denatured in dimethylsulfoxide in the
presence of a large excess of unlabeled and fragmented
HnRNA molecules (<20S). When the large ^{32}P-labeled
HnRNA molecules were separated after denaturation from
the small unlabeled HnRNA fragments by zone sedimentation, there was no reduction in the Poly A content of
the large ^{32}P-labeled RNA (Edmonds, Vaughan and Nakazato, 1971).

Poly A Sequences in the Cytoplasm

A major fraction of the Poly A sequences detected
in the cytoplasm of HeLa cells labeled for 1 hr with
^{32}PO$_4$ was found in the polyribosome portion of cytoplasmic extracts sedimented through sucrose gradients. Table 1 shows that almost 90% of the Poly A
sequences sedimented more slowly than polyribosomes
when the latter were disrupted with EDTA. Similar results were obtained with cytoplasmic extracts treated
with detergents to disrupt membranes. These data suggested an association of the Poly A sequences with
mRNA since for this labeling period it is the predominant labeled RNA species on polyribosomes which is
known to be displaced to the sedimentation region of
monosomes and subribosomal particles by EDTA treatment
(Penman, Vesco and Penman, 1968).

To increase the label in mRNA relative to ribosomal
RNA, a shorter labeling period of 30 min was done.
The ^{32}P distribution of the RNA isolated from these
polyribosomes, as sedimented in sucrose gradients is
shown in Figure 2. This distribution of RNA species
from 6 to 30S, with a maximum about 18S, is that commonly attributed to the mRNA of HeLa cells (Penman,

TABLE 1

Poly A sequences in cytoplasmic RNA

Treatment of cytoplasm	Polyribosomal RNA		Nonpolyribosomal RNA	
	CPM in Poly A	% Total CPM in Poly A	CPM in Poly A	% Total CPM in Poly A
None	1,700	3.2	1,720	0.35
+ EDTA	190	1.8	5,500	0.85

For Poly A analysis the gradient fractions were divided into a pool containing material sedimenting faster than 74S monoribosomes and a pool of material sedimenting at the 74S position plus all slower material. RNA preparation and analysis for Poly A was as described in Figures 1 and 2.

Vesco and Penman, 1968). The Poly A content is shown in Figure 2, as a percentage of the total RNA isolated from pooled fractions from four regions of the gradient. These quantities, ranging from 2.5 to 5% are in agreement with the Poly A content of the total polyribosomal RNA seen in Table 1. Higher values of about 15% are found if polyribosome-bound mRNA is hybridized directly to Poly T cellulose through its covalently bound Poly A sequence. Evidence for such covalent association is derived from the observation that such mRNA retains not only its original sedimentation properties (essentially those of Figure 2), but also an unaltered Poly A content of 15% (Nakazato and Edmonds, 1972).

The Size of Poly A Sequences

A similarity in the electrophoretic mobilities of the Poly A sequences derived from HnRNA species of different sedimentation velocities has already been noted. The mobilities on polyacrylamide gels of the Poly A sequences from polyribosomal RNA and from total nuclear RNA are compared with some well-characterized marker RNA species in Figure 3. The Poly A of both

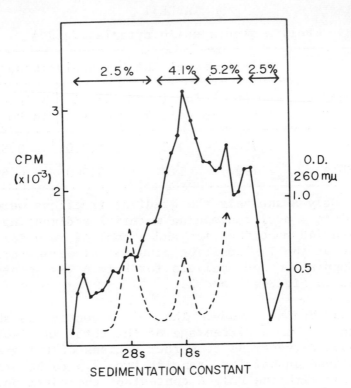

Figure 2.   Sedimentation and Poly A analysis of rapidly labeled polyribosomal RNA. A culture was labeled with ^{32}P-PO$_4$ for 30 min and polyribosomes were prepared as described by Girard *et al.* (1965). The RNA from polyribosomes was extracted, as in the legend to Figure 1, and run on a sucrose gradient in NETS in the SW40 rotor for 18.5 hr at 24,000 rpm, 18°. Aliquots of gradient fractions were taken for TCA precipitation and selected fractions were pooled for Poly A analysis. The Poly A content of the RNA fractions is shown above the UV absorption (---) and radioactivity profiles (-●-) from the gradient.

origins is seen to be relatively homogeneous and similar in length although the Poly A from polyribosomes did appear to move slightly faster. Lengths of 150 to

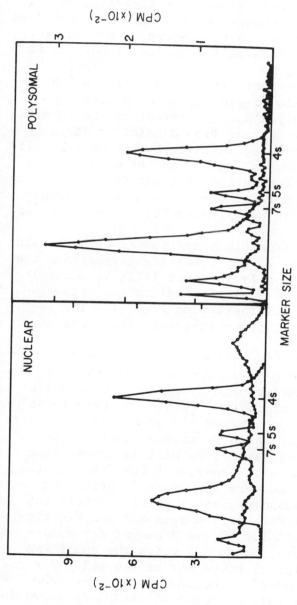

Figure 3. Acrylamide gel electrophoresis of Poly A isolated from labeled total nuclear RNA and rapidly labeled polyribosomal RNA. Poly A was prepared from nuclear RNA and polyribosomal RNA of cells labeled 60 min with ^{32}P PO_4, as in the legends to Figures 1 and 2. The Poly A was co-electrophoresed with ^3H-uridine labeled HeLa cell cytoplasmic RNA that had been heated to separate 7S rRNA from the 28S rRNA. Electrophoresis was for the equivalent of 17 hr at 35 volts on 14 cm gels with 10% cross-linkage. The approximate sedimentation constants of the marker RNA species are indicated on the abscissa. For details, see Edmonds, Vaughan and Nakazato (1971).

200 nucleotides would be estimated from plots of log molecular weight versus mobility constructed for the small well-characterized RNA markers (Loening, 1969).

This apparent discrepancy in size has been confirmed by co-electrophoresis of Poly A sequences from nuclear and cytoplasmic RNAs from cells labeled either for 60 min with $^{32}PO_4$ or for 90 min with ^{14}C-adenosine. Figure 4 shows electrophoretic patterns of the ^{32}P-labeled nuclear Poly A's co-electrophoresed with ^{14}C-labeled cytoplasmic Poly A's. Any effect on the lengths of these sequences resulting from dissimilar labeling of RNA molecules, produced in these experiments by different labeled precursors, was ruled out when results identical to those of Figure 4 were obtained by co-electrophoresis of nuclear Poly A from the ^{14}C-adenosine labeled cells and cytoplasmic Poly A from the ^{32}P-labeled cells.

Poly A sequences from all size classes of HnRNA are similar in size (Fig. 4, top), but are larger than the Poly A sequences from polyribosomes (Fig. 4, center) and from the RNA of more slowly sedimenting cytoplasmic components as well (bottom of Fig. 4). The sequences from both of these cytoplasmic fractions were indistinguishable.

Figure 4. (Opposite page.) Co-electrophoresis of Poly A's from the nucleus and cytoplasm. HeLa cells were labeled for 60 min with $^{32}PO_4$ in one experiment and for 90 min in another with ^{14}C-adenosine. Nuclear RNA isolated from each experiment was separated into molecules sedimenting faster, as well as slower than 45S. Cytoplasm from each experiment was fractionated into polyribosomes and more slowly sedimenting components (monosomes, subunits, etc.) prior to isolating RNA from each. The Poly A from each RNA was electrophoresed as described for Figure 3 except for a two-fold increase in duration during which the small RNA markers (seen in Fig. 3) moved out of the gel. For details of these procedures, see Edmonds, Vaughan and Nakazato (1971). ^{32}P-Poly A from > 45S HnRNA is shown in dotted line in all three panels. Top panel—^{14}C-Poly A from < 45S HnRNA. Center—^{14}C-Poly A from polysomal RNA. Bottom—^{14}C-Poly A from post-polysomal RNA.

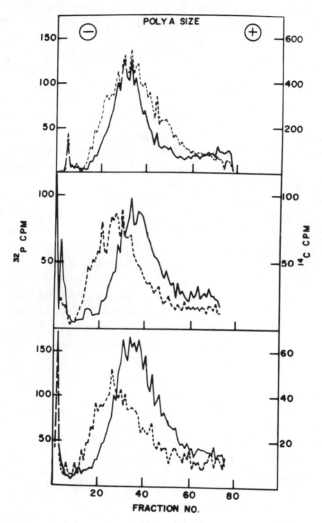

POLY A SEQUENCES IN ANIMAL RNA VIRUSES

The possibility of a translational function for
Poly A sequences suggested by their covalent association with HeLa mRNA led us to examine Poliovirus, a
picornavirus, and Eastern Equine Encephalitis (EEE), a
group A arbovirus, for Poly A sequences, since the
purified virion RNA of these two unrelated viruses is
infectious and can presumably function as mRNA in the
host cell cytoplasm where viral replication takes
place.

EEE virus (New Jersey strain) was prepared in chick embryo cell cultures infected at an input multiplicity of 0.1 plaque forming units per cell. After a 1 hr attachment period at 37°, 240 µC of carrier-free $^{32}PO_4$ was added to each 5 cm dish in phosphate-free Hank's salts solution containing 3% γ-globulin-free calf serum. Eighteen hours later virus was harvested from the supernatant fluid. The virus was concentrated and purified by elution from aluminum phosphate gel (Pfefferkorn and Hunter, 1963) and zone sedimentation on 5-20% sucrose gradients, followed by sedimentation to a pellet at 100,000 x g.

Poliovirus (type 1, Mahoney) was grown in suspension culture HeLa cells with an input multiplicity of 100 plaque forming units per cell. The cells were at a concentration of 4 x 10^6/ml in phosphate-free Eagle's minimal essential medium supplemented with 4 mM glutamine and actinomycin D (4 µg/ml). After 30 min, dialyzed calf serum (5%) and 18 mc of carrier-free $^{32}PO_4$ was added. The cells were harvested 6.5 hr later by centrifugation and were then subjected to four cycles of freezing and thawing to liberate virus. The virus was purified by banding at sedimentation equilibrium in a cesium chloride solution as previously described (Phillips, Summers and Maizel, 1968).

RNA was purified from both viral preparations by three successive hot phenol extractions of suspensions of virus in buffered sodium dodecyl-sulfate solution, as described in Figure 1. The size of the poliovirus RNA was analyzed by sucrose gradient sedimentation with HeLa cell rRNAs as sedimentation markers. A portion of poliovirus RNA was also sedimented on a sucrose gradient after being dissolved in a dimethyl sulfoxide (DMSO) solution together with an excess of small (<10S), unlabeled, synthetic Poly A. In this latter analysis the DMSO would have denatured any double-stranded RNA present (Strauss, Kelly and Sinsheimer, 1968), while the added unlabeled Poly A would prevent any contaminating small radioactive Poly A molecules from readhering to the viral RNA prior to sedimentation. These sedimentation analyses both indicated that the poliovirus RNA preparation was predominantly composed of intact 35S molecules. A portion

of RNA taken from the sedimentation peak of the poliovirus RNA dissolved in DMSO solution with unlabeled
Poly A was collected to provide a sample of viral RNA
which could be assumed to be free of any adventitiously
bound radioactive Poly A. The purified viral RNAs were
quantitatively analyzed for Poly A as described in
Figure 1.

Two different preparations of poliovirus RNA each
contained 0.75% Poly A, while a similar proportion of
Poly A, 0.69%, was found in EEE RNA. Table 2. We con-

TABLE 2

Poly A content of virus RNA

| Sample | % Of total cpm in Poly A[*] | |
	Polio	EEE
1	0.75	0.69
2	0.76, 0.75	—
2 DMSO-treated: re-isolated on Sucrose gradient	0.76	—

Nucleotide analysis of Poly A sequences

| | Polio | | EEE | |
	cpm	%	cpm	%
CMP	900	2.0	12	1.2
AMP	45,000	97.4	970	95.8
GMP	194	0.31	10	0.9
UMP	213	0.33	22	2.1

[*]The Poly A content of ^{32}P-labeled
virion RNA and the nucleotide composition of the Poly A were determined as
described by Edmonds, Vaughan and Nakazato (1971).

clude that the Poly A detected in poliovirus RNA is covalently part of that RNA sequence, since the proportion of Poly A was unchanged, 0.75%, when the RNA was further purified by zone sedimentation following solution in the presence of DMSO and unlabeled Poly A molecules.

The reported molecular weight of the poliovirus virion RNA is $2.4–2.5 \times 10^6$ daltons (Granboulan and Girard, 1969; Tannock, Gibbs and Cooper, 1970). From this value and the reported base composition of the RNA (Roy and Bishop, 1970), it can be calculated that the Poly A sequences in poliovirus RNA constitute about 53-55 nucleotides. Electrophoretic and compositional data indicate that poliovirus Poly A has a size of this order, so there must be one such sequence per virion RNA molecule. Accurate physical measurements of the size of EEE virion RNA are not available, but from its sedimentation rate in sucrose gradients, the molecular weight has been estimated to be 3.5×10^6 daltons (Armstrong, unpublished observations; Armstrong, Freeburg and Ho, 1971). We have determined the base composition of ^{32}P-labeled EEE virus virion RNA by conventional methods to be: AMP, 26.4%; UMP, 27.2%; GMP, 24.8%; CMP, 21.4%. From these data we calculate that the Poly A of EEE RNA includes approximately 70 nucleotides. Our electrophoretic analysis of EEE Poly A suggests that, as with poliovirus RNA, this Poly A is in one sequence per EEE virion RNA molecule.

The Poly A from each virus migrates in acrylamide gel electrophoresis as one peak (Fig. 5), with a mobility higher than that of the Poly A from the corresponding host cell RNA but lower than the mobility of tRNA. The relative mobility of Poly A from EEE RNA appears to be slightly less than that of Poly A from poliovirus RNA, indicating a small difference in length between the two. The fact that the viral Poly A's move more slowly in gels than tRNA is not incompatible with them being shorter in chain length, about 50 as opposed to 75-80 nucleotides, since molecules of Poly A have little secondary structure at neutrality, unlike the compactly folded tRNAs.

Analysis of a heavily radiolabeled sample of polio-

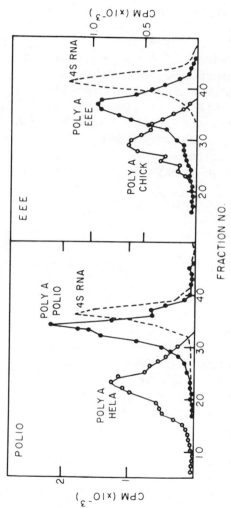

Figure 5. Acrylamide gel electrophoresis of Poly A isolated from poliovirus and EEE virus RNA and HeLa and chick cell RNA. Poly A was isolated from ^{32}P-labeled poliovirus and EEE virus RNA as described under Figure 1 for HeLa RNAs. ^{32}P-labeled Poly A from total nuclear RNA was from HeLa cells labeled 1 hr with ^{32}PO$_4$. ^{14}C-labeled chicken Poly A was from the polysome RNA of chick muscle culture cells labeled for 90 min with ^{14}C-adenosine. The growth of chick cells was described by Coleman and Coleman (1968), isolation of chick polysomes was described by Morse, Herrmann and Heywood (1971). Poly A samples were prepared for electrophoresis as described by Edmonds, Vaughan and Nakazato (1971). They were applied to 8.5 cm by 6 cm diameter 3.5% acrylamide gels cross-linked with 0.175% bis-acrylamide in an SDS-phosphate buffer system (pH 7). The gels were pre-electrophoresed for 45 min at 10 V/cm before sample application, after which they were run 1 hr 15 min at 10 V/cm and processed with a Maizel gel fractionator prior to scintillation counting of radioactivity.

239

virus Poly A gave the nucleotide composition shown in
Table 2. CMP was present as one nucleotide per Poly A
sequence of about 49 nucleotides, whereas GMP and UMP
were found only as traces. The Poly A available from
EEE RNA contained insufficient radioactivity for such
an accurate analysis, but the results obtained (Table
2) indicate that this Poly A also was essentially free
of nucleotides other than AMP.

The ratio of one CMP residue to 49 AMP residues in
isolated poliovirus Poly A probably results from
cleavage by RNase A at a CMP residue adjacent either
to the 3' end of a sequence [(AP)$_{48}$ApCp] within the
viral RNA or at its 5' terminus. Preliminary chemical
studies of poliovirus RNA by others (Wimmer, 1971)
suggest that the Poly A sequence is not the 3' terminus and is probably not within several nucleotides of
the 5' terminus either.

DISCUSSION

Two predictions can be made from our data on the
relative size and abundance of the covalently linked
Poly A sequences in the two classes of HeLa RNA and
the virion RNA of polio and EEE virus. First, many,
if not most of the HnRNA and mRNA molecules contain
Poly A sequences and second, the mRNAs of HeLa and the
RNA of the two viruses examined here contain only one
such sequence. The latter may also be true for HnRNA,
but for these very large RNAs it is difficult to ex-
clude the possibility of multiple Poly A sequences in
certain molecules which may be totally absent from
others.

This first prediction arises from the findings that
the Poly A content of the HnRNA molecules sedimenting
faster than 45S with molecular weights of 7-10 x 10^6
contained 0.46% Poly A (Fig. 1). One Poly A sequence
of 150-200 nucleotides in each molecule should constitute 0.4-0.6% of RNAs of this size. A similar
calculation for the total mRNA, assuming an average
molecular weight of 6 x 10^5 for the total mRNA fraction, and a slightly shorter Poly A sequence of perhaps 150 nucleotides should have 12-15% Poly A, con-

siderably higher than the 4 to 5% shown in Figure 2.
However, more recent measurements obtained by directly
binding HeLa mRNA to Poly T cellulose have now verified these predictions of approximately 15%.

That the mRNAs of HeLa contain a single Poly A sequence would be predicted from the inverse relationship of Poly A content and the size of mRNA molecules
seen in Figure 2. The exception is seen in the more
slowly sedimenting RNA molecules in this experiment
(i.e., 2.5% Poly A) which is labeled transfer RNA in
addition to some mRNA. The accepted molecular weights
for the RNAs of polio and EEE virus clearly allow only
one Poly A sequence per molecule of the size and in
the amounts found in these experiments.

The long sought evidence for a precursor-product
relationship between HnRNA and mRNA may have been
found in the similar Poly A sequences in each. Discussion of this question can be found in the paper of
Darnell *et al.* in this volume.

We have proposed that each molecule of HnRNA contains one sequence of potential mRNA adjacent to a
small Poly A sequence. The HnRNA molecule would be
rapidly degraded during or after transcription and
only a small portion of it including mRNA and an adjacent Poly A sequence would be conserved and exported
to the cytoplasm (Edmonds, Vaughan and Nakazato, 1971).
Recent awareness of the slightly greater lengths of
the nuclear Poly A sequences may require some refine-
ment of this model although, as noted earlier, the
shortened Poly A sequences in mRNA might arise during
its subsequent metabolism in the cytoplasm.

Although Poly A sequences may function as recognition signals for the processing of HnRNA as outlined
above, the existence of a Poly A sequence in cellular
mRNA as well as in the virion RNA of two viruses which
are essentially dissimilar and in the mRNAs synthesized by viral RNA polymerases of vaccinia (Kates,
1970) and vesicular stomatitis virions (Mudd and
Summers, 1970), all of which reproduce in cytoplasm,
makes it reasonable to hypothesize that Poly A sequences play some common role in the functioning of this
RNA. Although other possibilities might be considered,
the one process in which these mRNAs all function is

translation. The role that Poly A sequences may play in the translation process is now ready for investigation.

ACKNOWLEDGMENTS

This work was supported by grants from the USPHS National Institutes of Health (AM–10074, CA–10922, AI–02953, AI–08368). Dr. Maurice Vaughan is a Career Development awardee of the USPHS National Institutes of Health.

REFERENCES

Armstrong, J. A., Edmonds, M., Nakazato, H., Phillips, B. A., and Vaughan, M. H. (1972). *Science 176*, 526.

Armstrong, J. A., Freeburg, L. C., and Ho., M. (1971). *Proc. Soc. Exp. Biol. and Med. 137*, 13.

Coleman, J. R., and Coleman, A. W. (1968). *J. Cell. Physiol. 72, Suppl. 1*, 19.

Darnell, J. E., Wall, R., and Tushinski, R. J. (1971). *Proc. Nat. Acad. Sci. USA 68*, 1321.

Edmonds, M. (1971). *Procedures in Nucleic Acid Research*, G. L. Cantoni and D. R. Davies, eds., *Vol. II* (New York: Harper and Row), p. 629.

Edmonds, M., Vaughan, M. H., and Nakazato, H. (1971). *Proc. Nat. Acad. Sci. USA 68*, 1337.

Girard, M., Latham, H., Penman, S., and Darnell, J. E. (1965). *J. Mol. Biol. 11*, 187.

Granboulan, N., and Girard, M. (1969). *Virology 4*, 475.

Kates, J. (1970). *Cold Spring Harbor Symp. Quant. Biol. 35*, 743.

Lee, S. Y., Mendecki, J., and Brawerman, G. (1971). *Proc. Nat. Acad. Sci. USA 68*, 1331.

Loening, U. (1969). *Biochem. J. 113*, 131.

Morse, R. K., Herrmann, H., and Heywood, S. M. (1971). *Biochim. Biophys. Acta 232*, 403.

Mudd, J. A., and Summers, D. F. (1970). *Virology 42*, 958.

Nakazato, H., and Edmonds, M. (1972). *J. Biol. Chem. 247*, 3365.

Penman, S. Vesco, C., and Penman, M. (1968). *J. Mol. Biol. 34*, 49.

Pfefferkorn, E. R., and Hunter, H. S. (1963). *Virology 20*, 433.

Phillips, B. A., Summers, D. F., and Maizel, J. V. (1968). *Virology 35*, 216.

Roy, P., and Bishop, J. (1970). *Virology 6*, 604.

Strauss, J. H., Kelly, R. B., and Sinsheimer, R. L. (1968). *Biopolymers 6*, 793.

Tannock, G. A., Gibbs, A. J., and Cooper, P. D. (1970). *Biochem. Biophys. Res. Commun. 38*, 298.

Warner, J. R., Soeiro, R., Birnboim, H. C., and Darnell, J. E. (1966). *J. Mol. Biol. 19*, 349.

Properties of
RNA-Dependent
DNA Polymerases

David Baltimore

Department of Biology
Massachusetts Institute of Technology
Cambridge

The world of animal virology provides us with many
examples of unique transcription and replication systems (Baltimore, 1971a, b). The ones which are best
understood are found among those viruses which contain
RNA as their genetic material. These viruses utilize
RNA-dependent RNA synthesis for both transcription and
replication, and in one case, the RNA tumor viruses,
RNA-dependent DNA synthesis appears to be involved
during the initial phase of the growth cycle. *RNA-dependent RNA synthesis* is not one but a number of different processes; viruses like poliovirus, vesicular
stomatitis virus and reovirus, for instance, use
distinctly different modes of synthesis for their
transcriptional and replicational processes (Baltimore,
1971a, b).

These various modes of nucleic acid synthesis may
be specific to viruses. Their very elaborate nature,
however, suggests that they may not have evolved as

245

modes of synthesis for viruses. Since viruses most
probably have evolved from cells, it seems possible
that these specialized modes of nucleic acid synthesis
have also evolved from homologous processes in cells.
No such processes, however, have been demonstrated to
exist. Neither RNA-dependent RNA synthesis nor RNA-
dependent DNA synthesis has ever been convincingly implicated in any cellular process, although suggestions
of such activities have been made on occasion.

Possible places where these types of nucleic acid
synthesis might occur in cells are relatively easy to
imagine. The complex and mystifying process of differentiation offers innumerable places where one could
imagine that RNA-dependent RNA or DNA synthesis might
be involved. The complicated processes involved in
memory storage, in the immunologic system and even in
the basic molecular biological processes of cells also
offer many possibilities for conjecture. The very
ease with which one can imagine such processes under-
lines the necessity for stringent proof of their
existence.

The virus-specific mode of nucleic acid synthesis
which has drawn the most attention during 1971 is the
RNA-dependent synthesis of DNA (Baltimore, 1970; Temin
and Mizutani, 1970). It is instructive to examine the
present state of knowledge about this process in viruses in order to try to design criteria by which such
activities might be identifiable in cells. With this
background it will be possible to evaluate the published reports indicating that such activities do
exist in cells.

The RNA-Dependent DNA Polymerase of
RNA Tumor Viruses

RNA tumor viruses have been identified in many
species of animals (Schlom *et al.*, 1971). If purified
virions of any of these viruses are disrupted with a
nonionic detergent and then incubated in an appropriate cell-free system containing four deoxyribonucleo-
side triphosphates, synthesis of DNA occurs (Baltimore
1970; Temin and Mizutani, 1970; Schlom *et al.*, 1971).
Generally one of the nucleotides is labeled with a

radioactive atom and incorporation of radioactivity into acid-precipitable material is used as the criterion of synthesis. The reaction requires all four nucleotides and is inhibited by ribonuclease indicating that RNA is the template for DNA synthesis. Both double-stranded and single-stranded DNA molecules are found among the products of the reaction along with some DNA-RNA hybrids (Garapin et al., 1970; Fujinaga et al., 1970; Manly et al., 1971). Actinomycin D produces a partial inhibition of the DNA polymerase reaction by preventing synthesis of double-stranded DNA, but the drug does not affect the initial RNA-dependent synthesis (McDonnell et al., 1970; Manly et al., 1971). The 60-70S RNA species found in such virions will hybridize to the DNA product and therefore it appears that this RNA is the template (Rokutanda et al., 1970; Spiegelman et al., 1970a; Duesberg and Canaani, 1970). This reaction we have called the endogenous reaction.

If various polynucleotides, either deoxyribo- or ribo-, are added to disrupted virions, a stimulation of DNA synthesis is often observed and in some cases this can be 100- to 1000-fold (Spiegelman et al., 1970b; Mizutani et al., 1970; Spiegelman et al., 1970c Riman and Beaudreau, 1970). In such a reaction, the added polynucleotide is acting as template and we call this an exogenous reaction.

An enzyme can be purified from the virions of the RNA tumor viruses which will synthesize DNA on an added template. Only one major activity has been seen and this will copy both ribopolynucleotides and deoxyribopolynucleotides (Duesberg, Helm and Canaani, 1971; Kacian et al., 1971; Verma and Baltimore, unpublished results).

Primer Requirement

The RNA tumor virus DNA polymerase is a primer-dependent enzyme just as are other known DNA polymerases (Baltimore and Smoler, 1971; Smoler, Molineaux and Baltimore, 1971). This means that a single-stranded polynucleotide will only act as a template if a homologous primer is present. Table 1 illustrates the need for a primer to allow the avian myeloblastosis

TABLE 1

Various oligomers as primers for the AMV
DNA polymerase with poly(A) as a template

Additions	cpm [^{3}H]TMP incorporated
Poly(A)	234
Poly(A) + (dT)$_{14}$	15,800
Poly(A) + (dC)$_{14}$	505
Poly(A) + (dG)$_{14}$	315
Poly(A) + (dA)$_{14}$	140

Samples were incubated for 60 min at
37°. Standard reaction mixtures contained
18 nmole of [^{3}H]TTP (44 cpm/pmole), 210
pmole of poly(A), and either 300 pmole of
(dA)$_{14}$, 290 pmole of (dC)$_{14}$, 205 pmole of
(dG)$_{14}$, or 170 pmole of (dT)$_{14}$. See
Baltimore and Smoler (1971).

virus (AMV) DNA polymerase to copy poly(A). Only an
oligomer of dT will prime this synthesis, not oligomers of dC, dG or dA. A similar requirement for a
complementary oligomer can be demonstrated with other
templates (Baltimore and Smoler, 1971). Using oligo-
mer·template complexes the template specificities of
AMV DNA polymerase, mouse leukemia virus (MLV) DNA
polymerase and *E. coli* DNA Polymerase I have been determined (Table 2). The tumor virus polymerases show
a clear preference for ribopolynucleotide templates
while the *E. coli* enzyme prefers deoxyribopolynucleotide templates. However, there is ambiguity in that
poly(dC) is copied by the viral enzymes and poly(A) by
the *E. coli* enzyme.

The role of the primer has been assessed by utilizing the acid-soluble primer [5'-^{32}P] (dT)$_{10}$ (Smoler *et
al.*, 1971). If this primer is used with poly(A) as a
template, and ^{3}H-dTTP as a precursor, incorporation of
the primer into acid-precipitable material occurs in
parallel with ^{3}H-dTMP incorporation (Fig. 1; Table 3).

TABLE 2

Template specificity of DNA polymerases

Enzyme	Homopolymers							
	rA	rC	rI	rU	dA	dC	dI	dT
AMV DNA polymerase	+++	+++	++	0	0	++	0	0
MLV DNA polymerase	+++	++	+	++	0	+++	0	0
E. coli DNA Polymerase I	+	0	0	0	+++	++	+	++

Relative activities of different templates were estimated as given in Baltimore and Smoler (1971).
Polymers designated +++ or ++ were at least 100-fold
as active as templates designated 0 which showed no
detectable activity. All polymers were assayed in the
presence of complementary oligodeoxyribonucleotide
templates.

Incorporation of the primer requires poly(A) and dTTP.
The [5'-^{32}P] is not removed during the reaction. Also
it is not covered during synthesis because it remains
sensitive to alkaline phosphatase (Table 4). This indicated that synthesis proceeded by addition of mono-
mers onto the 3'-OH of the primer. The reaction is
diagramatically represented in Figure 2.

A further indication that synthesis occurs by addition to 3'-OH groups is the sensitivity of the reaction to (2',3') dideoxyTTP (ddTTP). This analog,
lacking both the 2'- and 3'-OH groups terminates synthesis when it is incorporated into growing chains and
can only be incorporated at 3'-OH ends of growing
chains, not at 5' ends (Atkinson et al., 1969). Both
the endogenous reaction and those exogenous reactions
which incorporate dTTP are sensitive to ddTTP (Table
5) indicating that all synthesis by the AMV DNA
polymerase involves addition at 3'-OH ends (Smoler et
al., 1971).

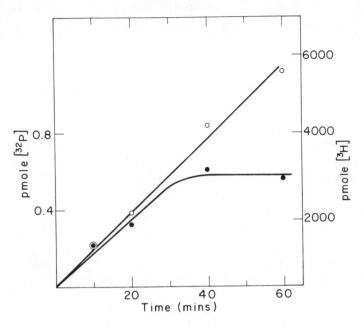

Figure 1. Time course of incorporation of [^{3}H]dTMP (O) and [$5'-^{32}$P](dT)$_{10}$ (●) by the AMV DNA polymerase. Reaction conditions were as described in Table 1. Reprinted from Smoler *et al*. (1971).

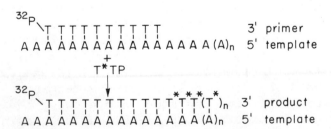

Figure 2. Schematic representation of primer function and direction of chain growth.

TABLE 3

Requirements for incorporation of $[5'-{}^{32}P](dT)_{10}$ by avian myeloblastosis virus DNA polymerase

	Incorporation	
	pmoles $[{}^{32}P]$	pmoles $[{}^{3}H]$
Complete	0.51	3,835
Minus $[{}^{3}H]dTTP$	< 0.01	——
Minus $[{}^{3}H]dTTP$ + ATP	< 0.01	——
Minus $[5'-{}^{32}P](dT)_{10}$	——	< 50
Minus poly(A)	0.01	< 50
Plus 17 nmoles $(dT)_{14}$	0.01	5,783

Complete reaction mixture contained 20 nmoles of poly(A), 75 nmoles $[{}^{3}H]dTTP$ (2.2 cpm/pmole), 8.1 pmoles $[5'-{}^{32}P](dT)_{10}$ (5500 cpm/pmole) and 3 μg AMV protein. See Smoler *et al.* (1971).

TABLE 4

Treatment of product containing $[5'-{}^{32}P](dT)_{10}$ and $[{}^{3}H]dTMP$

	pmoles $[{}^{32}P]$	pmoles $[{}^{3}H]$
Exp. 1 Product	0.43	3,725
Treated with NaOH	0.38	3,793
Exp. 2 Product	0.36	2,885
Treated with alkaline phosphatase	0.005	2,772

Products prepared as in Table 3 were purified by acid or ethanol precipitation and incubated either 18 hr at 37° in 0.5 N NaOH (Exp. 1) or 30 min at 70° with 2 μg of bacterial alkaline phosphatase (Exp. 2). See Smoler *et al.* (1971).

TABLE 5

Effect of ddTTP on endogenous and exogenous reactions
of the avian myeloblastosis virus DNA polymerase

Additions	ddTTP	pmoles [^{3}H]dTMP incorporation	% Of control
None	——	1.0	100
None	10^{-4} M	< 0.02	< 2
None	10^{-5} M	< 0.02	< 2
None	10^{-6} M	0.08	8
None	10^{-7} M	0.50	50
Poly(A) + (dT)$_{14}$	——	281	100
Poly(A) + (dT)$_{14}$	10^{-4} M	< 2	< 1

Complete reaction mixtures lacking dTTP but containing ddTTP where indicated were incubated for 20 min at
37°. [^{3}H]dTTP was then added and incubation continued
for 60 min (Exp. 1) or 120 min (Exp. 2). In Exp. 1,
135 pmoles of [^{3}H]dTTP, 6,000 cpm/pmole, was used. In
Exp. 2, 410 pmoles of poly(A), 170 pmoles of (dT)$_{14}$
and 19 nmoles of [^{3}H]dTTP, 42 cpm/pmole, were used.
See Smoler *et al.* (1971).

The requirement for a primer in the exogenous reaction raises the question of whether a primer might
also exist for the endogenous reaction. Recent evidence obtained by analysis of products on Cs$_2$SO$_4$ gradients indicates that a covalently bonded RNA–DNA
molecule is the initial product of the reaction and
suggests strongly that a small RNA in the virion is
acting as a primer in the endogenous reaction (Verma
et al., 1971).

*RNA-Dependent DNA Polymerases in
Cells*

The evidence accumulated thus far on the DNA polymerase of RNA tumor viruses suggests that a series of

criteria can be used to distinguish this enzyme from other known DNA polymerases and thus facilitate its identification in cells. These criteria involve both analysis of exogenous and endogenous reactions. Actually, the endogenous reaction is more diagnostic but the homopolymer templates are so active that they allow identification of much smaller amounts of polymerase activity. In evaluating these criteria it should be remembered that a putative cellular RNA-dependent DNA polymerase could have many properties different from those evidenced by the viral enzyme.

Table 6 shows in brief form some of the identifying

TABLE 6

Criteria for identifying an RNA-dependent DNA polymerase

I. *Enzymology*

    1. Homopolymer template specificity
    2. Ability to utilize short oligomers as primers
    3. Inhibition by single-stranded ribo-polynucleotides
    4. Utilization of natural RNAs as templates with preference for 60-70S tumor virus RNA

II. *Endogenous Reaction*

    1. Either in whole cells or in extracts
    2. Ribonuclease sensitivity
    3. Actinomycin D insensitivity
    4. Hybrid or covalently bonded RNA/DNA molecules identified on Cs_2SO_4 gradients or otherwise
    5. Hybridization to putative template

III. *Specific Inhibitor*

---

characteristics of the viral RNA-dependent DNA polymerase. Aside from satisfying the usual requirements for a DNA polymerase, the enzyme shows an unusual

pattern of template preferences in preparing ribopolynucleotide templates to deoxyribopolynucleotide templates (Table 2). This preference, however, is not
absolute. The ability to use short oligomeric primers
also serves to differentiate the viral polymerase from
known DNA-dependent DNA polymerases in mammalian cells
(Ross et al., 1971; Goodman and Spiegelman, 1971).
Inhibition of the viral polymerase by single-stranded
polynucleotides is also a characteristic of possible
diagnostic significance (Tuominen and Kenny, 1971).
Possibly the most characteristic property of the enzyme is its utilization of single-stranded natural RNA
templates and preference for 60-70S viral RNA over
other templates (Duesberg et al., 1971; Kacian et al.,
1971; Verma and Baltimore, unpublished results). This
preference probably relates more to the existence of
primer sites in the molecule than to any other property. A cellular enzyme might, especially in this regard, have very different template preferences.

The limited utility of template and primer preferences as diagnostic criteria is evident. Therefore,
analysis of an *endogenous reaction* either in whole
cells or in cell extracts will probably yield the
clearest evidence for the existence of an RNA-dependent DNA polymerase. The hallmarks of such an activity are ribonuclease-sensitivity and actinomycin-re-
sistance (Table 6). Furthermore, analysis of DNA
products on Cs_2SO_4 should clearly demonstrate DNA-RNA
hybrid molecules and/or DNA-RNA covalently bonded
molecules. Finally hybridization of the DNA to its
putative template should provide definitive evidence
of relatedness.

Actually, the best criterion would be sensitivity
or resistance to specific inhibitors. Actinomycin,
because it inhibits DNA-dependent but not RNA-dependent reactions, is diagnostic in a negative sense.
However, an inhibitor selective for RNA-dependent reactions (especially one that bound to RNA but not DNA)
would be most useful in screening cells for RNA-dependent DNA synthesis. Suggestive evidence that rifampi-
cin derivatives might be useful in this regard has
been published (Gurgo et al., 1971; Calvin et al.,
1971) and complete characterization of the inhibitory

spectrum of these compounds and their utility in cells is awaited.

At the present there are some suggestive reports of RNA-dependent DNA polymerases in various cell types, both malignant and normal (Gallo, Yang and Ting, 1970; Scolnick *et al.*, 1971; Stavrianopoulos, Karkas and Chargaff, 1971). However, a growing body of opinion and evidence favors the view that all of these reactions are the result of the use of nondiscriminating criteria and that actually no clear demonstration of an RNA-dependent DNA polymerase has been made (Ross *et al.*, 1971; Goodman and Spiegelman, 1971; Baltimore and Smoler, 1971). Certainly few of the criteria set for forth above (Table 6) have been met in any of the papers which have appeared.

A most intriguing suggestion about RNA-dependent DNA synthesis was made by Toccini-Valentini and Crippa (1971). They suggested that DNA amplificiation in frog oocytes may involve RNA-dependent DNA synthesis. Although they are far from proving this assertion, some hopeful evidence has been published.

Many laboratories are now searching in various cell types for evidence of RNA-dependent DNA synthesis and RNA-dependent RNA synthesis. It is clear that such unusual mechanisms of information transfer might be behind many biologic phenomena which are mysterious at present. However, we must be very rigorous in our criteria for defining such processes before we implicate them. It is too easy to solve mysteries by postulating unusual processes and then collecting a bit of vaguely suggestive evidence.

REFERENCES

Atkinson, M. R., Deutchser, M. P., Kornberg, A., Russel, A. F., and Moffatt, J. G. (1969). *Biochemistry 8*, 4897.

Baltimore, D. (1970). *Nature 226*, 1209.

Baltimore, D. (1971a). *Transactions of the New York Academy of Science 33*, 327.

Baltimore, D. (1971b). *Bact. Reviews 35*, 235.

Baltimore, D., and Smoler, D. (1971). *Proc. Nat. Acad. Sci. USA 68*, 1507.

Calvin, M., Joss, U. R., Hackett, A. J., and Owens, R. B. (1971). *Proc. Nat. Acad. Sci. USA 68*, 1441.

Duesberg, P. H., and Canaani, E. (1970). *Virology 42*, 783.

Duesberg, P., Helm, K. V. D., and Canaani, E. (1971). *Proc. Nat. Acad. Sci. USA 68*, 747.

Fujinaga, K., Parsons, J. T., Beard, J. W., Beard, D., and Green, M. (1970). *Proc. Nat. Acad. Sci. USA 67*, 1432.

Gallo, R. C., Yang, S. S., and Ting, R. C. (1970). *Nature 228*, 927.

Garapin, A-C., McDonnell, J. P., Levinson, W. E., Quintrell, N., Fanshier, L., and Bishop, J. M. (1970). *J. Virol. 6*, 589.

Goodman, N. C., and Spiegelman, S. (1971). *Proc. Nat. Acad. Sci. USA 68*, 2203.

Gurgo, C., Ray, R. K., Thiry, L., and Green, M. (1971). *Nature New Biology 229*, 111.

Kacian, D. L., Watson, K. F., Burny, A., and Spiegelman, S. (1971). *Biochim. Biophys. Acta 246*, 365.

McDonnell, J. P., Garapin, A., Levinson, W. E., Quintrell, N., Fanshier, L., and Bishop, J. M. (1970). *Nature (London) 228*, 433.

Manly, K., Smoler, D. F., Bromfeld, E., and Baltimore, D. (1971). *J. Virol. 7*, 106.

Mizutani, S., Boettiger, D., and Temin, H. M. (1970). *Nature 228*, 424.

Riman, J., and Beaudreau, G. S. (1970). *Nature 228*, 427.

Rokutanda, M., Rokutanda, H., Green, M., Fujinaga, K., Ray, R. K., and Gurgo, C. (1970). *Nature 227*, 1026.

Ross, J., Scolnick, E. M., Todaro, G. J., and Aaronson, S. A. (1971). *Nature New Biology 231*, 163.

Schlom, J., Harter, D. M., Burny, A., and Spiegelman, S. (1971). *Proc. Nat. Acad. Sci. USA 68*, 182.

Scolnick, E. M., Aaronson, S. A., Todaro, G. J., and Parks, W. P. (1971). *Nature 229*, 318.

Smoler, D., Molineaux, I., and Baltimore, D. (1971). *J. Biol. Chem. 246*, 7697.

Spiegelman, S., Burny, A., Das, M. R., Keydar, J., Schlom, J., Travnicek, M., and Watson, K. (1970a). *Nature 227*, 563.

Spiegelman, S., Burny, A., Das, M. R., Keydar, J.,
   Schlom, J., Travnicek, M., and Watson, K. (1970b).
   Nature *227*, 1029.
Spiegelman, S., Burny, A., Das., M. R., Keydar, J.,
   Schlom, J., Travnicek, M., and Watson, K. (1970c).
   Nature *228*, 430.
Stavrianopoulos, J. G., Karkas, J. D., and Chargaff,
   E. (1971). *Proc. Nat. Acad. Sci. USA 68*, 2207.
Temin, H., and Mizutani, S. (1970). *Nature 226*, 1211.
Toccini-Valentini, G. P., and Crippa, M. (1971). *The
   Biology of Oncogenic Viruses, 2nd. Lepetit Collo-
   quium on Biology and Medicine*, L. G. Sylvestri, ed.
   (Amsterdam: North-Holland Publ. Co.), p. 235.
Tuominen, F. W., and Kenny, F. T. (1971). *Proc. Nat.
   Acad. Sci. USA 68*, 2198.
Verma, I. M., Meuth, N. F., Bromfeld, E., Manly, K. F.,
   and Baltimore, D. (1971). *Nature New Biology 233*,
   131.

Programs for
Protein Synthesis

Introduction

The end result of differential activation of the genome is that proteins are synthesized (and some are degraded or secreted) in a precise temporal order and in precisely regulated amounts. One can imagine *a priori* that the controls might be exerted at the level of (1) initiation and termination of transcription, (2) processing of the transcripts, (3) their conveyance (in Eucaryotes) into the cytoplasm and association with polysomal complexes (a process that might require specific ribosomal subunits), (4) initiation of translation and, (5) polymerization of subunits. Studies of enzyme induction and repression in bacteria have demonstrated how, by careful biochemical characterization of the elements of the system, by appropriate external manipulation of the system to detect internal responses, and by judicious use of genetically altered strains, one can establish whether precon-

ceived sites of control really do operate and can
uncover previously unsuspected ones.

The same methodology is now being applied to the
study of developmental systems. A few notable examples
follow. Many other systems, equally intriguing and
provocative, might have served in their place including: bacterial sporulation, sporophore formation and
spore germination in *Blastocladiella*, erythropoiesis,
lens development, pancreatic secretory cell development, myogenesis and chondrogenesis, Lily anther development, embryoid development from somatic cells of
carrot, tobacco, etc.

The results of all these studies provide at least
one common conclusion: that the simple repressor-oper-
ator-coeffector regulatory circuits that appear to
control physiological modulations in bacteria cannot
be invoked in any straightforward way to account of
themselves for developmental regulation, particularly
in Eucaryotes.

Developmental Changes During Meiosis in Yeast

H. O. Halvorson, J. Haber, and S. Sogin

Rosenstiel Basic Medical Sciences Research Center
Brandeis University
Waltham, Massachusetts

Both translational and transcriptional regulation have been implicated to explain developmental regulation in various eucaryotic cells. Difficulty in determining the molecular basis of this regulation is due not only to the lack of genetic analysis but also to our limited understanding of the role of cell structures in regulating developmental processes. Understanding these complex relationships will undoubtedly be difficult and involve a multidisciplinary attack of the problem. For example, the ordered changes during development probably involve not only the formation of complex structures but also functions associated with these structures.

Among the simplest developmental systems in eucaryotic organisms is the process of meiosis and sporulation in the yeast *Saccharomyces cerevisiae*. This yeast is diploid in the vegetative phase (Fig. 1). Vegetative cells divide by budding. Given appropriate

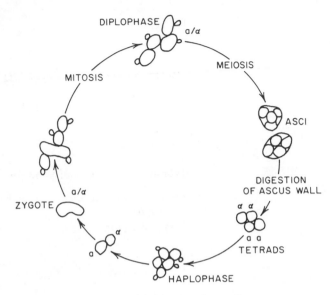

Figure 1.   Life cycle of yeast.

environmental conditions, meiosis occurs and precedes
sporulation.   Four haploid spores form within the
ascus.   Unless these spores are isolated they may fuse
with their opposite mating type neighbors (a and α) in
the ascus to restore the diploid state.

   Meiosis and sporulation in yeast is a process of
intracellular differentiation wherein cells stop
growth by budding (mitosis) and enter meiosis.   A particular physiological state must be established prior
to the induction of sporulation.   When cells are
transferred to sporulation medium (potassium acetate)
the ability to sporulate is dependent upon the physiological age of the culture, and as discussed below, of
the age of the cell in its cell cycle.   Wild-type
diploid cells in sporulation medium arrive at a point
of commitment to meiosis which may be observed by returning portions of sporulating cultures to glucose
medium in which cells ordinarily do not sporulate
(Sherman and Rome, 1963).   The point of commitment
coincides closely with the onset of the reductional
division.

Genetic Control

A number of genes have now been recognized which regulate sporulation and meiosis in yeast. In *Schizosaccharomyces pombe* Bresch and his coworkers (1968) have demonstrated that there are a number of distinct genes responsible for meiosis and sporulation and for the dissolution of the separate cell walls during regenerative fusion. Esposito and Esposito (1969) have isolated a number of temperature sensitive recessive mutants which are blocked in meiosis but undergo mitotic divisions. These results suggest that the two modes of nuclear division are under separate genetic control. Sporulation deficient mutants can be recognized by their inability to form viable asci detectable by microscopic examination. Some mutants permit meiosis to proceed through the first nuclear division while others proceed through the second meiotic nuclear division but do not complete ascus formation.

The availability of a number of temperature sensitive mutations of meiosis permits one to analyze the meiotic process by temperature shift experiments. Shifting cultures from a nonpermissive to the permissive temperature during meiosis defines the time of expression of a temperature sensitive function. The reverse experiment defines the end and thus the duration of the dependence on this gene product. Figure 2 shows a plot for three such temperature-sensitive mutants of meiosis in *Saccharomyces cerevisiae*.

The conditional mutants of meiosis, in addition to indicating discrete stages in the meiotic process, also provide a promising approach for studying unique biochemical events during development. However, little is known at the moment of the biochemical nature of these components.

Biochemical Events During Sporulation

Figure 3 shows the sequence of macromolecular synthesis occurring during sporulation. After cells are transferred to acetate medium, there is an immediate rise in dry weight followed 4 hr later (T_4) by a

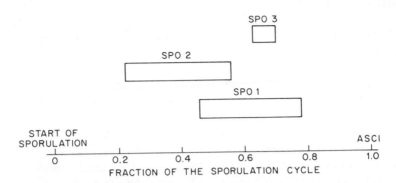

Figure 2. Physiological map of temperature-sensitive periods of three mutants. The duration and position of the temperature-sensitive period(s) of each mutant in the sporulation cycle are shown by horizontal bars. Data calculated from Esposito *et al.* (1970).

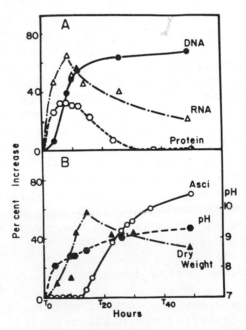

Figure 3. Biochemical changes during sporlation (data from Esposito *et al.*, 1969).

meiotic round of DNA replication. This DNA synthesis
achieves a percent increase equivalent to the total
ascus production. Accumulation of glycogen and lipid
commences in the early stages. RNA and protein content per cell increases to a maximum level at about T_8
and then subsequently declines. Two periods of active
protein synthesis occur: the first coincidental with
DNA synthesis and the other during ascus spore formation. Protein synthesis is required throughout sporulation and the overall process is accompanied by active RNA and protein turnover. Such a generalized
picture does not reveal much about the expression and
control of sporulation-specific events.

Our analysis of sporulating cultures has revealed
thus far one process which appears unique to sporulation. 20S RNA has been shown to accumulate in *Saccharomyces cerevisiae* under sporulation conditions (Kadowaki and Halvorson, 1971). It was found that 20S RNA
appeared only in strains which were diploid or pseudo-
diploid and thus able to sporulate. 20S RNA synthesized during sporulation was shown by RNA·DNA hybridization to contain sequences homologous with nuclear
but not mitochondrial DNA. The role of this apparently unique species during sporulation became apparent
only during the past year. One interesting feature of
the appearance of 20S RNA could be seen from the data
in Figure 4. Cells were labeled with tritiated adenine during presporulation growth, ^{32}P during sporulation for 18 hr, and then analyzed by polyacrylamide
gel electrophoresis. In addition to a peak in the 20S
region, it became apparent that the relative incorporation of ^{32}P into the 25S and 18S ribosomal RNA was
unequal, even though the two ribosomal RNAs are processed from a single precursor (Tabor and Vincent,
1969). Furthermore, the specific activity of the two
peaks became equal if the 20S peak counts were integrated with the 18S peak. The possibility that 20S
RNA was in fact a precursor of 18S RNA was tested more
directly by isolating 20S RNA from polyacrylamide gels
and carrying out an RNA:DNA hybridization competition
reaction with unlabeled ribosomal RNA from vegetative
cells (Fig. 5). The competition curves showed that
most (approximately 85%) of the labeled 20S RNA could

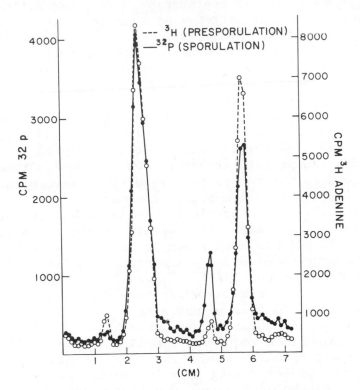

Figure 4. Polyacrylamide gel electrophoresis of RNA from sporulating yeast.
Cells were grown vegetatively to stationary phase in the presence of ^{3}H–adenine
and then placed in sporulation medium containing ^{32}P for 18 hr. The 20S RNA peak
(at ca. 4.6 cm) appears only during sporulation.

be competed off the DNA by ribosomal RNA. It is not
known why the 20S RNA accumulates during sporulation.
It thus appears resonable that the 20S RNA is similar,
if not identical to, the 20S ribosomal precursor
described by Udem and Warner (1971). Interestingly,
27S RNA, the precursor to 25S RNA does not appear to
accumulate at any time during sporulation.

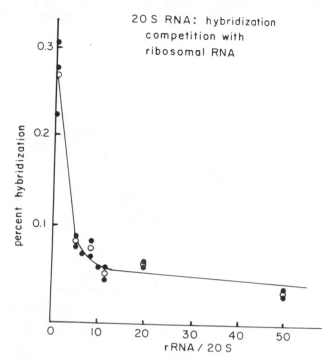

Figure 5. RNA·DNA hybridization competition curve for 20S RNA. ^{32}P 20S RNA was hybridized to DNA on filters in the presence of increasing concentrations of ribosomal RNA isolated from vegetative cells which do not contain 20S RNA. Approximately 85% of the 20S RNA is competed out by ribosomal RNA.

Cell Cycle Dependency of Sporulation

Both in examining the time and stage of arrest in mutants during the sporulation cycle and especially in characterizing the associated sensitive biochemical events during the sporulation cycle, a synchronously sporulating culture would be desirable. Such an approach has proven valuable with vegetative cultures to demonstrate the order of synthesis of a large number of enzymes (Halvorson, Carter and Tauro, 1971).

By the use of synchronized populations we have

found that the capacity of a cell to complete sporulation (Fig. 6) is highly dependent on its stage in the

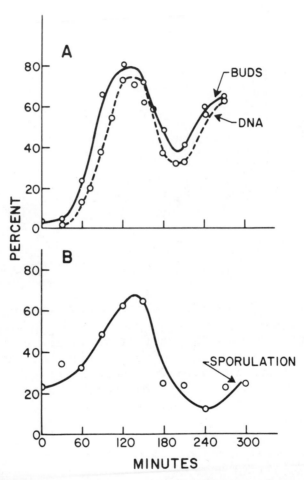

Figure 6. Cell cycle dependence of sporulation. A culture growing synchronously in
an acetate growth medium was assayed for
sporulation capacity at intervals over more
than one cell cycle. The degree of sporulation was low at the beginning of the
first cell scission. During the cell cycle
sporulation capacity increases, reaching a
maximum prior to cell scission.

vegetative cell cycle. In the early part of the cell cycle, up to the time of bud initiation and DNA replication, the intrinsic capacity of a cell to complete sporulation is low. Later in the cell cycle, however, the ability to complete sporulation rises significantly. The control of such periodic fluctuation of sporulation capacity is not known. It is possible that sporulation is simply limited by the availability of one or more essential compounds which either fluctuate periodically throughout a cell cycle or are simply present in greater amounts in cells of greater volume (i.e., later in the cell cycle). These results might also be explained by the existence of a control mechanism for the expression of sporulation-specific functions which would be inducible only during a limited period of the vegetative cell cycle.

A substantial portion of the decrease in sporulation capacity after cell scission can be attributed to a sharp difference in the sporulation capacity of newly formed daughter cells from that of mother cells. The observation that sporulation occurs in only one of the halves of a double cell about to divide 85% of the time, is in close agreement with the conclusions reached by Yanagita *et al.* (1970) in studying sporulation of mother and daughter cells from late stationary phase. Nevertheless, a significant dependency of sporulation activity on the cell cycle, for both mother and daughter cells, is still indicated by the extent of variation in sporulation capacity from the beginning to the end of the cell cycle.

The ability of *Saccharomyces cerevisiae* to undergo sporulation is predetermined by the stage in the vegetative cell cycle from which the cells were derived. In many respects this phenomenon is similar to the observation by Dawes, Kay and Mandelstam (1971) that in *Bacillus subtilis* sporulation appears to be initiated only during the period of the cell cycle during which DNA replication occurs. In *Saccharomyces cerevisiae*, Hartwell, Culotti and Reid (1970) have demonstrated that a number of mutants affecting different stages of the vegetative cell cycle are expressed much earlier in the cell cycle than the time of execution. In the case of sporulation, execution of sporulation events

also appears to depend on the expression of activities
prior to the exposure of the cells to the direct
stimuli for sporulation.

The development of homogeneous, synchronously sporulating cultures has made it possible to begin to examine the stage of sporulation in detail. From this
as well as the availability of temperature sensitive
meiotic mutants in yeast we hope to identify which
aspects of the normal vegetative cell cycle are maintained during sporulation and to examine in greater
detail the early regulatory events controlling the
transition of the vegetative cell to sporulation.

ACKNOWLEDGMENTS

The technical assistance of Miss Maria de los
Angeles Argomaniz was greatly appreciated. A portion
of this work was conducted in the Laboratory of Molecular Biology, University of Wisconsin, Madison,
Wisconsin. This work was supported by a Public Health
Service grant Al-1459, a National Science Foundation
grant B-1750, and by a National Science Foundation
fellowship held by one of us (J.E.H.).

REFERENCES

Bresch, C., Muller, G., and Egel, R. (1968). *Mol.
Gen. Genet. 102*, 301.

Dawes, I. W., Kay, D., and Mandelstam, J. (1971).
Nature 230, 567.

Esposito, M. S., Esposito, R. E., Arnaud, M., and
Halvorson, H. O. (1968). *J. Bacteriol. 100*, 180.

Esposito, M. S., Esposito, R. E., Arnaud, M., and
Halvorson, H. O. (1970). *J. Bacteriol. 104*, 202.

Esposito, M. S., and Esposito, R. E. (1969). *Genetics
61*, 79.

Halvorson, H. O., Carter, B. L. A., and Tauro, P.
(1971). *Advances in Microbial Physiology 5*, 47.

Hartwell, L. H., Culotti, J., and Reid, B. (1970).
Proc. Nat. Acad. Sci. USA 66, 352.

Kadowaki, K., and Halvorson, H. O. (1971). *J. Bacteriol.* *105*, 826, 831.

Sherman, F., and Roman, H. (1963). *Genetics 48*, 255.

Taber, R. L., and Vincent, W. S. (1969). *Biochim. Biophys. Acta 186*, 317.

Udem, S. A., and Warner, J. R. (1971). (in press).

Yanagita, T., Yagisawa, M., Oishi, S., Sando, N., and Suto, T. (1970). *J. Gen. Appl. Microbiol. 16*, 347.

Quantal Control

M. Sussman

Department of Biology
Brandeis University
Waltham, Massachusetts

P. C. Newell

Microbiology Unit
Department of Biochemistry
Oxford University
Oxford, England

INTRODUCTION

Any developmental program can ultimately be looked upon as a complex sequence of differential gene expression. Given a permissive environment and, where necessary, an external signal to trigger the start of the sequence, we imagine that the system thereafter generates its own cues and thereby ensures temporal control over the flow of developmental events. We guess, in some cases with increasingly good reason, that these cues emerge from protein actions and protein-protein interactions and that they serve to initiate the transcription of specific parts of the genome and to control the translation of the mRNA transcripts.

But control over the rate and the extent of gene expression is as much a hallmark of a developmental program as is its temporal precision. The formal scheme described above does not make provision even in principle for the fact that the amounts to which many gene products accumulate during development appear to be programmed. The work reported here bears on this aspect of the regulatory apparatus. For the past few years we have been engaged in studying the accumulation and disappearance of four functionally related enzymes during specific stages of cellular slime mold development. A common regulatory pattern has emerged which we have termed, Quantal Control. It may apply to other developmental systems as well, notably ciliogenesis during sea urchin cleavage (Stephens, 1972) and the sequential synthesis of enzymes during the cell growth and division cycle of the yeast *Saccharomyces cerevisiae* (Tauro and Halvorson, 1966).

THE DEVELOPMENTAL PROGRAM

Figure 1 summarizes the program of fruiting body construction in *Dictyostelium discoideum* as it is currently understood. The vegetative amoebae, after cessation of growth, collect together into multicellular aggregates. Depending on external conditions, each aggregate can either (a) transform itself directly into a fruiting body at the site of aggregation or (b) transform itself into a migrating slug and move away. The conditions favoring (a) include: low pH (below 6.5) and a buffer concentration sufficient to keep it there; high ionic strength in the substratum; absorption and removal of a volatile, alkaline material that is produced and excreted by the aggregating cells. (It is *not* NH_3.) The conditions favoring (b) include: high pH (7 or above) and the absence of buffer; low ionic strength in the substratum; the presence of the volatile, alkaline cell product at concentrations above threshold (Newell, Telser and Sussman, 1969).

As Figure 1 indicates, fruit construction can be completed within 24 hr at 22°. If, however, the aggregate transforms itself into a slug, it can continue to migrate for many days (Slifkin and Bonner, 1952). At any time, however, the slug can be induced to stop

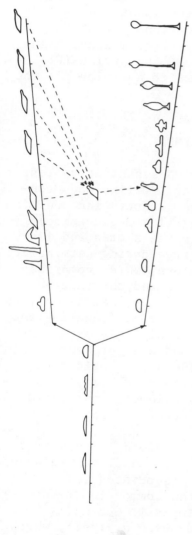

Figure 1. Alternative pathways to fruiting body construction in *Dictyostelium discoideum*. Vegetative cells are harvested from growth plates or liquid medium toward the end of the log phase. They are washed and dispensed in aliquots of 10^8 cells on 2 inch circles of Whatman #50 filter paper resting inside small petri dishes, each on an absorbant pad saturated with a solution (LPS) containing streptomycin, salts and .04 M phosphate buffer at pH 6.5. Cemented to the lid are two absorbant pads saturated with 1 M phosphate buffer, pH 6.0. Under these conditions, 10^8 cells form about 10^3 aggregates and these develop synchronously and directly into fruiting bodies. If the phosphate concentration in the LPS solution is reduced below .01 M, the salts are eliminated, and the pads in the lid are omitted, the aggregates develop into migrating slugs. If, however, the latter are transferred to fresh pads and exposed to the conditions which favor fruit construction as described above, they immediately revert to the fruiting mode.

migrating and to construct a fruiting body. This can
be done by shifting to the environmental conditions
which favor pathway (a). It can also be done by exposing the slug to omnidirectional light. In the dark,
the slug moves at random. In a horizontal light gradient it moves directly toward the light source. But
if exposed to overhead, omnidirectional light for as
little as 30 min, it immediately stops migrating and
constructs a fruiting body over a 7 hr period (Fig. 2).
The omnidirectional l ght appears to override all the
other environmental signals.

THE FOUR ENZYMES

   Polysaccharide and disaccharide synthesis represent
major metabolic activities during fruiting body construction by *D. discoideum*. The products include: the
disaccharide trehalose (glucose-1,1-α-D-glucoside),
which accumulates exclusively in the spores and is
used as a carbon and energy reserve during germination
(Clegg and Filosa, 1961); cellulose, which comprises
the rigid walls of the stalk cells and the tapered
cylindrical outer sheath that encloses them as well as
serving as a part of the spore covering (Gezelius and
Ranby, 1957; White and Sussman, 1961); glycogen, em-
bedded in the stalk integument and also present within
the cells (White and Sussman, 1963a); an acid mucopolysaccharide composed of galactose, *N*-acetylgala-
tosamine, and galacturonic acid uniquely associated
with the spores probably as an outer coating (White
and Sussman, 1963b).
   The enzymes under study are involved in this metabolic network. Figure 3 shows the functional relationships among them. They are associated in a
branched pathway leading from the common intermedi-
ates, G-1-P and UTP. The key logistic enzyme is *UDPG
pyrophosphorylase* (Ashworth and Sussman, 1967), which
supplies UDPG to at least four branches. One branch
leads to trehalose accumulation and involves the enzyme *Trehalose-6-P synthetase* (Roth and Sussman,
1968a). Another branch leads through UDPGalactose to
the synthesis of mucopolysaccharide and involves first

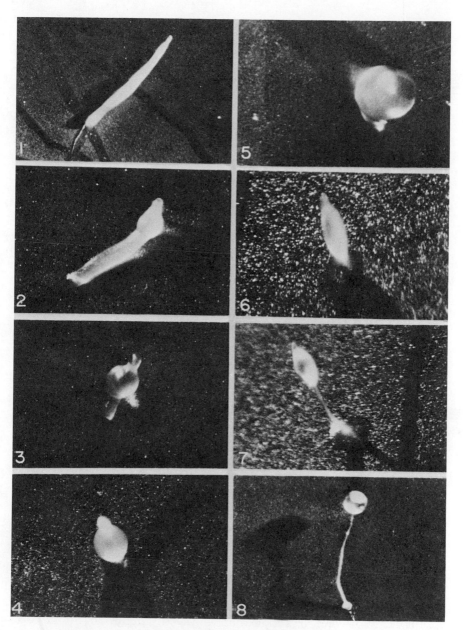

Figure 2. Sequence of fruit construction
by slugs exposed to overhead light (Newell,
Telser and Sussman, 1969).

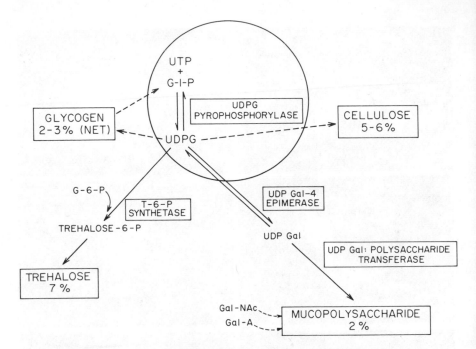

Figure 3. Pathways of polysaccharide and
disaccharide synthesis during fruiting
body construction in D. *discoideum.*

UDPGal-4-epimerase (Telser and Sussman, 1971) and then
UDPGal: polysaccharide galactosyl transferase (Sussman
and Osborn, 1964).

*The Pattern of Enzyme Accumulation
and Disappearance During Fruit
Construction*

Figure 4 summarizes the developmental kinetics of
the four enzymes during fruiting body construction.
At intervals, cells were harvested from filters,
broken, and the extracts were employed for measurements of specific enzyme activity. The patterns of
accumulation and disappearance were found to be keyed
precisely to particular morphogenetic stages. The
peak values attained by the population as a whole obviously must depend markedly upon the synchrony of

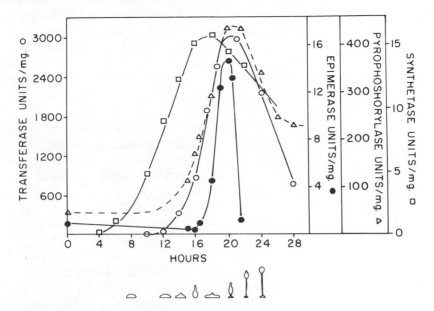

Figure 4. Developmental kinetics of the
four enzymes during fruit construction.

development. Routinely this is better than ± 0.5 hr
(out of the 24 hr required for fruit construction) and
the peak values varied by about ± 10% from experiment
to experiment.

In mutant strains which display morphogenetic de-
fects, these patterns of accumulation and disappearance are disturbed in a systematic way (Sussman and
Sussman, 1969). The disturbances are consistent with
the stage at which the development of the mutant becomes aberrant. In two mutant strains whose temporal
controls are deranged, the derangement was found to
extend also to the timing of enzyme synthesis and
destruction.

Does Accumulation Mean Synthesis?

Almost certainly yes. The accumulation of all four
enzymes has been shown to require concurrent protein
synthesis. Thus the levels of all are frozen within
minutes after exposure of the developing aggregates to

cycloheximide. The most detailed documentation exists
for UDPG pyrophosphorylase which has been purified to
apparent physical homogeneity (Franke and Sussman,
1971). Experiments involving the addition of known
amounts of the purified enzyme to crude extracts demonstrated that, at all stages of development, the assays
of activity in the extracts accurately reflect the concentration of the enzyme. Serological analyses revealed that the accumulation and disappearance of the
enzyme as a single antigenic determinant precisely
coincide with the changes in specific enzyme activity
and failed to demonstrate any cross reactive, enzymatically inactive cell component. Cells have been
pulsed with radioactive amino acids before, during,
and after pyrophosphorylase accumulation. The enzyme
has been precipitated from the crude extracts with
antiserum, and the precipitates dissolved in sodium
dodecyl sulfate and fractionated by SDS-acrylamide gel
electrophoresis in order to recover the enzyme subunit.
The pattern of radioactive incorporation is consistent
with the conclusion that the enzyme is synthesized *de
novo*, that there is no significant pool of inactive
subunits or precursor material and that there is no
significant level of pyrophosphorylase turnover prior to
the period of accumulation. Based upon the specific
activity of the purified enzyme, UDPG pyrophosphorylase
at its peak comprises about 0.5% of the cell protein.

*Does Each Enzyme Accumulate
Synchronously in All the Cells?*

Developing aggregates and migrating slugs have been
segmented into apex and base, front and back, core and
periphery, etc. and were employed for measurements of
UDPG pyrophosphorylase specific activity. At all
stages, the enzyme appeared to be homogeneously distributed (Newell, Ellingson and Sussman, 1968). Thus
if asynchrony of accumulation exists it would have to
be randomly distributed, i.e., limited to every other
cell for example. This highly unlikely eventuality
will be tested using labeled antibody preparations.
Even at its peak, UDPGal polysaccharide transferase
is absent from the stalk cells of the developing ag-

gregates and is present only in the prespore contingent (Newell, Ellingson and Sussman, 1968). (The mucopolysaccharide product is itself associated exclusively with the spores in the mature fruit.) However, nothing is known about the synchrony of accumulation within the prespore population.

Transcriptive Periods for Enzyme
Accumulation

Actinomycin D stops RNA synthesis in *D. discoideum*. Cells and cell aggregates exposed to it during fruit construction continue to develop at the normal rate for a short time. Morphogenesis continues for 1-3 hr but then stops at specific stages depending on the time of addition (Sussman *et al.*, 1967). Protein synthesis, whether measured by amino acid incorporation, polysome profiles, or accumulation of specific enzymes, appears to continue undiminished for several hours and then comes to a halt. By addition of actinomycin at appropriate times, the accumulation of each of the four enzymes and of at least seven others has been shown to require prior periods of RNA synthesis (Sussman and Sussman, 1965a; Roth, Ashworth and Sussman, 1968b; Sussman and Sussman, 1969).

Figure 5 delineates these transcriptive periods. The corresponding periods of actual enzyme accumulation are included for comparison. Some general conclusions can be made:

(1) The transcriptive periods occupy restricted portions of the developmental sequence and are staggered in time. *Hence, both the initiation and termination of transcription appear to be differentially regulated.*

(2) There are specific time lags of from 1-5 hr between the start of a transcriptive period and the appearance of the corresponding enzyme, and between the termination of the transcriptive period and the cessation of enzyme accumulation. These time lags can be altered drastically by appropriate shifts in the developmental program (see the following section). *Hence the program*

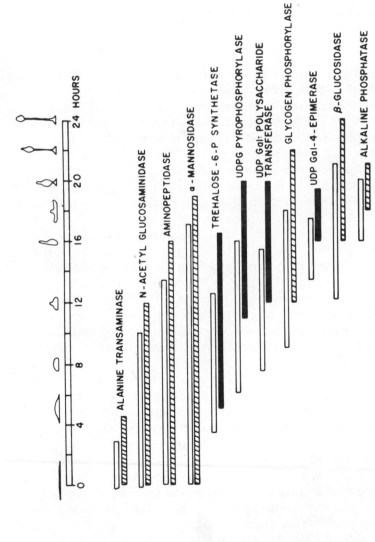

Figure 5. Transcriptive periods for enzyme synthesis. Open areas show transcriptive periods; solid areas show corresponding periods of enzyme synthesis. Data taken from Sussman and Sussman, 1965a; Roth and Sussman, 1968; Roth, Ashworth and Sussman, 1968; Loomis, 1969; Coston and Loomis, 1969; Telser and Sussman, 1970; Newell, Longlands and Sussman, 1971; Firtel and Bonner, 1972; Firtel and Brackenbury, 1972.

must include specific regulation of mRNA trans-
port from nucleus to the site of protein syn-
thesis and/or control of translation.

(3) There is a direct relation between the amount
of transcription allowed and the amount of enzyme that is synthesized. Thus if the developing cells are exposed to actinomycin during a
given transcriptive period, the later the actinomycin is added the greater is the amount of
the corresponding enzyme that subsequently
accumulates. *Hence the transcript, though*
stable for hours (considering the time lags be-
tween transcription and translation) is not in-
definitely stable. This stability appears to
be a function of utilization, not merely time.
Thus, a given amount of transcript, whether
translated immediately or delayed for several
hours by addition of cycloheximide, a reversible inhibitor of protein synthesis, yields
approximately the same amount of enzyme (Sussman, 1966).

The Pattern of Enzyme Synthesis
During Slug Migration

When developing aggregates elect to transform into
migrating slugs rather than to construct fruiting
bodies, the course of enzyme accumulation and disappearance is substantially altered as shown in Figure 6.
The pyrophosphorylase is synthesized very slowly, although it eventually reaches approximately the same
level; but none disappears. The transferase is synthesized at the usual rapid rate and then disappears
very slowly. The epimerase does not accumulate at all
(Newell and Sussman, 1970; Ellingson, Telser and
Sussman, 1971).

This pattern changes drastically whenever the migrating slugs are induced by exposure to omnidirectional light to stop migrating and start constructing
fruiting bodies (Figs. 7 and 8). If the light exposure occurs during the period of slow pyrophosphorylase accumulation the rate abruptly quickens and the
usual peak level of 450 units/mg is reached at the

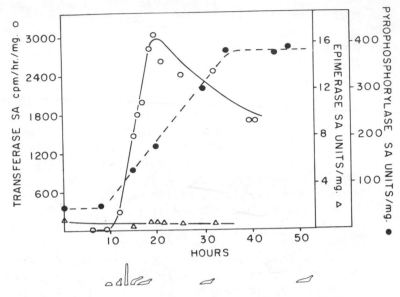

Figure 6. Developmental kinetics of three
enzymes during slug migration (Ellingson,
Telser and Sussman, 1971).

same morphogenetic stage as in aggregates that had
constructed fruits directly without ever having been
slugs. If the light exposure occurs after pyrophosphorylase accumulation has ceased, then a complete
second round of synthesis occurs and a second quantum
of pyrophosphorylase accumulates on top of the first!
It is accompanied by a delayed but otherwise normal
first round of epimerase synthesis. The performance
of UDPGal polysaccharide transferase is the same as
that of the pyrophosphorylase except for the fact that
first, the previously accumulated store of enzyme is
rapidly destroyed* and then the second quantum accumu-

---

*Both the transferase and the epimerase have been
shown to be preferentially released into the extracellular space toward the end of mucopolysaccharide
synthesis (Sussman and Lovgren, 1965b; Telser and
Sussman, 1971). This occurs at the time that the prespore cells extrude certain vacuoles (Hohl and Hamamoto, 1969), and these may in fact be the vehicles of

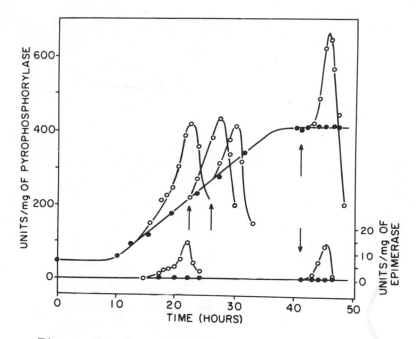

Figure 7. Pyrophosphorylase and epimerase synthesis in slugs which have been induced to fruit (O). The arrows refer to the times at which the slugs were exposed to omnidirectional light. Untreated controls (●). (Newell and Sussman, 1970).

lates. These rounds of enzyme synthesis display three noteworthy features:

(1) They require additional rounds of transcription which are initiated by the exposure of the slugs to omnidirectional light. Figure 9 shows data for the epimerase and transferase. The pyrophosphorylase gave the same results.
(2) Enzymes whose periods of accumulation are separated by as much as 5 hr (pyrophosphorylase vs. epimerase) in aggregates that fruit directly,

---

(* cont.) enzyme release. The abrupt disappearance of transferase seen in Figure 8 may conceivably also be preceded by vacuolar expulsion.

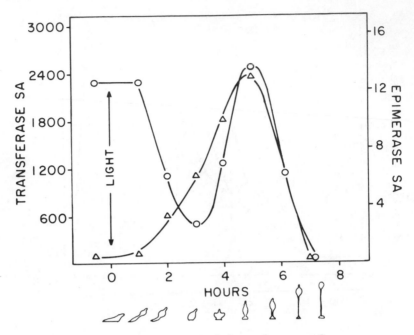

Figure 8.   Epimerase (Δ) and transferase
(O) synthesis in slugs which have been in-
duced to fruit. (Ellingson, Telser and
Sussman, 1971).

can begin simultaneously in slugs that are induced to fruit (see Figs. 7 and 8).   Furthermore, the transcriptive periods separated by as
much as 7 1/2 hr in the first condition (Fig. 6)
are initiated simultaneously in the second
(Fig. 9).

(3) In the normal round of pyrophosphorylase synthesis as it is carried out by aggregates that
construct fruits directly, there is a lag of
about 5 hr between the start of the transcriptive period and the synthesis of the enzyme.
In the second synthetic round (after slugs are
exposed to light), the lag is less than 1 hr.
This is true of the other enzymes also.

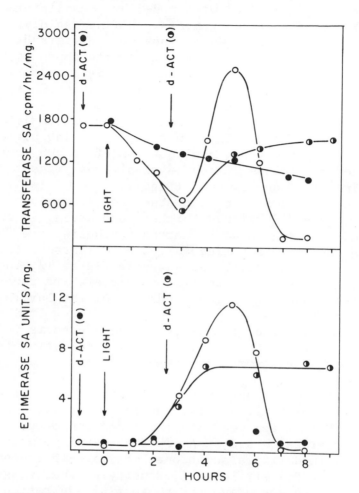

Figure 9. Effect of actinomycin on the
synthesis of transferase and epimerase
after light induction (Ellingson, Telser
and Sussman, 1971).

Enzyme Synthesis After Reaggregation
of Disaggregated Cells

Cell aggregates of *D. discoideum*, even after 18 hr
of development, can be rapidly dispersed to yield a
suspension of separated cells. If redeposited on filters at the original cell density, they reaggregate

almost immediately and within 2-3 hr recapitulate the
normal course of morphogenesis. Having rapidly re-
attained the stage at which they were dispersed, they
go on to construct fruiting bodies with approximately
normal timing (Newell, Longlands and Sussman, 1971;
Newell, Franke and Sussman, 1972). Figure 10 is a
schematic illustration of this process.

What happens to the course of enzyme accumulation
during this period? In each case, following aggregation, a complete additional quantum of enzyme is synthesized regardless of the level already present at
the time of disaggregation (Fig. 11).

What happens if once dispersed cells, having reaggregated and developed further, and having accomplished a second round of enzyme synthesis, are dispersed once more and redeposited on filters at the
original cell density? As shown in Figure 12, they
re-reaggregate and accomplish complete third rounds of
synthesis, once again independently of the levels of
enzyme activity already attained.

These additional rounds of enzyme synthesis induced
after reaggregation display several interesting features which are summarized below:

(1) If the disaggregated cells were redeposited at
    the original density but in the presence of
    EDTA, which prevents the formation of compact
    aggregates and stops further development, they
    formed little additional enzyme. If redepos-
    ited at very low cell density so that reaggre-
    gation was prevented by spatial separation,
    they formed none. Hence the resumption of nor-
    mal cell contacts must precede the acquisition
    of additional enzyme quanta.
(2) If the developing aggregates were exposed to
    actinomycin shortly before disaggregation, any
    further increase in enzyme levels was prevented
    (Fig. 13, left). (The reaggregating cells
    could recapitulate the previous sequence of
    morphogenesis, but then stopped when they
    reached the stage at which they had been dis-
    persed.) In contrast, undispersed cell aggre-
    gates exposed to actinomycin at the same time

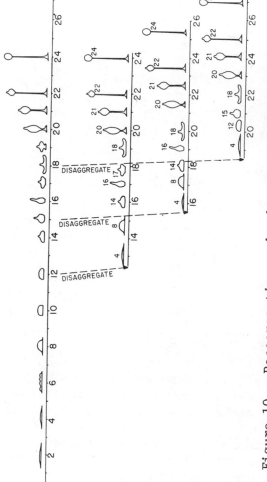

Figure 10. Reaggregation and subsequent development of disaggregated cells (Newell, Longlands and Sussman, 1971).

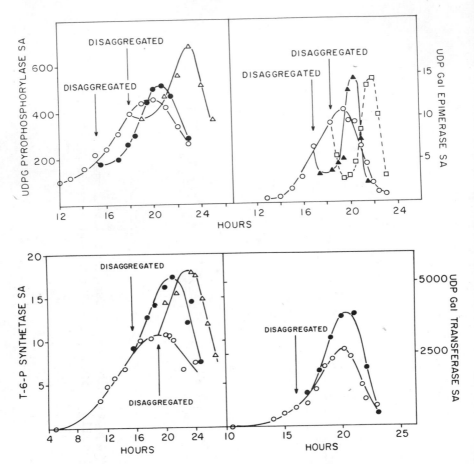

Figure 11. Enzyme synthesis during
reaggregation and subsequent development
(Newell, Longlands and Sussman, 1971).

continued to accumulate enzyme utilizing stored
transcript. Figure 13 (right) shows a corresponding discrepancy between disaggregated and
undisturbed cells when actinomycin was added
shortly after disaggregation. The data suggest
that additional rounds of synthesis require additional periods of transcription subsequent to
disaggregation. Further, the disaggregated
cells are unable to utilize stored transcripts
even after reaggregation.

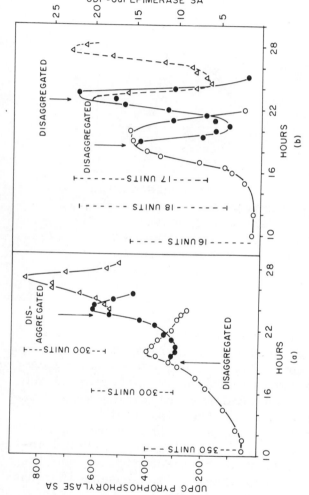

Figure 12 (a, b). Enzyme synthesis during two successive disaggregations and reaggregations. At each arrow, the developing aggregates were dispersed and the cells redeposited on fresh filters at the original cell density (Newell, Franke and Sussman, 1972).

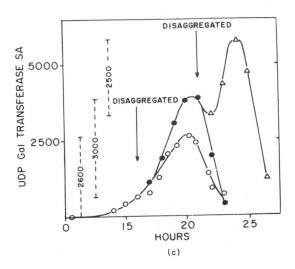

Figure 12(c). Enzyme synthesis during two
successive disaggregations and reaggregations. At each arrow the developing aggregates were dispersed and the cells
redeposited on fresh filters at the original cell density (Newell, Franke and
Sussman, 1972).

(3) During these additional rounds of synthesis,
the temporal relations amongst the four enzymes
and the lags between each transcriptive period
and the accumulation of the corresponding enzyme are altered in a manner analogous to the
alterations observed when migrating slugs are
induced to construct fruiting bodies. Thus
during the first round of UDPG pyrophosphorylase accumulation, a 5 hr lag was observed between the initiation of transcription and the
appearance of the enzyme. In subsequent rounds
the lag was an hour or less.

The data in Figure 14 indicate that disaggregation
and reaggregation do not lead to additional synthetic
rounds for all enzymes. As shown by Loomis (1969),
N-acetylglucosaminidase is present at low activity in
exponentially growing cells. It begins to accumulate

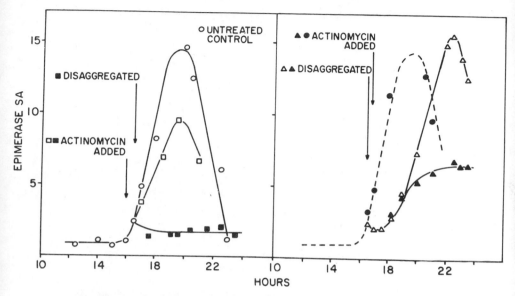

Figure 13. Effect of actinomycin on epimerase synthesis in reaggregated cells (Newell, Franke and Sussman, 1972).

as soon as the cells enter the stationary phase, hours before the onset of aggregation. Studies with mutants incapable of aggregation suggest that its synthesis is neither triggered nor modulated by the matrix of cell interactions that attends and follows aggregation. In addition it appears to have no functional relationship with the other four enzymes. As reported by Loomis, the specific activity of N-acetylglucosaminidase gradually rose over a 16 hr period from 60 units/mg protein to a plateau of about 200. Cells, disaggregated after 15 and 18 hr of development were redeposited on filters at the original cell density. In contrast with the results obtained with the other four enzymes, we find that no further accumulation of activity occurs.

Finally, it should be noted that Firtel and Bonner (1972) recently demonstrated that the enzyme glycogen phosphorylase also appears to be under quantal control in *D. discoideum* by doing disaggregation-reaggregation experiments.

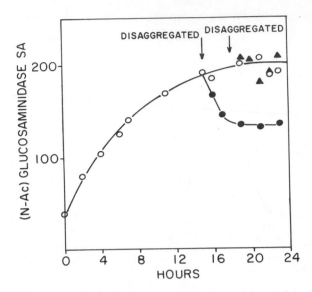

Figure 14.   The level of glucosaminidase in
reaggregated cells (Newell, Franke and
Sussman, 1972).

DISCUSSION

    As described previously, the four enzymes under ex-
amination constitute a functionally related group.
They play pivotal roles in a metabolic framework that
involves major expenditures of energy and raw materials and whose end products are essential to the developmental program of *Dictyostelium discoideum*.   They
can be assayed precisely and conveniently in crude extracts.   With degrees of rigor ranging from extremely
good to adequate, their specific activities have been
shown to be accurate reflections of their concentrations within the cells at any stage of development.
Hence they would seem to provide extremely useful
markers with which to explore various aspects of the
developmental program.   Here they have been used in an
attempt to deduce some of the rules that govern gene
epxression in these organisms.

*Some Statements About the Rules of
Gene Expression During Slime Mold
Development*

The following interpretations have been extracted
from the previously described results.

(1) *The accumulation of the four enzymes is in the
first instance the result of differential transcription.* The transcription of particular
genomic regions appears to be initiated and
terminated in precise accord with the flow of
morphogenetic events; events such as the establishment of cell contacts during aggregation
or the attainment of a particular stage of morphogenesis during slug migration or fruit construction, or in the shift from the one to the
other. However, this transcription does not
occur as if the genome were a tape unrolling
unidirectionally at an invariant rate. Depending upon shifts in the program as between migrating slugs and fruiting bodies, a given region may be transcribed now or later or not at
all. A region transcribed once, may be transcribed again as for example when, after one
round of enzyme synthesis, a slug is induced to
construct a fruiting body or when a cell aggregate is dispersed and caused to recapitulate the
morphogenetic sequence. In addition the temporal relations between the different transcriptive periods may be drastically altered by
shifts in the program so that transcriptive
periods that begin as much as 7 hr apart in one
mode may coincide in another. One begins to
suspect that feedback circuits may operate between the morphogenetic and transcriptive
events.

(2) *Specific time lags exist between the initiation
of transcription and the synthesis of the protein product.* It is possible that these lags
stem from the time required to process the
transcript and convey it to the cytoplasm or
from the operation of specific controls over

the initiation of its translation or from both.
Specific controls must operate here, too, since
nascent mRNA (and ribosomal subunits) have been
shown to appear in the cytoplasm within minutes
after fabrication (Kessin, 1971). Moreover,
these time lags are not invariant but depend
upon shifts in the developmental program.

(3) *The translative process is also keyed to morphogenetic events.* Thus, when, during the first
slow round of UDPG pyrophosphorylase synthesis
a migrating slug is induced to construct a
fruiting body, the rate of synthesis abruptly
increases. When developing aggregates in the
midst of enzyme synthesis employing stored reserves of transcript are dispersed, they immediately jettison these reserves and all further
synthesis of this particular group of enzymes
ceases until (and only until) reaggregation has
occurred and a new round of transcription has
commenced.

(4) *Whether the transcription of a particular portion of the genome is initiated now or later,
whether once or twice or even three times, the
response is always quantal* and a preset quantity of the protein product accumulates, i.e.,
300-350 units of UDPG pyrophosphorylase, 15
units of UDPGal-4-epimerase, etc. This result
implies that both the transcription of these
genes and the translation of the resulting mRNA
must be quantal processes.

(5) *Recapitulation of a part of the developmental
sequence induced by disaggregation does not
necessarily involve reactivation of all parts
of the genome that were initiated during the
first round of development.* Thus the accumulation of N-acetylglucosaminidase starts long before aggregation. Evidently the dispersal of a
developing aggregate does not provide the cue
to reactivate this portion of the genome and
the level of glucosaminidase is unaffected.

A Ticketing Theory of mRNA Translation:
One Possible Means of Quantal Control

The model illustrated in Figure 15 has recently
been proposed (Sussman, 1970). It is imagined that,

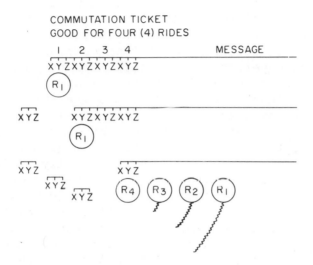

Figure 15. A ticketing model for control
of mRNA translation (Sussman, 1970).

in eukaryotes, each molecule of mRNA contains at its
5' hydroxyl end a set of tandemly repeated base sequences shown for convenience as triplets but probably
much longer than that. Each sequence serves as a
binding site for a ribosome and is usable only once.
It is postulated that only the terminal sequence is
sterically available for binding at any time. As the
bound ribosome moves along the mRNA, one of its proteins, an excising enzyme, would clip the terminal
sequence. That ribosome would pass by the remaining
repetitious sequences (the endonuclease would no
longer be in register for its attack), reach the initiating codon and begin translation. In the same way
a second, third, and fourth ribosome could attach successively and then excise the then terminal sequence.
After the last repetitious sequence has been removed,
no additional ribosomes can attach.

Two considerations follow from this model. First
it automatically provides for a quantal control of
translation. Whether translated now or later, rapidly
or slowly, the *functional* life span of a message would
be invariant. Second, mRNA molecules transcribed from
different genes having different numbers of tandemly
repeated sequences would have different functional
life spans and different levels of the corresponding
proteins would accumulate automatically. Mutations,
deletions, or duplications of the repetitious DNA sequences could permit natural selection to operate in
regulating protein levels in the cells.

The model can be expanded to provide for qualitative as well as quantitative control. Thus two or
more classes of mRNA molecules might exist, each with
a different tandemly repeated sequence occupying the
5' termini. These would serve as mutually exclusive
attachment sites for different classes of ribosomes.
In this way a cell might be potentially capable of
engaging in either of two alternative developmental
programs. The appropriate mRNA molecules might be
marked by one or the other repetitious sequence.
Though both classes are transcribed, only one
might be transported to the cytoplasm if only the
corresponding class of ribosomes were available for
attachment. The remaining class of mRNA would then
remain in the nucleus and be turned over.

ACKNOWLEDGMENTS

The results reported here were obtained by the com-
bined efforts of many individuals, notably including:
J. M. Ashworth, S. Cocucci, J. S. Ellingson, J. Franke,
W. F. Loomis, Jr., R. Roth, R. Sussman, and A. Telser.
This work was supported by National Science Foundation
grant GB5976X. M. S. was the recipient of a Career
Development Award from the National Institutes of
Health (K3-GM-1313). P. C. N. gratefully acknowledges
assistance from the Cancer Research Campaign.

REFERENCES

Ashworth, J. M., and Sussman, M. (1967). *J. Biol. Chem. 242*, 1696.

Clegg, J., and Filosa, M. (1961). *Nature 192*, 1077.

Coston, M. B., and Loomis, W. F., Jr. (1969). *J. Bacteriol. 100*, 1208.

Ellingson, J. S., Telser, A., and Sussman, M. (1971). *Biochim. Biophys. Acta 244*, 388.

Firtel, R., and Bonner, J. (1972). *Develop. Biol.* (in press).

Firtel, R., and Brackenbury, R. (1972). *Develop. Biol.* (in press).

Franke, J., and Sussman, M. (1971). *J. Biol. Chem. 246*, 6381.

Gezelius, K., and Ranby, B. (1957). *Exp. Cell Res. 12*, 265.

Hohl, H. R., and Hamamoto, S. T. (1969). *J. Ultrastructure Res. 26*, 442.

Kessin, R. (1971). Ph.D. Thesis. Brandeis University.

Loomis, W. F., Jr. (1969). *J. Bacteriol. 97*, 1149.

Newell, P. C., Ellingson, J. S., and Sussman, M. (1968). *Biochim. Biophys. Acta 177*, 610.

Newell, P. C., Franke, J., and Sussman, M. (1972). *J. Mol. Biol.* (in press).

Newell, P. C., Longlands, M., and Sussman, M. (1971). *J. Mol. Biol. 58*, 541.

Newell, P. C., and Sussman, M. (1970). *J. Mol. Biol. 49*, 627.

Newell, P. C., Telser, A., and Sussman, M. (1969). *J. Bacteriol. 100*, 763.

Roth, R. M., Ashworth, J. M., and Sussman, M. (1968). *Proc. Nat. Acad. Sci. USA 59*, 1235.

Roth, R. M., and Sussman, M. (1968). *J. Biol. Chem. 243*, 5081.

Slifkin, M., and Bonner, J. T. (1952). *Biol. Bull. 102*, 273.

Stephens, R. E. (1972). This volume, p. 343.

Sussman, M. (1966). *Proc. Nat. Acad. Sci. USA 55*, 813.

Sussman, M. (1970). *Nature 225*, 1245.

Sussman, M., Loomis, W. F., Jr., Ashworth, J. M., and
    Sussman, R. R. (1967). *Biochem. Biophys. Res. Comm.*
    26, 353.
Sussman, M., and Lovgren, N. (1965). *Exp. Cell Res.*
    38, 97.
Sussman, M., and Osborn, M. J. (1964). *Proc. Nat. Acad.*
    Sci. USA 52, 81.
Sussman, M., and Sussman, R. R. (1965). *Biochim. Bio-*
    phys. Acta 108, 463.
Sussman, M., and Sussman, R. R. (1969). *Symp. Soc. for*
    Gen. Microbiol. XIX, 403.
Tauro, P., and Halvorson, H. O. (1966). *J. Bacteriol.*
    92, 652.
Telser, A., and Sussman, M. (1971). *J. Biol. Chem. 246*,
    2252.
White, G. J., and Sussman, M. (1961). *Biochim. Biophys.*
    Acta 53, 285.
White, G. J., and Sussman, M. (1963a, b). *Biochim. Bio-*
    phys. Acta 74, 173, 179.

Control Aspects of Glutamine Synthetase Induction in Embryonic Neural Retina

P. K. Sarkar, A. W. Wiens, M. Moscona, and A. A. Moscona

Department of Biology
The University of Chicago
Chicago

INTRODUCTION

The induction of tissue-specific enzymes in embryonic cells represents an important aspect of differentiation and therefore the underlying mechanisms merit detailed analysis. To facilitate studies on enzyme induction in embryonic vertebrate development, experimental systems are required which meet the following conditions: (1) A well-defined population of embryonic cells that can be isolated in sufficient amounts for biochemical work in different phases of embryonic development, and in which a tissue-characteristic enzyme increases sharply at a specific stage of differentiation; (2) a chemically defined inducer which can elicit in these cells the induction of this enzyme precociously (i.e., at a much earlier embryonic age than that at which the enzyme increases in normal embryogenesis); such precocious induction is especially

important because, in order to achieve a true understanding of the mechanisms which control differentiation, it is essential to be able to temporarily modify
developmental processes by eliciting the expression of
specific genes before the normal time in ontogeny.
The prococious induction of *glutamine synthetase* in
the embryonic chick neural retina (Moscona, 1971) by
11-β hydroxycorticosteroids meets these conditions and
provides a model system suitable for studying mechanisms which control specific macromolecular synthesis
in the differentiation of higher vertebrate embryonic
cells. Analysis of this system, together with studies
on enzyme patterns in slime mold life cycles (Sussman,
1966; Gerisch, 1968), insect metamorphosis (Wyatt,
1968; Wigglesworth, 1966), and on enzyme induction in
neoplastic cell cultures (Tomkins, 1969) have already
contributed significantly towards elucidation of factors and processes in differentiation.

Glutamine Synthetase

   Glutamine synthetase (GS) [1-glutamate:ammonia li-
gase (ADP), EC 6.3.1.2], catalyzes the conversion of
1-glutamate to glutamine. Glutamine is, of course,
important in many biosynthetic processes, including
those for nucleotides and polysaccharides. In neural
cells, GS is of special interest because of the postulated role of glutamine and its metabolites in membranes and synapses (Salganicoff and DeRobertis, 1965)
and the importance of glutamine as a donor of NH_3 in
the biosynthesis of pyridine nucleotide coenzymes.
   GS purified from sheep brain (Tate and Meister,
1971; Ronzio *et al.*, 1969) was reported to have a molecular weight of 392,000 and to consist of eight apparently identical subunits. GS from chicken neural
retina was recently purified in our laboratory and
determined to have a subunit molecular weight of
42,000; the subunits appear to be identical and have a
relatively high content of aspartic acid (Sarkar and
Moscona, 1972). Routinely, the specific activity of
GS is assayed in tissue sonicates or homogenates by
the glutamyl-transferase reaction (Kirk and Moscona,
1963; Moscona, Moscona and Saenz, 1968). For radio-

immunochemical measurements of the enzyme, antisera prepared against purified GS were used (Alescio and Moscona, 1969; Alescio, Moscona and Moscona, 1970).

The Normal Developmental Pattern of Glutamine Synthetase in Embryonic Neural Retina

During early embryonic development of the chick, the level of GS specific activity in the neural retina is low and increases very slowly; however, on about the 16th day of embryogenesis retinal GS activity begins to rise very sharply and increases approximately 100-fold over the next six days, then plateaus at this high level. In the chick embryo this pattern of changes in GS activity is characteristic for the developmental program of the neural retina and is not seen in other tissues (with the exception of certain brain regions associated with vision) (Shimada, Piddington and Moscona, 1967; Piddington, 1971). It is particularly important that the increase of retinal GS activity coincides closely with the functional differentiation and maturation of the neural retina (Piddington and Moscona, 1965). A similar corresondence exists also in the development of the mammalian eye, in which retinal GS activity increases sharply at the time of functional maturation of the retina (Chader, 1971). The specific metabolic reasons for the very marked increase of GS during retina maturation remain to be determined.

The sharp increase in GS activity in the neural retina is not due to cell proliferation; it takes place at a time at which there is very little overall growth in the retina. In the chick embryo neural retina, cell number and total protein increase up to about the 12th day of embryonic development, but thereafter growth of this tissue ceases almost completely (Fig. 1); the overall rate of RNA synthesis in the neural retina (determined as the incorporation of labeled uridine into total cellular RNA) declines by about 60% between the 8th and the 12th day of embryonic development (Barbara M. Goldman, unpublished data) and remains essentially unchanged through the 18th day,

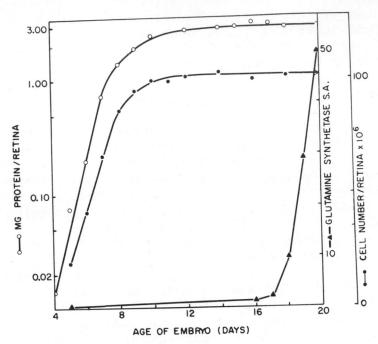

Figure 1. Glutamine synthetase activity in
the neural retina of the chick embryo increases sharply after the 16th day of embryonic development. Cell number and total
protein content in the neural retina increase until the 11-12th day of embryogenesis and do not change conspicuously thereafter (Moscona, 1971).

coincident with the decline in the rate of growth in
the retina during this period. Thus, on the 16th day
—i.e., at the time when GS activity begins to rise
sharply—in the embryo the neural retina is essentially a nongrowing tissue engaged predominantly in differentiation; the rapid increase of GS activity is not
accompanied by measurable increases in the rate of
total RNA, DNA or protein synthesis.
    There is evidence that the sharp rise of GS in the
retina starting on the 16th day of embryonic development is triggered by a corticosteroid which becomes
available in the embryo at this time (Piddington, 1970)

and serves as the natural inducer of the enzyme, possibly in conjunction with a fetal serum factor (Moscona and Kirk, 1965; Moscona, 1971).

*Precocious Induction of Retinal
Glutamine Synthetase by
Glucocorticoids*

The rapid increase of GS activity in the neural retina can be elicited precociously, several days before the regular time, by making the steroid inducer available prematurely to the retina. This induction is obtainable both *in vitro* and *in vivo* by treatment with certain 11-β hybroxycorticosteroids (Fig. 2), in-

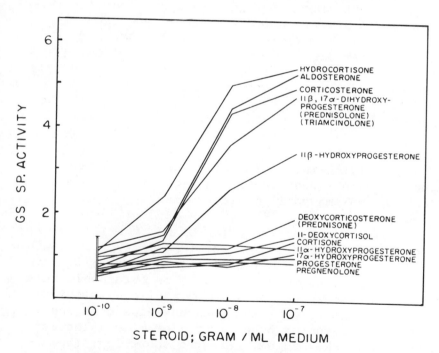

Figure 2. Precocious hormonal induction of GS isolated neural retina from 12-day chick embryos: dose-response effects of various steroids during 24 hr in culture (Moscona, 1971).

cluding hydrocortisone, corticosterone, aldosterone
and related synthetic steroids (Moscona, 1971; Moscona
and Piddington, 1966; Reif-Lehrer and Amos, 1968).
For obvious technical reasons the tissue culture system (actually "organ culture") is more suitable for
detailed analysis on the mechanism of induction, and
most of the studies were carried out in Erlenmeyer
flask cultures of isolated retina tissue obtained from
12-day chick embryos, using hydrocortisone (HC) as the
inducer (Moscona, Moscona and Saenz, 1968).

GS induction by HC is selective for the neural
retina in that the steroid does not similarly induce
GS in other chick embryonic tissues, (with the exception of certain brain regions associated with vision,
Piddington, 1971). The inductive effect is limited
to this particular group of 11-β hydroxycorticosteroids and no other hormone yet tested was found to
induce GS in the neural retina; cyclic AMP and its
dibutyryl derivative do not induce retinal GS; nor do
they enhance the inductive effect of HC (Moscona, 1971;
Moscona and Jones, in preparation).

Mechanism of Glutamine Synthetase
Induction

In the following we summarize some of the experimental findings concerning the precocious induction
of GS in the embryonic neural retina.

The precocious induction of GS in vitro is eli-
cited by adding hydrocortisone (HC; 0.3 µg/ml) to cultures of retina tissue freshly isolated from 12-day
chick embryos; GS induction by HC is obtainable in a
culture medium which consists only of physiological
salt solution (Tyrode) and does not depend on the
presence of serum or other macro- or micromolecular
additives in the medium. After a short lag, the enzyme activity begins to rise sharply and in 24 hr
reaches values 8 to 15 times higher than the uninduced
level (Moscona, Moscona and Saenz, 1968; Moscona, 1971).

The induced increase of GS activity requires con-
tinuous protein synthesis and was demonstrated immuno-

chemically to be due to increased rate of synthesis of the enzyme and to accumulation (Fig. 3) of the newly

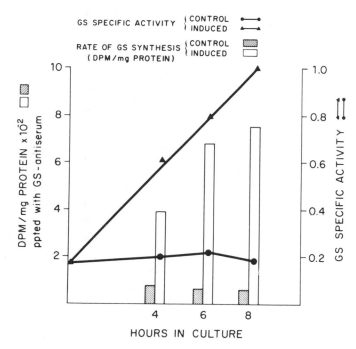

Figure 3. Induction of GS in embryonic neural retina: changes in the rate of enzyme synthesis measured by immunotitration of the radioactively labeled enzyme. Comparisons of GS specific activities and rates of GS synthesis, in induced and control retinas. (Moscona, Frenkel and Moscona, 1972.)

made GS, and not to activation of preformed enzyme precursors (Alescio and Moscona, 1969; Alescio, Moscona and Moscona, 1970; Moscona, Frenkel and Moscona, 1972). Immunological tests revealed no disparities between the enzyme which is responsible for GS activity in the uninduced retina, that induced by HC, and that present in mature retina; this, and other evidence (Kirk and Moscona, 1963) strongly indicate that in GS induction

the steroid elicits increased synthesis and accumulation of the same enzyme which is present at low level
in the uninduced retina, and that we are not dealing
here with different enzyme proteins that have similar
enzymic activities.

*Although the induction of GS represents increased
synthesis of the enzyme, this enhancement is differential* and does not involve measurable changes in the
overall rate of protein synthesis, or in the total
protein content of the retina. Unlike in other systems (Manchester, 1970; Tata, 1971) in the retina the
steroid inducer does not elicit measurable changes in
the rate or overall pattern of RNA synthesis. Therefore, retinal GS induction represents a differential
or selective biosynthetic response of the retina cells
to the steroid inducer (Moscona, Frenkel and Moscona,
1972).

*A close correspondence exists in GS induction be-
tween the increase of enzyme activity and the accumulation of newly made enzyme.* Using radioimmunochemical
methodology it was established (Moscona, Frenkel and
Moscona, 1972) that the low level of GS in the uninduced neural retina represents a balance between a low
rate of enzyme synthesis and continuous enzyme degradation; and that the HC-induced accumulation of GS in
the retina is due predominantly to a progressive increase in the rate of enzyme synthesis, and not to
cessation of enzyme turnover.

*Concurrent DNA synthesis in the retina is not
essential for GS induction;* complete inhibition of DNA
synthesis simultaneously with the addition of the
steroid inducer, or preceding addition of the inducer,
does not prevent the induction of GS (Moscona, Moscona
and Jones, 1970; Moscona, 1972). Moreover, prolonged
inhibition of DNA synthesis before the addition of the
inducer does not measurably reduce the inducibility of
the retina. In the experiments summarized in Figure 4
retinas from 9-day chick embryos were treated in culture for 24 hr with 10^{-5} M cytosine arabinoside which
rapidly and completely stops thymidine incorporation

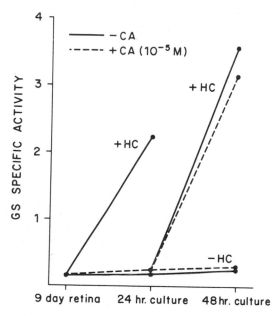

Figure 4.   Effect of continuous inhibition
of DNA synthesis in embryonic neural retina
on inducibility of GS.   Retinas isolated
from 9-day chick embryos were cultured in
the presence of 10^{-5} M cytosine arabinoside
(CA) and without it; to some of the cultures, hydrocortisone was added at zero hr
$(0.3 \times 10^{-7}$ g/ml); to others it was added
after 24 hr.   Controls received no steroid.
The data show that continuous inhibition
for 48 hr of DNA synthesis did measurably
reduce the responsiveness of the retina to
GS induction (Moscona, 1972).

into retina DNA; the retinas were then exposed to the
inducer (HC) for the ensuing 24 hr in the continuous
presence of the inhibitor.   In spite of this persis-
tent inhibition of DNA synthesis for 48 hr the inducibility of GS was not significantly reduced as shown by
the fact that, following the addition of HC, enzyme
activity rose to levels similar to those induced in
retinas that were not treated with the inhibitor.   It

is therefore clear, that in this system the inducer effect does not depend, directly or indirectly, on DNA replication, either ongoing at the time of inducer addition, or during the preceding 24 hr. Although no direct evidence is available concerning gene amplification in GS induction, the above findings make this possibility unlikely.

GS induction in the retina requires RNA synthesis: complete inhibition of transcription [with 1 to 10 μg/ml of actinomycin D (Act D)] at the time of inducer addition prevents the increase of GS. However, if inhibition of RNA synthesis is delayed until various times after the beginning of induction, continued accumulation of GS becomes progressively less sensitive to this inhibition; thus, blocking transcription after 4 hr of induction does not stop GS synthesis, and the amount of enzyme continues to increase.

Analysis of these and related findings indicates that HC elicits the accumulation of RNA templates for GS synthesis; these templates are stable in that, following their accumulation GS continues to be made even though further transcription is completely halted with Act D. This conclusion was substantiated by immunochemical measurements of the rates of GS synthesis which showed that, if transcription is stopped with Act D after 4 hr of induction, GS synthesis continues but the *rate* of enzyme synthesis remains at the 4 hr level for several hours; subsequently it declines and stops, presumably due to depletion of components or precursors necessary for GS synthesis. In spite of the arrested and declining rate of GS synthesis, the enzyme accumulates in this situation to levels similar to those in normal induction, although in the latter case the rate of enzyme synthesis continues to increase with time; the reason for this is that Act D stops the degradation of GS, while in normal induction degradation continues, resulting in similar enzyme levels in both cases (Moscona, Frenkel and Moscona, 1972).

The induced accumulation of active and stable RNA templates for GS synthesis is not dependent on con-

current protein synthesis. This was shown by experiments in which cycloheximide (at concentrations inhibitory to protein synthesis) was added to retina cultures together with HC, or 1 hr earlier; after 4 hr of incubation the inhibitor was washed out to reinitiate protein synthesis, and simultaneously Act D (10 µg/ml) was added to halt all further transcription. During further incubation, GS activity increased significantly, practically without a lag period. Therefore, (a) uninduced cells contain the proteins necessary for transcribing and processing the RNA required for GS synthesis; (b) the critical inductive effect of HC does not take place at the level of translational processes but at the level of transcription, or between transcription and translation. Thus, the overall evidence favors the conclusion that the inducer elicits accumulation of transcripts for GS synthesis (Moscona, Moscona and Saenz, 1968; Moscona, Frenkel and Moscona, 1972); it is possible that HC causes increased transcription of mRNA for GS; or that it causes preservation of transcripts that would otherwise turn over (Moscona, 1972).

The continuous presence of HC is required to maintain GS synthesis at the induced rate (Moscona, Frenkel and Moscona, 1972). If after 4 hr of induction, the inducer is thoroughly washed out from the cultures, the rate of GS synthesis slows down and eventually declines almost to that of uninduced controls (Fig. 5). Since, after 4 hr of induction stable templates for GS are present, we conclude that their expression (or persistence) depends on some processes for which the constant availability of the inducer is essential. However, if simultaneously with the withdrawal of the inducer, RNA synthesis is completely stopped (Act D; 10 µg/ml), the preformed templates for GS continue to make the enzyme for at least several hours at the 4-hr rate (Fig. 5).

Further analysis of these effects suggests that the translational expression of the structural transcripts for GS synthesis is controlled by labile products of regulatory genes: a "suppressor," and a "desuppressor"

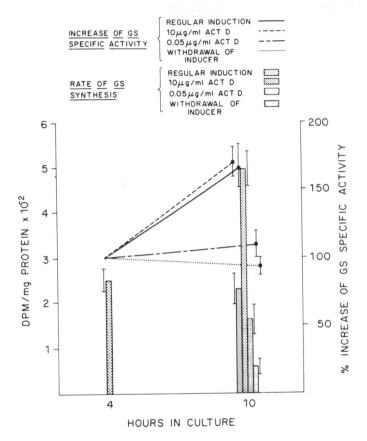

Figure 5. Effects of additions of Act D (high and low doses) and of inducer withdrawal, after 4 hr of induction on GS specific activity and on the rate of GS synthesis. The rate of synthesis was measured by immunoprecipitation (Moscona, Frenkel and Moscona, 1972).

which counteracts the effect of the suppressor. The suppressor is being continuously produced; the desuppressor is elicited by the steroid inducer and can be inhibited by a low dose of Act D (Fig. 5). Withdrawal of the inducer "releases" the suppressor and GS synthesis declines. Complete inhibition of transcription in the induced retina stops both these regulators and allows for continued expression of the preformed templates. Thus, the function of the steroid inducer in this system is not only to elicit the accumulation of stable mRNA for GS, but also to stimulate transcription for the labile desuppressor whose presence is essential for the continued expression of GS templates at the induced rate (Moscona, Frenkel and Moscona, 1972; Moscona, 1972).

A working hypothesis and a model for the mechanism of GS induction, based on the above and related findings was described in detail elsewhere (Moscona, Frenkel and Moscona, 1972; Moscona, 1972).

In order to define in further detail the molecular events in GS induction, we have pursued the following lines of work: (a) analysis of GS synthesis at the polysomal level; (b) inhibition of GS induction by proflavine. Some of the results are summarized here.

Analysis of Glutamine Synthetase
Synthesis at the Polysomal Level

The experiments directed towards the detection and localization of GS synthesis at the level of polysomes from HC-induced retina were designed on the basis of the following information. (1) During the first few hours of induction, stable templates for GS accumulate which have a half-life of several hours. (2) The fact that GS is rich in aspartic acid enabled us to use ^{14}C-Asp for preferential labeling of the enzyme. (3) The molecular weight of the subunit of retinal GS was determined to be 42,000 (Sarkar and Moscona, 1972).

The detailed conditions and experimental procedures for labeling of GS in these studies were described elsewhere (Sarkar and Moscona, 1971). Short labeling periods were used to reduce incorporation into ribosomal proteins and into nucleic acids. Total incorporation of ^{14}C-Asp into proteins in the HC-induced retina was found to be approximately two-fold higher than in the control (uninduced) tissue. To examine these differences at the polysomal level, 15,000 g supernatants from control and induced retinas labeled with ^{14}C-Asp were fractionated in sucrose gradients. The polysomal profile from the induced retina displayed

a generally higher incorporation of radioactivity into
proteins, and an apparent peak in the region of 12-14
ribosome polymers (Fig. 6). It can be calculated that

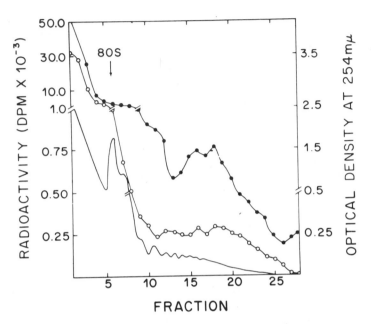

Figure 6. Sucrose-gradient (15-40%) analysis of 15,000 x *g* supernatants from control
and hydrocortisone-induced retinas labeled
in culture for 15 min with ^{14}C-Asp (3.3
μC/ml). Absorbance profiles (solid line)
were identical in both cases. (O), radioactivity (control); (●), radioactivity
(induced) (Sarkar and Moscona, 1971).

mRNA attached to 12-14 ribosome polymers corresponds
to peptide chains of molecular weights in the range of
35,000 to 45,000; this encompasses the 42,000 molecular weight of the subunit of retinal GS.
The increased incorporation of ^{14}C-Asp into retina
proteins was demonstrated to be closely related with
the induction of GS by HC; it was not detected in
other tissues, in which HC does not induce GS. In the
neural retina the increased incorporation of ^{14}C-Asp

was seen only in conjunction with GS induction, and was not present under conditions when the enzyme is not induced, for example in monolayer cultures of retina cells (Sarkar and Moscona, 1971).

The close correlation between increased ^{14}C-Asp incorporation and GS induction was further substantiated by showing that selective inhibition of GS induction by proflavine (see below) prevents the increased incorporation of ^{14}C-Asp. The induction of GS by HC is completely prevented by the addition of 8 μM (2.54 μg/ml) proflavine to the culture medium simultaneously with the addition of HC. Although this concentration of the drug does not measurably affect total protein synthesis, it completely prevents the increased ^{14}C-Asp incorporation characteristic of GS induction in the retina and reduces it to the uninduced control level (Table 1).

To examine the distribution of rapidly labeled RNA in the polysomal profile, control and HC-induced retinas were labeled for 15 min with ^{3}H-uridine and chased for 2 hr in the presence of 10 μg/ml Act D. Analysis of the polysomal profile showed an increased amount of ^{3}H-uridine-labeled RNA in the zone of high ^{14}C-Asp incorporation (Sarkar and Moscona, 1971). These findings raise the possibility that this RNA may represent mRNA for GS and this problem is currently under investigation.

Inhibition of Glutamine Synthetase
Induction by Proflavine

In screening for agents which might selectively inhibit GS induction while affecting only minimally overall protein or RNA synthesis, we found that proflavine (3,6-diaminoacridine), at 8 μM concentration, added together with HC to cultures of neural retina (from 10- or 12-day chick embryos) totally blocked the normal induction of GS. Examination of the inhibition of GS induction by 8 μM proflavine showed that:

   (1) Ongoing DNA synthesis is not required for proflavine inhibition of GS induction by HC.

TABLE 1

Effect of proflavine on the incorporation of ^{14}C-Asp
into proteins in the embryonic neural retina and its
correlation to glutamine synthetase induction

Tissue	^{14}C-Asp incorporation (dpm/mg protein)		GS specific activity
Uninduced controls	(a)	17,337	0.3 -0.5
	(b)	9,400	
	(c)	19,832	
HC-induced	(a)	28,098	1.3 -1.5
	(b)	22,800	
	(c)	28,177	
HC and 8 µM proflavine	(a)	10,760	0.25-0.5
	(b)	9,480	
	(c)	14,531	

(a), (b) and (c) are data from different sets of
experiments. The increase in ^{14}C-Asp incorporation in
the HC-induced retina is 1.5- to 2.5-fold over control.
In earlier experiments, 2- to 3.4-fold increases were
observed (Sarkar and Moscona, 1971); these differences
may be due to the use of different batches of fetal
bovine serum in the culture medium. GS specific activity is expressed as µM of glutamyl hydroxamate
formed per hr per mg of protein (Kirk and Moscona,
1963).

(2) The inhibition is at least partially reversible;
if, after 6 hr in culture medium with 8 µM proflavine, the retina tissue was thoroughly
washed to remove proflavine and further incubated in the presence of HC for 18 hr, up to
60% of normal induction was obtained.

(3) There is no evidence that 8 µM proflavine
blocks GS induction by stopping the access of
HC to the cells; the uptake of labeled HC by
the neural retina tissue is not measurably affected in the presence of this concentration of
proflavine.

(4) Treatment with 8 µM proflavine reduces overall
RNA synthesis in the retina by only 10-15%;
sucrose density gradient fractionation of labeled RNA shows that this reduction affects
tRNA synthesis somewhat more than that of heterogeneous and ribosomal RNA. No evidence was
found that this concentration of proflavine
affects the processing and transfer of RNA from
the nucleus to the cytoplasm in the retina.
The presently available data are consistent
with the possibility that proflavine blocks GS
induction by inhibiting preferentially the formation of transcripts essential for GS synthesis. This is supported by the fact that proflavine does not interfere with the translation
of GS on preformed, stable templates which accumulate during the first few hours of induction.

(5) Protein synthesis continues at a normal rate
for at least 24 hr in retina tissue cultured in
the presence of 8 µM proflavine; hence, the
blocking of GS induction is not due to a general inhibition of the mechanisms essential for
overall protein synthesis. However, as
described above, the increased incorporation of
^{14}C-Asp which is characteristic of GS induction
in the retina by HC, is prevented by proflavine.

(6) The blocking by proflavine of the HC-induced
increase of GS activity is due to inhibition of
GS synthesis and not to the formation of inactive GS; this was demonstrated definitively by
radioimmunoprecipitation assays of GS using
specific GS-antiserum (Table 2). This conclusion was confirmed by electrophoresis of the
precipitates obtained with anti-GS serum on
polyacrylamide-SDS gels; it was found that, in
the region of the gel corresponding to the GS
subunit, there was a high level of radioactivity in preparations from HC-induced retina, but
no radioactivity in the preparations from
retinas in which GS induction by HC was prevented by proflavine.

TABLE 2

Inhibition of glutamine synthetase synthesis
by proflavine

Tissue	Counts precipitated by anti-GS serum (DPM/mg protein)
Uninduced control	7,760
HC-induced	17,062
HC and 8 μM proflavine	5,960

Neural retinas from 12-day embryos were cultured
(in 3 ml Tyrode, 20% fetal bovine serum and 1% peni-
cillin-streptomycin mixture) with the additions as indicated in the table. After 1 hr incubation 0.5 μC/ml
^{14}C-amino acid mixture was added per culture. Following further incubation for 23 hr, the tissues were
washed, sonicated in phosphate buffer and 100,000 g
supernatant was obtained, with which the immunopreci-
pitations were performed using anti-GS serum, with
normal serum as control. The precipitated material
was washed, taken up in 0.1 M NaOH, reprecipitated
with 15% TCA, collected on Millipore filters, washed,
dried and counted. The data represent the difference
between the counts precipitated by antiserum and
normal serum.

A detailed interpretation of the above results with
proflavine requires further analysis of the detailed
mechanism of proflavine action in this system. It is
of interest that in T4 bacteriophage the mutagenic
effects of acridines (including proflavine) differ
markedly from those of 5-bromodeoxyuridine (Orgel and
Brenner, 1961) and that the latter, contrary to proflavine, does not inhibit GS induction in the retina
(Moscona, 1972; Moscona, in preparation).

SUMMARY

In this article we have (1) described the features
of the retinal GS induction system; (2) summarized

some of the experimental results concerned with the analysis of molecular events in GS induction; and (3) described briefly two current lines of work in this laboratory directed towards further analysis of this system.

ACKNOWLEDGMENTS

This work was supported by research grant HD 01253 from the National Institutes of Health (to A. A. M.), by a research fund from the American Cancer Society Institutional Grants IN-41J and K to the University of Chicago (P. K. S.), and by a postdoctoral stipend from the National Institutes of Health Training Grant TO 1-HD00297 to P. K. S. and A. W. W.

REFERENCES

Alescio, T., and Moscona, A. A. (1969). *Biochem. Biophys. Res. Commun. 34*, 176.

Alescio, T., Moscona, M., and Moscona, A. A. (1970). *Exptl. Cell Res. 61*, 342.

Chader, G. J. (1971). *Arch. Biochem. Biophys. 144*, 657.

Gerisch, G. (1968). In *Current Topics in Developmental Biology*, A. A. Moscona and A. Monroy, eds. (New York: Academic Press).

Kirk, D. L., and Moscona, A. A. (1963). *Develop. Biol. 8*, 341.

Manchester, K. L. (1970). In *Mammalian Protein Metabolism*, H. N. Munro, ed., *Vol. 4*, (New York: Academic Press), p. 229.

Moscona, A. A. (1971). In *Hormones in Development*, M. Hamburgh and E. J. W. Barrington, eds. (New York: Appleton-Century-Crofts), p. 169.

Moscona, A. A. (1972). In *Symposium on Biochemistry of Cell Differentiation: 7th Mt . Fed. Eur. Biochem. Societies, Varna, Bulgaria 1971*, (London: Academic Press).

Moscona, A. A., and Kirk, D. L. (1965). *Science 148*, 519.

Moscona, A. A. and Piddington, R. (1966). *Biochim. Biophys. Acta 121*, 409.

Moscona, A. A., Moscona, M., and Jones, R. E. (1970). *Biochem. Biophys. Res. Commun. 39*, 943.

Moscona, A. A., Moscona, M., and Saenz, N. (1968). *Proc. Nat. Acad. Sci. USA 61*, 160.

Moscona, M. H., Frenkel, N., and Moscona, A. A. (1972). *Develop. Biol.* (in press).

Orgel, A., and Brenner, S. (1961). *J. Mol. Biol. 3*, 762.

Piddington, R. (1970). *J. Embryol. Exp. Morph. 23*, 729.

Piddington, R. (1971). *J. Exp. Zool. 177*, 219.

Piddington, R., and Moscona, A. A. (1965). *J. Cell Biol. 27*, 247.

Reif-Lehrer, L., and Amos, H. (1968). *Biochem. J. 106*, 425.

Ronzio, R. A., Rowe, W. B., Wilk, S., and Meister, A. (1969). *Biochemistry 8*, 2670.

Salganicoff, L., and DeRobertis, E. (1965). *J. Neurochem. 12*, 287.

Sarkar, P. K., and Moscona, A. A. (1971). *Proc. Nat. Acad. Sci. USA 68*, 2308.

Shimada, Y., Piddington, R., and Moscona, A. A. (1967). *Exptl. Cell Res. 48*, 240.

Sussman, M. (1966). In *Current Topics in Developmental Biology*, A. A. Moscona and A. Monroy, eds., *Vol. 1*, (New York: Academic Press), p. 61.

Tata, J. R. (1971). In *Current Topics in Developmental Biology*, A. A. Moscona and A. Monroy, eds., *Vol. 6*, (New York: Academic Press), p. 79.

Tate, S. S., and Meister, A. (1971). *J. Biol. Chem. 68*, 781.

Tomkins, G. M. (1969). In *Problems in Biology: RNA in Development*, E. W. Hanly, ed. (Salt Lake City: University of Utah Press), p. 145.

Wigglesworth, V. B. (1966). In *Cellular Differentiation and Morphogenesis*, W. Beerman *et al.*, ed. (Amsterdam: North Holland Publ. Co.), p. 180.

Wyatt, G. R. (1968). In *Metamorphosis*, W. Etkin and L. Gilbert, eds. (New York: Appleton-Century-Crofts), p. 143.

Protein Synthesis During Cleavage

Paul R. Gross

Department of Biology
Massachusetts Institute of Technology
Cambridge
and
Marine Biological Laboratory
Woods Hole, Massachusetts

Fertilization of an egg is followed by a period during which the most obvious events are cell divisions, usually with very short intermitotic intervals. The outcome of this is the production of a number of cells, usually a large number, from the initially large single zygotic cell, and sometimes, but not always, their arrangement as a single-layered hollow ball called a blastula. The dramatic visible movements of cells and groups of cells relative to one another comes later, just before and during gastrulation. Like many other conclusions of classical embryology, those concerned with the functions and accomplishments of cleavage depend heavily upon the visible, and hence it is not surprising that the main purpose of cleavage should traditionally have been described as mere subdivision, or, somewhat more subtly, as a re-establishment of normal, somatic balance between nuclear and cytoplasmic masses. The implications of such a de-

323

scription of cleavage for analysis of gene expression during development are clear. They are that not much of importance happens in this regard, the whole process having been in some way programmed during oogenesis, and definitive gene "action" is supposed to begin at about the time of gastrulation.

This view is incorrect, and its correction is in itself important. Much more important, however, is the need to understand the nature of gene expression during cleavage, for it is now abundantly clear, upon functional as well as biochemical grounds, that gene expression is continuous from the moment of fertilization (and from before fertilization in eggs that are fertilizable before the maturation divisions are completed). This expression of information encoded in DNA takes the form of protein synthesis (translation) in all embryos examined to date, and of RNA synthesis (transcription) in most of them. The outcomes of these processes are crucial for further development: if they are stopped or experimentally altered, development cannot proceed normally. It is therefore of utmost importance in establishing the biochemical plan of development to analyze gene expression at these two levels (at least), and to attempt to understand how events at the two levels influence one another.

Our purpose here is to review a number of key points that have emerged from the biochemical study of animal embryo cleavage during the past few years. To this end, primary emphasis will be given protein synthesis and the origins of the information, in the form of messenger RNA (mRNA) molecules, that control it. Because echinoderm embryos, and particularly those of sea urchins, have been the most thoroughly studied in this regard, the review deals explicitly with results obtained in those forms, but they have been selected for presentation with a view to generality. To the best of my knowledge, the major conclusions offered here are very generally, if not universally applicable. In order to facilitate the synthesis of a large body of information and to focus attention on the main points, the results to be discussed have been selected from the recent literature or from among data (of my own colleagues or associates) to be published shortly.

This will permit the reader to judge for himself the value of the conclusions, by reference to published experimental details.

The general position to be proposed is simply summarized: Protein synthesis and RNA synthesis are normal and essential concomitants of cleavage, and they go on in all cells of the embryo. DNA synthesis does as well. Hence gene action at every level defined as such is a part of cleavage. The patterns of gene action during cleavage establish an essential precondition for subsequent morphogenesis and the later, more visible differentiations of cells. The precondition includes the conversion of old proteins to new ones, via breakdown and resynthesis from amino acids, and the stepwise divergence of function among the daughters of early mitoses, so that even in the most classically "regulative" of embryos, products of early cleavage have restricted potency and characteristic behavior *in situ* and in culture (Hynes, Raff and Gross, 1972).

The protein synthesis of cleavage is under the control of two main classes (and several subclasses) of mRNA. One of these is a product of transcription long before translation. These so-called "maternal" mRNA molecules are produced during oogenesis, stored untranslated in the egg cytoplasm, and utilized selectively during cleavage and afterward. The other arises from transcription of the rapidly-increasing mass of embryonic genomes, and some, if not all of these messengers are translated immediately. Because chromosomal proteins are among the products of the latter class, it is possible and indeed likely that divergence of gene function among cleavage cells is brought about via divergence of protein synthetic patterns among those cells. Some of the evidence upon which this position is based will now be taken up.

Protein Synthesis in General

In all embryos whose cleavage periods have been studied thus far with adequate radiotracer methods, protein synthesis from a pool of free amino acids is easily detected (Gross, 1967a, b for review). This synthesis is essential for the continuance of cleavage,

and of course, for all subsequent development. Hultin
(1961) first showed that inhibition of protein synthesis with puromycin brings cell division rapidly to a
halt. Later analysis of the mechanism of action of
puromycin revealed that puromycin peptides are produced, and it was suggested that these might be cyto-
toxic in certain circumstances. Hultin's observation
indicated a rather significant specificity of the
effect, in the sense that an ongoing mitotic division
could be blocked by the drug if it were applied prior
to metaphase, but if the drug was added during or after
metaphase, it was the next division that was inhibited.
Such an outcome of general cytotoxicity would have
been surprising, and it is entirely ruled out by more
recent studies with other inhibitors.

Hogan and Gross (1971), for example, employed the
ipecac alkaloid emetine as a protein synthesis inhibitor in sea urchin embryos. This drug acts in a manner
quite unrelated to that of puromycin. Emetine slows
or stops the movement of ribosomes along the mRNA in
polyribosomes, apparently by inhibition of translocase
activity. It operates, therefore, in the same way as
the better-known (structurally related) cycloheximide.
It has no *immediate* effect on the metabolism of other
macromolecules nor on energy metabolism, and yet, in
the experiments referred to, the effect on cell division and development was identical with that of puromycin.

What particular products of protein synthesis are
indispensable for cleavage and later development is
not known, but structural proteins of the mitotic apparatus (see below) and perhaps enzymes required for
DNA synthesis (Hogan and Gross, 1971) are among the
candidates.

In eggs that are fertilized and fertilizable only
as oötids (e.g., sea urchin eggs), the spawned oötid
synthesizes proteins at a very low rate, and the rate
increases greatly within a very few minutes after
fertilization. The absolute rates have been measured,
in sea urchins at least, over the whole period of development. Rising by thirty to seventy-fold from the
egg base rate during the first cleavage cycle, they
continue to increase up to the early blastula stage

and then remain more or less constant at about 3×10^{-8} micromoles of amino acid (equivalent) per minute per embryo. These rate changes are matched quite closely by the changes in the polyribosome fraction, i.e., the fraction of all ribosomes actually engaged upon the translation of mRNA in polyribisomes. A sample of such data, obtained by calculations based on incorporation and instantaneous amino acid pool specific activities, is shown in Figure 1 (from Fry and Gross, 1972). These rates, converted to protein mass, indi-

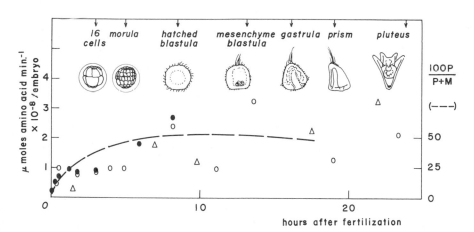

Figure 1. Absolute rates of amino acid incorporation into proteins, calculated (Fry and Gross, 1970b) from instantaneous uptake (pool plus proteins), incorporation (proteins alone), and measured amino acid pool sizes, as a function of stage of development in *Arbacia punctulata*. The different symbols represent separate experimental series carried out at different times. The dashed line represents the percentage of all ribosomes functioning in protein synthesis on polyribosomes. These data were collected from the literature and, in certain instances, from the same experiments as were used for the rate estimates. The figure is from Fry and Gross, 1972.

cate that the sea urchin embryo converts most of its yolk to amino acids and thence to other proteins during during development to the pluteus (feeding larva) stage.

The patterns of synthesis (i.e., kinds and amounts of different proteins made) change continuously and characteristically with stage (Gross, 1967). In embryos of other forms, the timing of protein synthetic activation associated with development is different. In some cases, it is the onset of maturation itself, in others it is fertilization during maturation (sometimes reinitiating an arrested meiosis), and in still others, it appears to be a complex of hormonal stimuli within the maternal body that switches on the accelerating protein synthesis and the pattern changes characteristic of cleavage and later stages (see, e.g., Ecker and Smith, 1971, for *Rana pipiens*). In all cases, however, specific and characteristic protein syntheses, at rates very much greater than those of the fully-grown but resting oocyte, are an indispensable normal concomitant of cleavage. To the extent that translation represents a late step in the expression (action) of genes, gene action is therefore continuous during cleavage (see also Fry and Gross, 1970a, b).

Cyclic variations in protein synthesis, coupled to the mitotic cycles of cleavage stage embryos, have been reported by Mano (1970), but we have not been able to confirm the claim that these alterations reflect true rate changes in the three sea urchin species available to us. There *are* cyclic variations in the permeability to labeled precursors and hence in pool specific activities, however. In our experience, therefore, the 30-60 min cell cycles of cleavage do *not* include a pronounced metaphase depression of protein synthesis and polyribosome content of the kind that takes place in cell cultures. It would be surprising, considering the 5-min duration of metaphase in sea urchin blastomeres along with the 2-5 min transit time for a mean polypeptide chain (Fry and Gross, 1972), if there were such a depression.

"Maternal mRNA"

The proposal that egg cytoplasm contains a store of
untranslated mRNA for use after fertilization, was
made originally on the basis of experiments with the
transcription inhibitor actinomycin D (Gross and
Cousineau, 1963). The key findings were that even in
the presence of sufficient drug to reduce RNA synthesis by more than 90 percent, a near-normal increase of
protein synthesis followed fertilization and was maintained through and often beyond the blastula stage.
These studies were extended and refined by Gross,
Malkin and Moyer (1964) *in vivo* and by Stavy and Gross
(1969a, b) for *in vitro* protein synthetic systems,
with the result that the hypothesis was strengthened.
Experiments on protein synthesis in activated enucleate
merogones (egg fragments) by Denny and Tyler (1964)
and by Brachet, Ficq and Tencer (1963) led to the same
conclusions, but the possibility of cytoplasmic transcription, later confirmed (see below), made this a
weaker demonstration of the action of stored mRNA.
    Despite occasional arguments concerned with the
timing or mechanism of action of actinomycin (e.g.,
Thaler, Cox and Villee, 1969), the drug has been
proven to act as was originally proposed (Greenhouse,
Hynes and Gross, 1971), and the conclusions drawn from
the behavior of embryos exposed to it are valid (Gross,
1967a b; Davidson, 1968).
    There exists evidence, in any case, for the pres-
ence of "maternal" mRNA in eggs, obtained from experiments that do not involve the use of antimetabolites
or inhibitors. This evidence is of several kinds,
but the most impressive categories are those concerned
with analysis of the postfertilization "switch" in
protein synthesis rates and with the direct observation of messenger-like RNA in the egg via molecular
hybridization.
    A careful analysis of the basal protein synthesis
in unfertilized eggs indicates that there is no important translational lesion, i.e., that the rate of
ribosome transit over a mean mRNA is the same as
after fertilization (Fry and Gross, 1972; Humphreys,
1969; Kedes *et al.*, 1969). We have not been able to

confirm in detail the observation by Metafora, Felicetti and Gambino (1971) of a translational inhibitor
in the egg, but even if their proposals are correct,
the differences found between unfertilized egg in *in
vitro* systems and those from embryos are entirely insufficient to account for the rate change at fertilization *in vivo*. The sizes and other properties of
polyribosomes in unfertilized eggs compared with those
in the zygote rule out any simple and general initiation lesion (Kedes *et al.*, 1969; Denny and Reback,
1970; Humphreys, 1971). Yet a near-normal acceleration
of protein synthesis occurs at fertilization and afterward even if there is little or no transcription
allowed. The latter indicates that mRNA is present in
the egg cytoplasm (not the pronucleus, because the
same observations can be made on enucleates, see
below), but is unavailable for translation in the egg.
　　The egg contains RNA that hybridizes with DNA to
the extent of more than 3% of the genome. This RNA is
neither ribosomal nor tRNA. The 3% figure is a minimal
one for sea urchins, and includes only repetitive sequences, or sequences transcribed from repetitive DNA,
some of which, however, function as mRNA—the mRNA for
histones is an example (Hynes and Gross, 1972; Kedes
and Birnstiel, 1971). Davidson and his colleagues
have shown, even more impressively, that the amphibian
egg contains a very large amount of informational RNA
transcribed from unique DNA—RNA that is virtually
certain to contain a large proportion of functional,
or potentially functional mRNA (e.g., Davidson and
Hough, 1971; Britten and Davidson, pp. 5-27). Preliminary results of similar studies in sea urchin eggs
and embryos yield the same conclusions. Thus it is
not only true that eggs contain RNA capable of direct-
ing protein synthesis after fertilization, but the
functional studies referred to above indicate that the
RNA is in some way mainly untranslatable in the egg,
i.e., "masked." What is the nature of this mask is
not known, and it remains a major problem for cell
biology as well as for embryology, since it is now
clear that mRNA can exist free of the translational
machinery in the cytoplasm of all eukaryotic cells, to
be selected for translation upon receipt of an appro-

priate signal (Martin, Tomkins and Granner, 1969; Fan
and Penman, 1970).

The relevance of this information to the topic of
this volume should be clear: Not only does the ongoing
and essential protein synthesis of cleavage represent
a process of gene expression, but a part of the translation is done on primary gene products, mRNA molecules, made much earlier (months to years), during
oogenesis. Thus the particular gene actions are the
physical basis (or one of the bases) of the "programming" of eggs. They are gene actions taken in the
first instance while the oocyte is being prepared in
the maternal body, and in the second through transla-
tion-level controls of a so far undiscovered mechanism,
once definitive developmental life has begun.

The program for protein synthesis inscribed upon
the maternal mRNA population is not, however, the only
one that functions during cleavage, as we shall now
see.

mRNA Synthesis During Cleavage

Synthesis of new RNA begins in the normal zygote
before the first division is complete, and continues
thereafter throughout development. There have been
some arguments expressed concerning precisely how
early nuclear transcription begins, but autoradiographic and biochemical evidence recently obtained
proves that it occurs during the first cell cycle
(Selvig, Greenhouse and Gross, 1972), with a notable
acceleration at about the eight-cell stage (Wilt,
1970). It is furthermore evident that some of the
new transcription products emerge from the nucleus *ab
initio* to function in polyribosomes as mRNA (Rinaldi
and Monroy, 1969) and do so without detectable delay
(Kedes and Gross, 1969) and without the necessity for
simultaneously synthesized protein to serve a transport function (Hogan and Gross, 1971). It is possible
that nuclear RNA synthesis does begin rather later in
amphibian cleavage—perhaps as late as the very early
blastula (Gurdon and Woodland, 1969), but as failure
to label is not entirely reassuring evidence for the
absence of synthesis, this question remains somewhat

unsettled. Suffice it to say that the maternal RNA
program is very soon joined, in all appropriately
studied cases, by a program for protein synthesis
originating upon the chromosomes of cleavage cells.

The new RNA appears to be produced in *all* cells of
the embryo. Autoradiographic methods sometimes give
misleading results, owing to the high frequency of
mitotic figures (and hence absent nuclei) among blastomeres, but a thorough study of cleavage cells as
early as the 16-cell stage, *in situ* and isolated in
culture, provides convincing evidence that all three
kinds of blastomeres make RNA and do so according to
the characteristic and changing patterns seen as an
average in the whole embryo (Hynes and Gross, 1970;
Hynes, Greenhouse, Minkoff and Gross, 1972).

During cleavage, the products of transcription include mRNA, little or no ribosomal RNA, and virtually
no transfer RNA, although the latter becomes radioactive through end-labeling of the pCpCpA sequence
(Gross, Kraemer and Malkin, 1965; Greenhouse, Hynes
and Gross, 1971). New RNA in the nuclei is mainly in
the form of very large polynucleotides, with modal
sedimentation constants of about 40S; and new RNA of
the cytoplasm is found mainly in association with
polyribosomes, having sedimentation constants of 5 to
20S (Hogan and Gross, 1972). The precursors to ribosomal RNA, which can be labeled with moderate specificity by the use of tracer methionine (since these
polynucleotides become methylated during processing),
are first detectable, and then barely so, at the late
blastula stage (Hogan and Gross, 1972). Most of the
RNA made during cleavage is therefore heterogeneous in
sedimentation, and its base competition is DNA-like.

The question arises as to the novelty of the sequences, i.e., the extent to which RNA, and particularly mRNA, made during cleavage duplicates the pro-
grammatic content of the maternal mRNA. On this score
evidence is difficult to obtain, since a complete
analysis would require competition-hybridization with
RNA transcribed from unique DNA, a technique not yet
perfected. There are, nevertheless, convincing indications as to the nature of the answer. Competition-
hybridization between populations of maternal mRNA

molecules and radioactive mRNA produced during cleavage
show that for polynucleotides copied from repetitive
DNA, at least, and some of these *are* mRNA molecules,
sequence identity cannot exceed 30%, and is probably
much less (Hynes and Gross, 1972). The program inscribed on embryonic mRNA is, in other words, a different program from that stored in the egg cytoplasm
during oogenesis. Thus the expression of gene action
during cleavage, to the point of translation at least,
is an expression of two separate programs read out
from the genes at two very different periods of the
embryo's life-history.

There is RNA synthesis on cytoplasmic genomes
throughout development (Hartman and Comb, 1969), and
it can even be detected in activated, enucleate merogones (Chamberlain, 1970; Craig, 1970; Selvig, Gross
and Hunter, 1970). One would wish to know what is the
origin of this RNA, and also whether or not it contributes to the translatable polynucleotides used during
cleavage. The solution to this problem is now available. Evidence from molecular hybridization experiments suggested strongly, but could not prove, that
cytoplasmically transcribed RNA is made in the mitochondria, whose DNA content is about 16 times the
haploid nuclear amount in sea urchin eggs and thousands of times the haploid nuclear amount in amphibian
eggs. While in amphibians mitochondrial transcription
appears not to begin early (Dawid *et al.*, 1971), it
surely does in the sea urchin. This has been proven
by a number of independent means, including selective
sensitivity to drugs such as ethidium bromide, selective insensitivity to other drugs such as colchicine,
and direct autoradiographic localization in stratified
mitochondria (Selvig, Greenhouse and Gross, 1972).
The product appears to be made up of a mixture of two
stable and discrete mitochondrial RNA species (sedimenting at 12 and 21S) with some heterogeneous
material.

Is it, then, translatable? The answer is no! The
work of Selvig (Selvig, 1971; Selvig, Greenhouse and
Gross, 1972) makes it quite clear that RNA transcribed
on mitochondrial genomes does not under any circumstances associate in detectable quantities with normal

cytoplasmic polyribosomes when these are properly prepared (Raff *et al.*, 1972). Hence if the mitochondrial RNA is translated, it must be on a machinery quite
separate from the normal one, and it cannot contribute
more than a few percent, at most, to the total protein
synthesis.* In virtue thereof, cytoplasmically transcribed RNA cannot provide any alternative to the maternal mRNA hypothesis, and cannot, specifically,
account for the ability of activated enucleate merogones to display the normal pattern of protein synthesis increase (see also Craig and Piatigorsky, 1971).

A final point in regard to new, or embryonic mRNA
must now be dealt with. I have indicated that tran-
scription-blocked embryos show a near-normal pattern
of increasing protein synthesis following fertilization, and in fact, they show normal or supernormal
incorporation of certain tracer amino acids into proteins for many hours thereafter. How can this be re-
conciled with the proposal just made that new mRNA
joins the maternal program on the polyribosomes early

---

*
There are several grounds for this latter conclusion. The mitochondria of zygotes can be stratified
centrifugally, for example, to form a distinct layer.
If populations of fertilized eggs are given brief exposures to radioactive amino acids and then stratified
in this way, the material can be fixed, sectioned, and
autoradiographed. The result of such experiments is
that all of the radioactive protein (ca. 50% of which
is nascent if the labeling periods are shorter than 2
min) is found in the cytoplasmic matrix, as evidenced
by silver grain distribution in the overlying emulsion
(Kedes, Gross, Cognetti, and Hunter, 1969; and also
Gross, 1967c). With longer periods of labeling, the
nucleus becomes radioactive, but the mitochondrial
layer never shows a significant localization of radioactive (i.e., new) proteins. Also, in the preparation
of *in vitro* protein synthetic systems from eggs and
embryos (e.g., Stavy and Gross, 1969a, b), homogenate
fractions display activity that is strictly proportional to their content of ribosomes or surviving
ordinary polyribosomes, never to their content of
intact or fragmented mitochondria.

on, and contributes importantly to the translation products throughout cleavage?

The solution to the problem is fortunately a trivial one. New mRNA contributes to a negligible fraction of the protein synthesis during the first few cleavages, but by the 16-cell stage it begins to make a significant impact on the total rate. By mid-cleavage (ca. 200 cells), it accounts for nearly 30% of the ongoing protein synthesis (Kedes and Gross, 1969a). The reason why the expected differential incorporation as between normal and transcription-inhibited embryos is not seen with certain tracer amino acids is that the amino acid pool sizes are coupled in some way to transcription. The leucine pool, in particular, shows a large increase in the normal embryo at about mid-cleavage, but in the presence of actinomycin the pool changes very little.* Hence, in experiments with tracer leucine, the pools achieve a higher specific activity at the morula stage in the presence of actinomycin (permeabilities being unaltered) than they do normally. When *absolute* rates of synthesis are measured with the aid of pool specific activities and incorporation data, it is clear that the synthesis rate is always correlated with the relative proportion of polyribosomes (see also Fig. 1) whether actinomycin is present or not. Conversely, the absence of the embryonic program makes itself felt as a predictable loss in total protein synthesis, i.e., about 30% during cleavage, a large part of which is chromosomal proteins (see below). The resolution of this rate paradox is due to B. Fry (Fry, 1970; Fry and Gross, 1972).

---

*The free leucine pool, for example, is about 50 nm per embryo until about the 16-32 cell stage, whether or not actinomycin is present. Normally, it then rises steadily and levels off at about 200 nm by the time of hatching (ca. 8 hr). In the presence of actinomycin, the rise is smaller, leading to a plateau of about 110 nm per embryo, or less. As indicated in the text, rates of entry into the embryo by tracer amino acids (and, in fact, by tracer nucleosides(are not changed by the drug.

Examples of "Maternally" and
"Embryonically" Programmed Proteins

There remains, now that the general form of the dual programming mechanism has been presented, the perhaps heuristic task of fixing the main points by recourse to examples. To this end I shall report an example of a specific protein product (actually two) encoded on maternal mRNA, together with an example of the protein synthetic role played by embryonic, or new mRNA.

Recently, the introduction of vinblastine precipitation as a method for purification of microtubule proteins (especially the A and B tubulins) provided an opportunity to test the notion that these proteins are among the products of cleavage-stage synthesis. Raff and his colleagues (1971) established that such synthesis does indeed occur in sea urchin embryos. They found, furthermore, that a soluble pool of the tubulins is maintained at all times and at constant size.[*] On this basis and with the aid of pulse- and chase-data, they proposed that the pool is used continuously through development, and especially in cleavage, for the production of ordered and insoluble structure, and is replenished by new proteins assembled from the amino acid pool.

In the course of their work, Raff *et al.* examined the sensitivity of tubulin synthesis to transcription-block with actinomycin. The observation was that this

[*]This point was examined by estimating the content per embryo, at various stages, of colchicine-binding protein sedimenting at about 6S, it having been demonstrated separately that such activity is proportional to vinblastine-precipitable radioactivity in separated tubulins (Raff *et al.*, 1971). Both the colchicine-binding assay and the vinblastine-precipitation reaction have a good deal of inherent variability, however, so that "constant size" really means "roughly constant." This point has been investigated by alternate methods, however, and the results presented at this Symposium by R. A. Stephens confirm the conclusions as to tubulin pools and synthesis very elegantly.

synthesis, in marked contrast to others (see below), is insensitive to blockade of RNA production after fertilization. This led to the suggestion that microtubule monomer proteins are among those encoded on maternal mRNA.

The maternal mRNA hypothesis has been established almost beyond doubt by a number of means, as described earlier, but proof that a *particular* protein is translated on such mRNA required attention to alternative explanations of the actinomycin effect (or lack thereof), alternatives that do not apply in general. Among these were, for example, (1) insensitivity to actinomycin of this, and only this class of mRNA, (2) origin of the tubulin mRNA on cytoplasmic genomes after fertilization, (3) synthesis of tubulins on mRNA produced *just* before fertilization and surviving for some time thereafter, (4) labeling of tubulins by enzymatic end-addition of tracer amino acid, rather than via a normal ribosome-mediated net synthesis on polyribosomes.

All of these possibilities have been tested directly (Raff *et al.*, 1972), and none of them applies. Tubulins are made on ordinary polyribosomes and their synthesis is completely stopped by translocase inhibitors such as emetine. They are synthesized normally in enucleate activated merogones. The synthesis is insensitive to ethidium bromide, which blocks mitochondrial transcription. It either starts or accelerates greatly at fertilization and increases selectively, over general protein synthesis on maternal mRNA (in actinomycin) during early cleavage. Thus the monomeric proteins of microtubules, important structural components of mitotic spindles and asters, cilia, cell cortices and perhaps even membranes, are among those for which the program is built into the egg cytoplasm well in advance of the exigencies of cleavage and early development.

The best example of proteins encoded upon mRNA transcribed during cleavage proper, and translated at once, is that of certain chromosomal histones. This subject has been ventilated a good deal of late, and therefore we need not make a detailed argument here.

Infante and Nemer (1967) were the first to notice

that a class of light polyribosomes (modal polymer number five) becomes very prominent during cleavage in sea urchin embryos, and they pointed out that this class fails to appear if transcription is blocked. They argued correctly that the bulk of cleavage polyribosomes, which appear and function whether transcription is permitted or not, must be translating maternal mRNA, while the light ones might well contain new mRNA. This suggestion was confirmed in due course by Nemer and his colleagues and by others, using labeled uridine to trace new mRNA in polyribosomes. They speculated originally that the light polyribosomes might be inactive, but this idea was abandoned when it became evident that all polyribosomes are functional (Kedes and Gross, 1969a) and more important, when indirect evidence began to accumulate in favor of a role for light polyribosomes in the synthesis of chromosomal histones (Borun, Scharff and Robbins, 1967). Eventually, evidence of many kinds led to the firm conclusion that histones are synthesized throughout cleavage, maximum rates being observed coincidentally with the time of maximum mitotic activity at mid-cleavage, and that the histones are synthesized on light polyribosomes. These polyribosomes also contain a discrete family of mRNA molecules synthesized concurrently, with modal sedimentation constant of 9S and an associated (perhaps derivative) family sedimenting at 20S (Kedes, Gross, Cognetti and Hunter, 1969; Nemer and Lindsay, 1969; Kedes and Gross, 1969b).

More recent unpublished work has uncovered some reassuring properties of these mRNAs, including their power to stimulate protein synthesis, and probably histone synthesis, *in vitro*, the release of histones from polyribosomes containing them *in vitro*, the sensitivity of their production to interference with DNA synthesis, etc. Kedes and Birnstiel (1971) have shown, furthermore, that these mRNAs (of which there appear to be 3-5 discrete species) are transcribed from moderately iterated sequences in the DNA, a finding which, if confirmed and explained, will add greatly to our understanding of eukaryotic genome organization. The only matter of doubt comes from the finding that some histone synthesis, apparently a small frac-

of the normal, continues in the presence of actinomycin, when the light polyribosomes with new 9S RNA do
not exist (J. Rudermann, unpublished data). It is not
clear whether the apparently maternal histone mRNAs
bear the same sequences as the embryonically-tran-
scribed ones. This doubt does not in any way mitigate
the strength of the examples of proteins programmed
upon embryonic messages—it merely leaves unsettled
the issue of uniqueness with respect to the maternally
provided program. Hopefully, the uncertainty will be
alleviated in the near future by quantitative studies
on embryonic histone synthesis both *in vivo* and *in
vitro*, coupled with improved methods for separation
of the several types.

ACKNOWLEDGMENTS

In Symposia like this one, contributors are expected
to refer to the work of their own laboratories, but
there is here what may appear as an excessive density
of personal citations. It would be incorrect to see
them as such, however. The numerous references to the
work of one group reflects what has been for several
years its large size and high quality. My ability to
assemble such a review statement as this means, therefore, that I have had the good fortune to work at a
place to which outstanding young people have come for
graduate study and postdoctoral research. What is
reported here is thus in large part a summary of *their*
progress, and I acknowledge that with pleasure and
gratitude.
I acknowledge the keen sportsmanship of my friend
and colleague Professor M. Sussman, who, exclusive of
occasional threats and cries of "murder!" tolerated
my delays in completing this manuscript with characteristic charm and good humor.
Our work has been supported by research grants from
the National Science Foundation, the National Institute of General Medical Sciences, and the National
Cancer Institute. Several of my young colleagues have
been National Institutes of Health trainees or fellows.

REFERENCES

Borun, T. W., Scharff, M. D., and Robbins, E. (1967). *Proc. Nat. Acad. Sci. USA 58*, 1977.

Brachet, J., Ficq, A., and Tencer, R. (1963). *Exp. Cell Res. 32*, 168.

Chamberlain, J. P. (1970). *Biochim. Biophys. Acta 213*, 183.

Craig, S. P. (1970). *J. Mol. Biol. 47*, 615.

Craig, S. P., and Piatigorsky, J. (1971). *Devel. Biol. 24*, 214.

Davidson, E. H. (1968). *Gene Activity in Early Development* (New York: Academic Press).

Davidson, E. H., and Hough, B. R. (1971). *J. Mol. Biol. 56*, 491.

Dawid, I., Swanson, R. F., Chase, J. W., and Rebbert, M. (1971). *Carnegie Inst. Wash. Yearbook 69*, 575.

Denny, P. C., and Reback, P. (1970). *J. Exp. Zool. 175*, 133.

Denny, P. C., and Tyler, A. (1964). *Biochem. Biophys. Res. Commun. 14*, 245.

Ecker, R. E., and Smith, L. D. (1971). *Devel. Biol. 24*, 559.

Fan, H., and Penman, S. (1970). *J. Mol. Biol. 50*, 655.

Fry, B. J., (1970). Ph.D. Thesis, Mass. Institute of Technology.

Fry, B. J., and Gross, P. R. (1970a). *Devel. Biol. 21*, 105.

Fry, B. J., and Gross, P. R. (1970b). *Devel. Biol. 21*, 125.

Fry, B. J., and Gross, P. R. (1972). *Devel. Biol.* (in press).

Greenhouse, G. A., Hynes, R. O., and Gross, P. R. (1971). *Science 171*, 686.

Gross, P. R. (1967a). *Curr. Topics Devel. Biol. 2*, 1.

Gross, P. R. (1967b). *New Eng. J. Med. 276*, 1230, 1297.

Gross, P. R. (1967c). *Canadian Cancer Conference 7*, (London: Pergamon Press).

Gross, P. R., and Cousinaeu, G. H. (1963). *Biochem. Biophys. Res. Commun. 10*, 321.

Gross, P. R., Kraemer, K., and Malkin, L. I. (1965). *Biochem. Biophys. Res. Commun. 18*, 569.

Gross, P. R., Malkin, L. I., and Moyer, W. A. (1964).
  Proc. Nat. Acad. Sci. USA 51, 407.
Gurdon, J. B., and Woodland, H. R. (1969). *Proc. Roy.
  Soc. Ser. B 173*, 99.
Hartman, J. F., and Comb, D. G. (1969). *J. Mol. Biol.
  41*, 155.
Hogan, B., and Gross, P. R. (1971). *J. Cell Biol. 49*,
  692.
Hogan, B., and Gross, P. R. (1972). *Exp. Cell Res.* (in
  press).
Hultin, T. (1961). *Experientia 17*, 410.
Humphreys, T. (1969). *Devel. Biol. 20*, 435.
Humphreys, T. (1971). *Devel. Biol. 26*, 201.
Hynes, R. O., and Gross, P. R. (1970). *Devel. Biol.
  21*, 383.
Hynes, R. O., and Gross, P. R. (1972). *Biochim. Bio-
  phys. Acta 259*, 104.
Hynes, R. O., Greenhouse, G. A., Minkoff, R., and
  Gross, P. R. (1972). *Devel. Biol.* (in press).
Hynes, R. O., Raff, R. A., and Gross, P. R. (1972).
  Devel. Biol. (in press).
Infante, A. A., and Nemer, M. (1967). *Proc. Nat. Acad.
  Sci. USA 58*, 681.
Kedes, L. H., and Birnstiel, M. L. (1971). *Nature 230*,
  165.
Kedes, L. H., and Gross, P. R. (1969a). *J. Mol. Biol.
  42*, 559.
Kedes, L. H., and Gross, P. R. (1969b). *Nature 223*,
  1335.
Kedes, L. H., Gross, P. R., Cognetti, G., and Hunter,
  A. L. (1969). *J. Mol. Biol. 45*, 337.
Kedes, L. H., Hogan, B., Cognetti, G., Selvig, S. E.,
  Yanover, P., and Gross, P. R. (1969). *Cold Spring
  Harbor Symp. Quant. Biol. 34*, 717.
Mano, Y. (1970). *Devel. Biol. 22*, 433.
Martin, D., Tomkins, G., and Granner, D. (1969). *Proc.
  Nat. Acad. Sci. USA 62*, 248.
Metafora, S., Felicetti, L., and Gambino, R. (1971).
  Proc. Nat. Acad. Sci. USA 68, 600.
Raff, R. A., Colot, H. V., Selvig, S. E., and Gross,
  P. R. (1972). *Nature 235*, 211.
Raff, R. A., Greenhouse, G. S., Gross, K. W., and
  Gross, P. R. (1971). *J. Cell Biol. 50*, 520.

Rinaldi, A. M., and Monroy, A. (1969). *Devel. Biol.*
    19, 73.
Selvig, S. E. (1971). Ph.D. Thesis, Mass. Institute of
    Technology.
Selvig, S. E., Greenhouse, G. A., and Gross, P. R.
    (1972). *Cell Differentiation* (in press).
Selvig, S. E., Gross, P. R., and Hunter, A. L. (1970).
    Devel. Biol. 22, 343.
Stavy, L., and Gross, P. R. (1969a). *Biochim. Biophys.*
    Acta 182, 193.
Stavy, L., and Gross, P. R. (1969b). *Biochim. Biophys.*
    Acta 182, 203.
Thaler, M. M., Cox, M. C. L., and Villee, C. A. (1969).
    Science 164, 832.
Wilt, F. H. (1970). *Devel. Biol. 23*, 444.

Ciliary Protein Synthesis in Sea Urchin Embryos

R. E. Stephens

Department of Biology
Brandeis University
Waltham, Massachusetts
and
Marine Biological Laboratory
Woods Hole, Massachusetts

In terms of building a specific structure, cilia
formation at the blastula stage represents the first
true morphological step for the developing embryo.
This event is under maternal control, apparently the
result of long-lived messenger RNA, since partheno-
genetically activated eggs will develop quite normally
(Harvey, 1940) and cilia will form in the absence of
DNA-dependent RNA synthesis (cf. Gross, 1964). Evidence exists that ciliary proteins are synthesized at
a constant rate well in advance of ciliogenesis
(Auclair and Meismer, 1965) and that a large pool of
ciliary precursor is formed, since multiple regeneration of cilia can take place in the absence of protein
synthesis (Auclair and Siegel, 1966). Given that the
bulk of ciliary protein pre-exists prior to actual
cilia formation, what triggers the morphogenetic process? Is some minor but critical component synthesized *de novo*, or does some "activation" event take
place?

It is the object of this study to determine the synthetic sequence of specific ciliary proteins by pulse-labeling embryos at various times prior to and during ciliogenesis, isolating cilia after hatching, fractionating the cilia electrophoretically into known structural components, and finally determining the relative specific activities of these proteins at each relevant time point in development. What quantity of which protein is made when, and can this be related to morphogenesis?

METHODS

Fertilized eggs of the artic boreal sea urchin *Strongylocentrotus droebachiensis* were grown through hatching at 7.5°. At this temperature, first division is reached at 3 hr, cilia formation begins at 26-27 hr, and hatching takes place at 33 hr (Stephens, 1972a). During successive points in development, aliquots containing 1 ml of cells were given 3-hr pulses with 20 μC of ^{14}C-leucine, chased with 1% cold leucine in sea water, and allowed to form cilia and hatch. Cilia were then isolated from the embryos by hypertonic sea water treatment (Auclair and Siegel, 1966) and the axonemes ("9 + 2" structures minus membranes, matrix, and cell debris) were prepared by Triton X-100 treatment (Stephens and Linck, 1969).

The ciliary axonemes were subjected to sodium dodecyl sulfate (SDS) polyacrylamide gel electrophoresis (Shapiro, Vinuela and Maizel, 1967) to separate the constituent proteins by molecular weight. Known standards were run in parallel. The gels were quantitatively stained with Fast Green (Gorovsky, Carlson and Rosenbaum, 1970), recorded microdensitometrically, and then sliced longitudinally, dried, and analyzed autoradiographically by the methods of Fairbanks, Levinthal and Reeder (1965).

Thus 11 successive time periods during early development could be investigated on a comparative semi-quantitative basis. Identical numbers of embryos were treated with the same amount of isotope for identical lengths of time; all were allowed to develop cilia in

parallel and the constituent proteins of the isolated
cilia were analyzed under identical conditions. *Only
the time point of the pulse was varied.*

In addition, these same embryos were permitted to
regenerate cilia for 6 hr, the regenerated cilia were
isolated and analyzed as above, and then compared with
the virgin cilia. A decrease in specific activity of
any given component would indicate a discrete pool,
replenished by cold leucine incorporation after decil-
iation, while the absence of any component in the regenerates would indicate that only enough of that component was synthesized for one generation. No change
in specific activity would imply a very large pool of
ciliary precursor.

RESULTS

Between the second division and the onset of ciliogenesis, the isolated ciliary axonemes showed virtually identical specific activities of 2700 cpm/µg total
protein, demonstrating a constant level of bulk protein synthesis. The pulse period during the first
division, where total synthesis is always lower, had
a value roughly 3/4 that of the later, constant time
points. During the actual process of ciliogenesis,
the specific activity increased by 25% and 55% for
the 27—30-hr and 30—33-hr pulses, respectively. Since
the same number of cilia are represented in all cases,
this substantial increase in ^{14}C-leucine incorporation
would reflect either increased or *de novo* synthesis,
or both, during active ciliogenesis. The bulk of the
ciliary proteins clearly *are* made continuously prior
to ciliogenesis, but a clear increase in synthetic
rate takes effect during the morphogenetic process.
To what components can this be attributed?

Analysis of individual proteins of the ciliary
axoneme reveals that these proteins fall into four
general categories: those which pre-exist in the unfertilized egg; those which are synthesized constantly
prior to ciliogenesis; those which appear to decrease
with time; and those which are synthesized *de novo*
only during morphogenesis. Figure 1 represents an op-

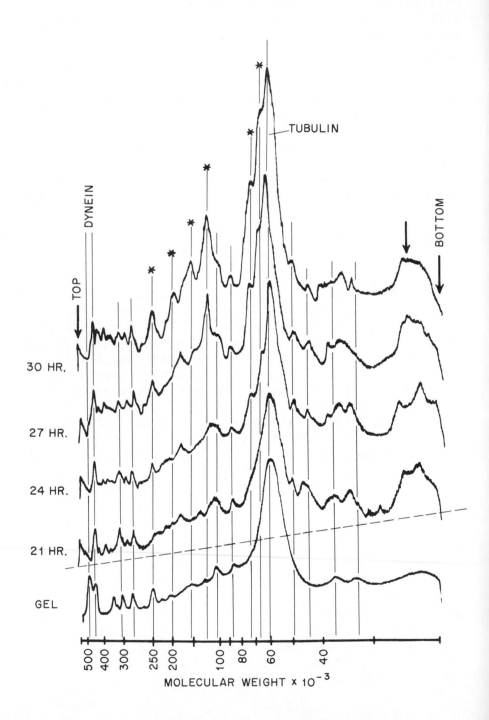

tical gel scan (bottom) representative of all of the
gels analyzed, an autoradiographic scan of a 21—24-hr
pulse period representative also of any earlier pulse
period, and autoradiograph scans of the 24—27-hr,
27—30-hr, and 30—33-hr pulse periods, representing
initiation, early growth, and final growth of cilia,
respectively. Comparison of the 21-hr scan (before
ciliogenesis) with the 30-hr scan (after ciliogenesis)
will illustrate best the significant changes that take
place in the protein synthetic profile during morphogenesis; all profiles prior to the 21—24-hr pulse
period are indistinguishable from it, except that the
0—3-hr period is slightly reduced overall.

Pre-Existing Proteins

The ciliary ATPase *dynein* (Gibbons, 1965) has two
subunits with molecular weights of approximately
500,000 and 470,000 (Linck, 1971). Quite significantly, only the lower molecular weight subunit shows any
label during embryogenesis. Since the higher molecular weight subunit is quite prominent in the gel scans,
this subunit must pre-exist in the unfertilized egg,
to be used in later development, for ciliogenesis.
Consistent with this view is the finding by Weisenberg and Taylor (1968) of a 13S dynein-like ATPase in
both unfertilized eggs and in the mitotic apparatus,
coupled with the later discovery of only one dynein
band in SDS-gels of unfertilized eggs (Stephens, 1972b).
This higher molecular weight subunit is found in at
least 10-fold excess over that required for cilia for-

Figure 1. (Opposite page.) Microdensitometric traces
of Fast Green stained SDS-gel of ciliary axoneme, representative of all time points studied (bottom trace)
and microdensitometric traces of autoradiograms of
ciliary SDS-gels from samples pulse-labeled at 21—24
hr, 24—27 hr, 27—30 hr, and 30—33 hr (upper four traces). The unmarked arrow indicates the tracking dye,
stars indicate proteins synthesized *de novo* during
ciliogenesis. The molecular weight scale is approximate, serving as a reference point. The broken line
signifies the film base line.

mation, lending credence to its possible mitotic function.

Constant or Continuous Synthesis

The lower molecular weight dynein subunit, many "background" proteins, adenylate kinase (MW = 32,000), and tubulin are all made uniformly during early development. Since these constitute the bulk of the axonemal protein, it becomes obvious why isolated cilia appear to have the same specific activity regardless of when the embryo was pulse-labeled. Removal of membranes, matrix, and accidental cell debris, leaving only the "9 + 2" structure, emphasizes the increase in specific activity during active ciliogenesis, but of these continuously synthesized proteins only tubulin shows an increase in synthesis during this morphogenetic period, amounting to 30-50%.

Apparent Decreased Synthesis

A protein component with a molecular weight of about 310,000 is prominent in the gel scans and in all autoradiograms prior to and including the 21—24-hr pulse period. During active ciliogenesis, the specific activity of this protein appears to decrease twofold or more, suggesting a limiting role for its synthesis. A major difficulty in substantiating this point is the marked increase in other minor proteins in this same molecular weight region during the later time periods, obscuring this particular component.

Synthesis de novo

The 27—30-hr and 30—33-hr pulses demonstrate a great increase in the specific activity of some six protein bands previously undetectable in the 21—24-hr or prior pulse periods. These apparent de novo proteins, having molecular weights of approximately 250,000, 200,000, 165,000, 130,000, 73,000, and 68,000, are marked with stars in the figure. These represent proteins of the linkage-spoke complex, responsible for geometrical symmetry in the "9 + 2" microtubule pattern. One of these proteins, *nexin* (Stephens, 1970b), has been isolated; it ties together the nine outer doublets and commonly exists as a dimer with molecular weight of 165,000. It is possible that the approximate 73,000 and 250,000 molecular weight proteins could be monomer and trimer, respectively, of this linkage protein. Regardless of identity, it is perhaps significant that the *de novo* synthesis of such architectural protein components should mark the initiation of ciliogenesis.

Regeneration

Upon autoradiographic analysis of regenerated cilia, it was found that the specific activities of all continuously synthesized proteins, except tubulin, was essentially unchanged, implying a rather large precursor pool for these components. The specific activity of tubulin was decreased by 1/4 to 1/3 in the regenerates, indicating that a pool three to four times larger than that needed for one generation was synthesized at any given time point prior to ciliogenesis. The pool formed during ciliogenesis, already about 50% higher in specific activity than that of tubulin formed earlier, showed only a 5-10% dilution effect upon regeneration. Thus the claim of previous workers that the bulk ciliary proteins are derived from a substantial pool is readily demonstrated for the separated structural components.

Most importantly, the regenerates from embryos labeled during ciliogenesis showed no significant amount of label in the *de novo* proteins, a finding that is consistent with a single "round" of synthesis in the amount needed for one generation only. This is *not* inconsistent with the multiple regeneration of cilia in the absence of protein synthesis (Auclair and Siegel, 1966) since multiple "rounds" of these proteins would have accumulated by late gastrula, when such regeneration studies were performed. In addition, it is a debatable point as to whether puromycin totally inhibits protein synthesis or whether some synthesis does take place. Based on total cellular protein synthesis, the amount of linkage-spoke complex needed to be synthesized for multiple regeneration would be small indeed.

DISCUSSION

The structural and enzymatic proteins of embryonic cilia are clearly synthesized differentially. One subunit of dynein pre-exists in the unfertilized egg, while the other subunit only, is synthesized, continuously, after fertilization. Whether prefertilization tubulin is utilized for later cilia formation has yet to be established. Like the lower molecular weight dynein subunit, the bulk of the ciliary proteins—consisting chiefly of tubulin and adenylate kinase—are also synthesized uniformly throughout early development. During ciliogenesis, only tubulin increases significantly in specific activity while a 310,000 molecular weight component appears to decrease somewhat. The morphogenetic event is particularly marked by the *de novo* appearance of linkage and spoke proteins, necessary for the geometrical architecture of the basic "9 + 2" array. During this period the latter protein complement is made only in an amount sufficient for one round of ciliogenesis, since regenerated cilia show no significant label in any of these components. However, analysis of the regenerated cilia indicates a substantial pool of the remaining ciliarly precursors. Of these, tubulin appears to be limiting, but even in this case there is enough protein present for about four generations, based upon a 1/4 to 1/3 dilution of the pool upon regeneration, and assuming that all can be utilized.

Previous studies by Robbins and Shelanski (1970) with HeLa cells and by Raff *et al.* (1971) with sea urchin embryos have also demonstrated the continuous and relatively constant synthesis of total tubulin during the cell cycle. The latter study, and the present one, find no evidence of differential synthesis for the A- and B-subunits of tubulin (cf. Stephens, 1970a). How much tubulin, if any, can be used by both the mitotic apparatus and by cilia cannot be determined at the present time.

The *de novo* synthesis of architectural elements would be a reasonable way to initiate organelle formation. Synthesis of a limited amount of such components during any given period would serve to limit growth or number. The production of quantal amounts of protein has its parallel in enzyme formation in the developing cellular slime mold *Dictyostelium discoideum* (Newell, Longlands and Sussman, 1971). The mode of control of such a process is clearly of great developmental interest.

The "activation" of pre-existing proteins cannot be eliminated by the data given here. In fact, there is mounting evidence that mitotic tubulin can be differentially activated or mobilized during a discrete period of the cell cycle (Stephens, 1972b). *In vitro* evidence indicates that binding of GTP, the formation of an intramolecular disulfide bond, and the presence of divalent cations are of critical importance to the polymerization of flagellular B-tubulin (Stephens, in preparation). Similar parameters may influence ciliary tubule formation.

In this same connection, why do doublet microtubules not arise from mitotic centrioles since both tubulin and dynein are present well prior to ciliogenesis? Perhaps "activation" of the centriole to serve as a "crystallization center" for the pre-existing and *de novo* proteins is a requisite step in basal body function.

Questions regarding the formation and utilization of the tubulin pool or pools, the nature of the linkage and spoke proteins and their interactions with ciliary microtubules, the control of synthesis of limiting structural components, the possible collateral role of dynein in mitosis, and the potential activation of critical structural components—proteins or basal bodies—remain unanswered. Their solution is essential to a thorough understanding of the mechanics of the important morphogenetic event of ciliary assembly.

ACKNOWLEDGMENTS

The author is grateful to Dr. Melvin Spiegel of Dartmouth College for suggesting the basic experimental scheme some eight years ago. This work was supported by United States Public Health Service grants GM 15,500 and GM 265.

REFERENCES

Auclair, W., and Meismer, D. M. (1965). *Biol. Bull.* *129*, 397.

Auclair, W., and Siegel, B. W. (1966). *Science 154*, 913.

Fairbanks, G., Levinthal, C., and Reeder, R. H. (1965). *Biochem. Biophys. Res. Commun. 20*, 393.

Gibbons, I. R. (1965). *Arch. Biol. Liège 76*, 317.

Gorovsky, M. A., Carlson, K., and Rosenbaum, J. L. (1970). *Anal. Biochem. 35*, 359.

Gross, P. R. (1964). *J. Exp. Zool. 157*, 21.

Harvey, E. B. (1940). *Biol. Bull. 79*, 166.

Linck, R. W. (1971). Ph.D. Thesis, Brandeis University.

Newell, P. C., Longlands, M., and Sussman, M. (1971). *J. Mol. Biol. 58*, 581.

Raff, R. A., Greenhouse, G., Gross, K. W., and Gross, P. R. (1971). *J. Cell Biol. 50*, 516.

Shapiro, A. L., Vinuela, E., and Maizel, J. V. (1967). *Biochem. Biophys. Res. Commun. 28*, 815.

Stephens, R. E. (1970a). *J. Mol. Biol. 47*, 353.

Stephens, R. E. (1970b). *Biol. Bull. 139*, 438.

Stephens, R. E. (1971). *Biological Macromolecules*, S. N. Timasheff and G. D. Fasman, eds., *Vol. V*, (Marcel Dekker, Inc.), p. 355.

Stephens, R. E. (1972a). *Biol. Bull. 132* (in press).

Stephens, R. E. (1972b). *Biol. Bull. 132* (in press).

Stephens, R. E., and Linck, R. W. (1969). *J. Mol. Biol. 40*, 497.

Weisenberg, R., and Taylor, E. W. (1968). *Exp. Cell Res. 53*, 372.

Transient Controls of Organ-Specific Functions in Pituitary Cells in Culture

Armen H. Tashjian, Jr.

Pharmacology Department
Harvard School of Dental Medicine
Boston

Richard F. Hoyt, Jr.

Department of Pharmacology
Harvard Medical School
Boston

INTRODUCTION

The preceding papers in this symposium have served by a variety of means to illustrate the enormity of information contained within the genome of eucaryotic cells and to underline the complexity by which expression of this genetic heritage is controlled. An understanding of the control of gene expression in eucaryotic cells is made particularly difficult when one considers the influence of interactions between cells in the Metazoa. While a multicellular organism is essentially a clone of cells possessing an identical intranuclear genome, the patterns of gene expression differ markedly from one group of cells to another within the organism and appear to be heritable and exquisitely controlled phenomena.

Control mechanisms which govern genetic expression may not be the same in all animal cells. Signals

triggering the expression of certain cell- or organ-
specific functions may arise within the target cell
itself; others come from closely adjacent cells, and
still others, for example neural and humoral agents,
arise from distant sites within the organism. Each cell
is exposed directly or indirectly to a variety of signals, and the biologist must be very clever indeed to
design experiments in which these multiple signals and
the individual responses to them can be analyzed mean-
ingfully in the intact organism. The purpose of this
report is to describe an experimental system that may
permit more precise studies of the mechanisms controlling the expression of cell-specific functions in
eucaryotic cells.

We reasoned that the regulation of cell-specific
functions could be studied profitably using a homogeneous population of healthy cells capable of carrying out several, unique, differentiated functions
which were responsive to one or more experimentally
controllable, chemical signals. In mammals, the endocrine system offers examples of cells that are adap-
table to this type of experimental approach. Endocrine secretory cells epitomize cells which both
generate and respond to highly specialized signals.
The fact that the generated signals are exported from
the cell offers certain practical advantages, for
example, less attention need be paid to intracellular
protein turnover than is the case in studies of intracellular enzymes.

In recent years, several clonal strains of rat
pituitary tumor cells have been established in tissue
culture, and these cultures have two properties which
make them of special interest in the study of genetic
expression in eucaryotic cells. First, they produce
two exportable protein hormones and an intracellular
enzyme, the production of each of which can be controlled by a variety of experimentally introduced
signals. Second, because the cells produce specific
markers and are neoplastic, they offer the opportunity
to examine the control of the expression of cell-specific function in tumor cells and to compare these
mechanisms with those in the fully differentiated non-
malignant cell. One of the two hormones produced by

these cells in culture appears to represent the inappropriate expression of genetic information, since it is believed that in the normal pituitary gland, each of these hormones is produced by a separate, distinct cell type. Thus, studies of the control of production of these two proteins might lead eventually to an understanding of how the cancer cell is able to open a portion of its genetic file that usually is kept tightly closed and unread in the fully differentiated cell. The corollary of this is, of course, an understanding of the manner in which normal cells differentiate and dedifferentiate.

MATERIALS AND METHODS

Origin of the Clonal Strains of Pituitary Cells

During the past six years, in collaboration with Sato, Yasamura and Bancroft, a variety of clonal strains of functional rat pituitary tumor cells have been established in culture. The establishment of the original three strains (GH$_1$, GH$_1$2C$_1$ and GH$_3$) has been described in detail (Tashjian *et al.*, 1968), and the history of several of the more recent lines was reviewed in 1970 by Sonnenschein, Richardson and Tashjian (1970), and Tashjian, Bancroft and Levine (1970). Clones derived from a single cell (Puck, Marcus and Ciecura, 1956) have been made at least once and often as many as 10 times from primary cultures or serially propagated cells and each clonal strain synthesizes and secretes rat growth hormones (GH)*. In addition, most strains also produce rat prolactin. The hormones secreted in culture are indistinguishable, both biologically and immunologically, from authentic rat pi-

---

*Abbreviations used are: GH = growth hormone; HC = hydrocortisone sodium succinate; db-cAMP = N^6,0^2'-dibutyryl adenosine 3',5'-cyclic monophosphate; ergocryptine = the methanesulfonate of 2-Br-α-ergocryptine; TRH = thyrotropin releasing hormone; and EGTA = [ethylenebis(oxyethylenenitrilo)] tetraacetic acid.

tuitary GH and prolactin. Most strains have shown no
evidence of senescence, such as a decrease in growth
rate or hormone production, during serial propagation
for periods as long as three to four years.

Figure 1 summarizes the relationship of the cells

HYPOTHALAMUS

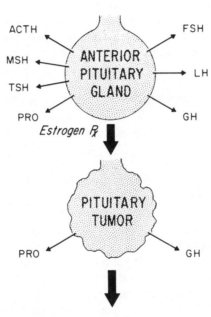

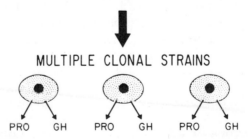

Figure 1. Relationship between the anterior pituitary gland and clonal strains of
hormone-secreting rat pituitary tumor
cells.

in culture to the rat pituitary gland. The anterior
lobe of the pituitary secretes at least seven hormones.
The secretion (or release), and possibly also the synthesis, of these hormones is under the control of factors which are produced in various hypothalamic neu-
rones and which reach the pituitary directly through a
portal circulation between the hypothalamus and the
pituitary gland (McCann and Porter, 1969). Accumulating evidence supports the current view that each of
the pituitary hormones is synthesized and secreted by
a specific cell type and that, with the exception of
the gonadotrophins (FSH and LH), each hormone may have
its own hypothalamic releasing and inhibiting factors
(McCann and Porter, 1969; Meites, 1970).

Chronic estrogen treatment in the rat may give rise
to pituitary tumors and these tumors, which can be
transplanted between rats of the same inbred strain,
often secrete more than one hormone (Takemoto *et al.*,
1962). It was from such a multihormone producing
tumor (MtT/W_5) that all the cell strains used in these
experiments were derived. They are referred to col-
lectively as GH cells, specific strains having appropriate subscript- or letter-designations, such as GH
or $GH_1 2C_1$.

Method of Culture of Growth
Hormone Cells

The method of culture of GH cells in monolayer has
been described previously (Tashjian *et al.*, 1968). In
brief, cultures were grown in plastic tissue culture
dishes containing 3 ml of Ham's F 10 medium (Ham,
1963) supplemented with 15% horse serum and 2.5% fetal
calf serum. Dishes were incubated at 37° in a humidi-
fied atmosphere of 5% CO_2 and 95% air. GH cells (GH_3)
can also be grown in suspension cultures (Bancroft and
Tashjian, 1971), but all of the experiments described
in this report were performed on monolayers.

Measurement of Growth Hormone
and Prolactin

GH and prolactin were assayed immunologically in
culture medium or in cell homogenates by the method of
microcomplement fixation (Wasserman and Levine, 1961).
Each assay method is specific, and no immunological
cross-reactions between GH and prolactin have been
detected (Tashjian, Bancroft and Levine, 1970; Tashjian, Levine and Wilhelmi, 1968). The 95% confidence
limits of a single determination are ± 20-25%; in most
of the experiments reported at least duplicate culture
dishes were used for each experimental point.

Measurement of Glutamine Synthetase

After incubation, the medium was removed and the
cells were washed four times with 3-4 ml of 0.15 M
NaCl. The cells were then scraped mechanically into
distilled water (3 dishes/2.0 ml) and ground thoroughly in an all glass homogenizer. An aliquot was removed for protein determination and the remaining homogenate was frozen for assay of glutamine synthetase
activity using the colorimetric method reviewed by
Waelsch (1955). One unit of enzyme activity is defined as 1 μmole L-glutamic acid γ-monohydroxamate
formed per mg protein per hr.

Measurement of Cell Protein

Cell protein was determined by the method of Lowry
et al. (1951).

Incorporation of Labeled Amino Acids
into Cell Protein

One to two μCi of ^{14}C-reconstituted protein hydro-
lysate were added to the medium of each dish and incorporation of ^{14}C-amino acids into trichloroacetic
acid-insoluble material was determined as previously
described (Bancroft, Levine and Tashjian, 1969).

Preparation of Tissue Extracts

Crude acid extracts of acetone powders of bovine hypothalamus, cerebral cortex, kidney and liver were prepared using 0.1 N HCl as described previously (Tashjian, Bancroft and Levine, 1970). The extracts were adjusted to pH 7.5, the precipitates which formed were removed by centrifugation, and the neutral supernatant solutions were designated as the "crude extracts." They were characterized further by gel filtration, dialysis and heating as described by Bancroft and Tashjian (1970).

Scanning Electron Microscopy

Cells were grown on glass coverslips in plastic Petri dishes. At the end of the experiment the coverslips were washed four times in F 10 medium without serum. The cells were then fixed for 15 min at room temperature in 6% glutaraldehyde in F 10 medium without serum. The coverslips were washed again in saline and were prepared for microscopy according to the method of Fujita, Tokunga and Inone (1971) using acetone as the drying solvent. Cultures were coated with carbon and with gold-palladium alloy using a Kinney high vacuum evaporator and were examined with a Jeolco JSMU-3 scanning electron microscope.

Materials

Tissue culture dishes (60 x 15 mm) were purchased from Falcon Plastics. Ham's F 10 medium (with and without glucose), horse serum and fetal calf serum were obtained from Grand Island Biological Co. Hydrocortisone sodium succinate was from the Upjohn Co. The following agents were purchased from Nutritional Biochemicals Corp. (17β-estradiol, testosterone, 17α-methyltestosterone, 2-deoxy-D-glucose, cycloheximide and puromycin dihydrochloride) or Sigma Chemical Co. (corticosterone;5β,3α-tetrahydrocortisol, cortisol and the sodium salt of $N^6,0^{2'}$-dibutyryl adenosine 3',5'-cyclic monophosphoric acid). Sandoz Ltd. donated the 2-Br-α-ergocryptine methanesulfonate, and the synthetic thyrotropin releasing hormone (Lot No. 842-553) was the gift of Abbott Laboratories.

RESULTS AND DISCUSSION

General Characteristics of Growth Hormone Cells in Culture

Most of the experiments reported here were performed with the GH_3 strain of rat pituitary cells. Unless otherwise stated the features to be described are those of GH_3 cells, but they are characteristic of all GH cells with the exception of the GH_12C_1 strain and of two subclones of GH_3-spinner cells, none of which produce prolactin.

At least 12 serial clones of GH_3 cells, each derived from a single cell of the preceeding clone, all produce[*] both GH and prolactin. This finding suggests strongly that individual cells are able to synthesize and secrete both hormones. However, since we have not yet measured the production of both proteins by a single cell, it is possible, but unlikely, that all mass cultures are made up of two types of cells further differentiated from a single parental cell: one producing GH and the other prolactin.

When dispersed with Viokase and plated sparsely, GH_3 cells, after a lag period of about 36 hr, grow exponentially with a population doubling time of about 50-60 hr. The cells eventually reach a state, which we have called the early stationary phase (Bancroft, Levine and Tashjian, 1969), in which cell protein stops increasing exponentially and levels off or in-

---

[*]Production of either GH or prolactin is defined as the amount of hormone which accumulates in the medium during a defined period of time (the hormones being highly stable in medium), divided by the cell protein at the time of collection (Tashjian, Bancroft and Levine, 1970; Bancroft, Levine and Tashjian, 1969). Thus, each reported value for production represents the average specific rate of appearance of the hormone in the medium.

creases at a greatly reduced rate. The cells never
attain a confluent monolayer. GH and prolactin are
produced during both the exponential and early stationary phases of growth. Most of the experiments reported
here were performed in exponentially growing cells.

   The ratio of immunologically active hormone stored
in the cells to that secreted into the culture medium
is low. In the case of GH, there is only approximately a 15- to 30-min supply of hormone in the cells
(Bancroft, Levine and Tashjian, 1969), while in the
same cells intracellular prolactin is equal to that
appearing in the medium in about 1-2 hr (Tashjian,
Bancroft and Levine, 1970). We have not as yet been
able to detect by complement-fixation inhibition
(Wasserman and Levine, 1961) an immunologically active
precursor molecule for either hormone. The low concentration of intracellular hormone may reflect rapid
secretion into the medium of newly synthesized proteins or a rate of intracellular GH and prolactin
turnover considerably in excess of the figure of 1%
per hr proposed by Eagle *et al.* (1959) as the average
rate of turnover of cell protein. We have not conclu-
sively demonstrated which of these possibilities is
predominant or whether both are true. Based on the
assumption that there is little or no intracellular
turnover of these exportable proteins (Bancroft,
Levine and Tashjian, 1969), the calculated values for
the fraction of the total protein synthesized by GH
cells represented by GH is about 2% for unstimulated
and 14% for HC-stimulated cells. Comparable values
for prolactin are about 2% for unstimulated and as
high as 25% for cells maximally stimulated by organ
extracts (Tashjian, Bancroft and Levine, 1970; Bancroft and Tashjian, 1970). If correct, these high
values of specific protein synthesis should make it
possible to use the stimulated GH cells as a source
of protein-specific polysomes bearing specific mRNA.

   Lastly, it should be emphasized that the GH cells
are aneuploid (Sonnenschein, Richardson and Tashjian,
1970) with modal numbers of chromosomes among different strains ranging from 69 to 75 per cell (normal 2n
for the rat is 42 chromosomes). These findings show
clearly that a markedly abnormal karyotype does not by

itself preclude organ-specific function in mammalian cells in culture; conversely, differentiated function of cells in culture does not necessarily imply the existence of a normal or even near-normal karyotype.

Effects of Hydrocortisone on Growth Hormone and Prolactin Production

Addition of HC to the medium of GH_3 cells stimulates the rate of production of GH 4-8 times that observed in control cells (Bancroft, Levine and Tashjian, 1969). The stimulation has a lag period of 24-36 hr, which can be reduced to 12-18 hr by frequent medium changes (Bancroft and Tashjian, 1970), reaches a maximum at 70-100 hr, and is observed at an HC concentration as low as 5×10^{-8} M. Cells maximally stimulated with 3×10^{-6} M HC produce 50-160 µg GH/mg cell protein per 24 hr. Removal of HC from the medium causes a return of the rate of GH production to that in control cells (Bancroft, Levine and Tashjian, 1969). That the effect of HC on GH_3 cells is not merely a general stimulation of cell metabolism is shown by several findings. Addition of HC to the medium of cells growing exponentially causes a 40-50% decrease in the rate of cell growth and a 30-40% decrease in the relative rate of amino acid incorporation into acid-insoluble material (Bancroft and Tashjian, 1970). In addition, the stimulation of GH production occurs equally well in medium containing or lacking glucose, as a carbon or energy source, and in medium containing 2-deoxyglucose, an inhibitor of intracellular glucose utilization (Fig. 2). Further evidence for the specificity of the effect of HC is noted when the relative rates of production of prolactin and GH are compared. While HC stimulates the production of GH about 8-fold, it depresses the rate of prolactin production to less than 25% of that in control cultures (Fig. 3). The dose-response relationships betweeen HC and the production of GH and prolactin are essentially mirror images of each other (Tashjian, Bancroft and Levine, 1970); the dose of HC which just stimulates GH production also just suppresses prolactin, and intermediate and maximum effects on the production of both hormones occur

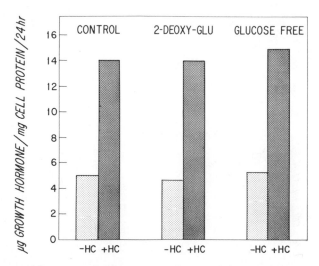

Figure 2. Effects of HC on GH production by GH₃ cells in control F 10 medium, in medium containing 2-deoxy-D-glucose (0.20 mg/ml), and in medium lacking glucose and supplemented with dialyzed serum ("glucose free"). All cultures were established in control medium. At zero time, fresh medium (either control, plus deoxyglucose, or glucose free) containing HC (3×10^{-6} M) or lacking HC was added to duplicate dishes. The medium was collected after 72 hr for hormone assay and the dishes were washed and frozen for cell protein. Total cell protein was decreased below levels in control medium by about 10% in medium with deoxyglucose and about 20% in glucose free medium.

at similar concentrations of HC.

We have found recently that GH cells produce the enzyme glutamine synthetase as well as GH and prolactin, and that the activity of this enzyme is stumulated in GH₃ cells by HC (Fig. 4), much as it is in explanted chick embryo retinal cells (Reif-Lehrer, 1968; Morris and Moscona, 1970). The presence of glutamine synthetase adds another type of controllable

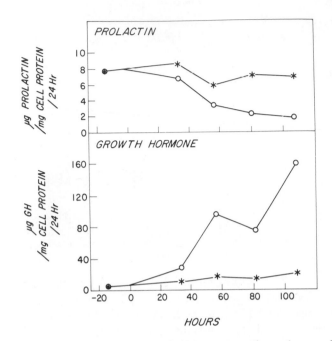

Figure 3. Effects of HC on prolactin and GH production. Duplicate dishes were used for each point. At zero time, fresh medium either containing HC (3 x 10^{-6} M) or lacking HC was added to each dish. Medium was collected at intervals from HC-treated (o) and control (*) dishes and frozen for hormone assay. These dishes were washed and frozen for determination of cell protein. [Reproduced from Tashjian, Bancroft and Levine (1970).]

marker, an inducible intracellular enzyme, to the pituitary cell system and should prove particularly useful in future studies of the mechanisms of the HC effects in these cells and in cell hybridization studies (Sonnenschein, Richardson and Tashjian, in press).

This leads us now to a brief consideration of the mechanisms whereby HC affects GH cells. As described earlier, there is little intracellular GH or prolactin in relation to the quantity of either hormone appear-

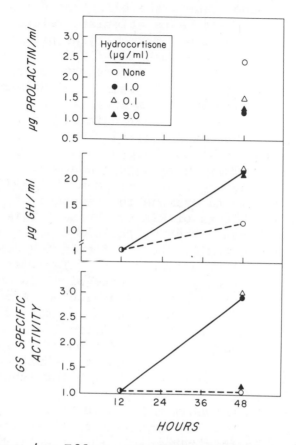

Figure 4. Effects of three concentrations of HC on prolactin, GH and glutamine synthetase (GS) activity in GH₃ cells. At zero time fresh medium either containing HC (●, Δ or ▲) or lacking HC (O) was added to each dish. Medium and cells were collected at 12 and 48 hr for hormone and enzyme assays. HC did not affect hormone or enzyme levels at 12 hr. At 48 hr, all three dose levels of HC caused the characteristic decrease in prolactin and increase in GH appearance in medium, while 0.1 and 1.0 µg/ml of HC stimulated GS activity about 3-fold, and the high dose of HC (9.0 µg/ml) had no effect on GS in this experiment.

ing in the medium. Thus, the effect of HC on GH production cannot merely be to stimulate release of stored
hormone into the medium. Moreover, as previously shown
by Bancroft, Levine and Tashjian (1969), cycloheximide
and puromycin each suppress incorporation of labeled
amino acids into cellular protein by 93-98% and suppress GH production by both stimulated and control
cells by at least 94%. Such findings, still however,
do not tell us whether HC stimulates GH production by
a direct action on hormone biosynthesis or by inhibition of intracellular hormone degradation. In either
case, the long lag period, a minimum of 12-18 hr, has
made it difficult to examine the response of GH cells
to certain inhibitors of RNA synthesis. This difficulty arises because the GH_3 cells are particularly
sensitive to the toxic effects of even low dose levels
(0.01-0.10 µg/ml) of actinomycin D. Thus, it has not
been possible to study the effects of HC in cells
whose protein synthesis has been blocked reversibly
with cycloheximide and in which RNA synthesis either
is or is not inhibited by actinomycin D as has been
done by Tomkins and his colleagues on HTC cells (Tomkins *et al.*, 1969). In such an experiment the GH
cells die or are clearly unhealthy before the effect
of HC can be observed. It may be possible to assess
the role of RNA metabolism in the HC effect by using
other inhibitors of RNA synthesis or to shorten the
lag period. In this regard, it is noteworthy that GH_3
cells show fewer toxic manifestations to cordycepin
(3'-deoxyadenosine) than they do to actinomycin D.
Should the effect of cordycepin be the same in GH_3
cells as it is in HeLa cells (Penman, Rosbash and Penman, 1970), namely to inhibit the labeling of mRNA but
not the transport of mRNA nor the labeling of nuclear
heterogeneous RNA, the drug would be useful in examining the possible role of newly transcribed mRNA in the
effect of HC on hormone production.

Steroid hormones are active as metabolic signals
in a variety of target tissues both *in vivo* and *in
vitro*. At present, binding of the steroid molecule to
receptor sites on a specific cytoplasmic protein is
considered to be an important, early step in hormone
action (O'Malley, 1971; Williams-Ashman and Reddi,

1971). The events which follow from this supposedly
primary interaction are often diverse and are, for
many tissues, as yet uncertain. In several systems
steroid receptors have been partially separated from
other cytoplasmic proteins and the binding properties
and steroid specificity have been examined. It has
been postulated, for example, in HTC cells (Samuels
and Tomkins, 1970) that steroids can be classified
into groups as inducers, anti-inducers and inactive
steroids. In the GH3 cell system, HC and corticosterone, the major circulating glucocorticoid in the rat,
have similar potencies in their effects on GH and prolactin production (Fig. 5). At the dose levels tested,
cortisone and tetrahydrocortisol did not stimulate GH
production, and they were considerably less potent
than the other two steroids in diminishing the production of prolactin. It is of interest to note that HC
and corticosterone were both optimal inducers of tyrosine aminotransferase in HTC cells, that tetrahydrocortisol was an inactive steroid and that cortisone
as well as testosterone and methyltestosterone were
anti-inducers (Samuels and Tomkins, 1970). We have
tested testosterone and 17-methyltestosterone alone
and together with HC in GH3 cells (Fig. 6). Neither
steroid by itself had effects on GH production, but
each was able to reduce the effect of HC. Methyltestosterone was a more potent antagonist of the HC effect
than was testosterone, as it also was in HTC cells
(Samuels and Tomkins, 1970). Although not presented,
testosterone and methyltestosterone also inhibited the
effect of HC on suppressing prolactin production while
neither of the inhibiting steroids had an effect of
its own. Additional steroids must be tested, the receptor isolated and characterized, and the relationship of steroid binding to its effect on each function
of the cell examined in detail before it will be possible to assess the functional significance which
steroid receptor molecules may play in GH cells.

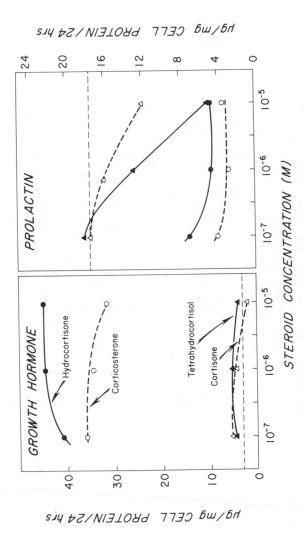

Figure 5. Effects of 4 steroids on GH and prolactin production in GH₃ cells. Replicate cultures were grown to moderate density in the absence of added steroids. At zero time, aliquots of fresh medium containing each of one of the 4 steroids listed in the figure or lacking added steroids were added to culture dishes. The dishes were incubated for 4 days, the medium was collected, and fresh medium either with or without the appropriate steroid was added for an additional 3 days. The data shown are for the collection period 4 to 7 days. The dashed horizontal lines give the levels of GH and prolactin production in control cultures not treated with added steroids. In addition to results shown, pre-disone (Δ^1-cortisone) had little effect on either GH or prolactin production at 10^{-5} M while dexamethasone, a synthetic highly potent glucocorticoid, was ~10 X as active as HC.

368

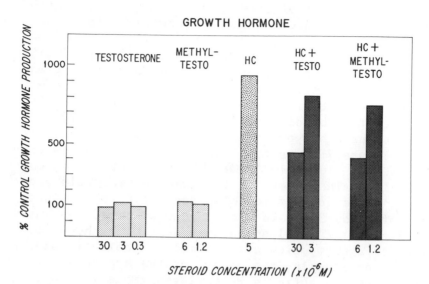

Figure 6. Effects on GH production of
testosterone and 17α-methyltestosterone
alone or in combination with HC. Replicate cultures of GH₃ cells were grown to
moderate density in the absence of added
steroids. At zero time, fresh medium containing either testosterone, 17α-methyl-
testosterone, HC (3 x 10⁻⁶ M), testosterone
+ HC, 17α-methyltestosterone + HC, or no
steroid was added. Cultures were incubated
4 days, medium was collected for hormone
assays and cells were washed and frozen for
protein determination. The results are expressed as GH production in treated cultures as a percentage of that in control
cultures (100%).

Effects of 17β-Estradiol on Growth Hormone and Prolactin Production

The results of the experiments described above were
concerned with an agent, HC, which increases GH production and decreases the production of prolactin. In
order to illustrate the flexibility of the GH cell
system, we should like now to present data pertaining

to the reciprocal of this effect, namely stimulation of prolactin production and a decrease in the production of GH.

There is evidence for the presence of estradiol receptors in both the hypothalamus and pituitary gland (Kahwanago, Heinrichs and Herrmann, 1970; Notides, 1970), but whether estrogenic steroids affect prolactin secretion by direct action on the pituitary as well as through hypothalamic effects is uncertain as yet. Figure 7 shows that estradiol, at very low concentrations (10^{-11}–10^{-10} M), has a direct action on GH cells. The effect is to increase prolactin production by about 2-fold while decreasing the production of GH by 30%. In a more general sense, we have noted that estradiol appears to sensitize the GH cells to other agents which also stimulate prolactin production. For example, the effect of the hypothalamic tripeptide, thryotropin releasing hormone (*vide infra*), on GH_3 cells is greater in the presence of estradiol than in the absence of the steroid. That estradiol receptors are present in the pituitary gland (Kahwanago, Heinrichs and Herrmann, 1970; Notides, 1970) and that the steroid can affect prolactin production directly through an action on pituitary cells in culture, make it likely that the effects of estrogen on pituitary prolactin *in vivo* are due to actions on both the hypothalamus and pituitary gland. A direct effect of estrogen on explanted anterior pituitary fragments has also recently been reported by Lu, Koch and Meites (1971).

Effects of Hypothalamic and Other
Tissue Extracts on Growth Hormone
Cells

The findings that GH_3 cells produced prolactin under basal culture conditions and that the rate of prolactin production could be either raised or lowered at will suggested to us that these cells might prove useful as a system for the assay of hypothalamic factors which either stimulate or inhibit prolactin release and for the study of their actions as well. Previously available techniques, both *in vivo* and *in*

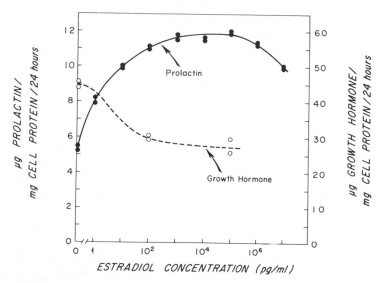

Figure 7. Effects of 17 β-estradiol on prolactin and GH production by GH$_3$ cells. Replicate cultures were grown to moderate density in the absence of estradiol. At zero time, fresh medium either containing estradiol (at the concentrations shown) or lacking estradiol was added to duplicate dishes. The dishes were incubated for 3 days, the medium was collected, and fresh medium either with or without estradiol was added for an additional 4 days. The data shown are for the collection period 3-7 days. Essentially the same results were observed in the 0-3 day collection period. There was was no effect of estradiol, at any dose level tested, on cell growth.

vitro with explanted pituitary fragments, possessed limitations (Meites, 1970) which often made experiments difficult to perform and difficult to control.

In most mammals, it has been thought that the predominant hypothalamic influence on pituitary prolactin release is inhibitory (McCann and Porter, 1969). It was, therefore, a surprise to us when crude extracts of bovine (and rat) hypothalamic tissue caused a

marked, 6- to 10-fold increase in prolactin production
by GH cells, accompanied by a decrease in GH production (Tashjian, Bancroft and Levine, 1970). No inhibitory effects on prolactin production were observed
as early as 2 or 4 hr after adding the extracts to the
culture medium, and significant stimulation was detected between 4 and 7 hr. There were no concomitant
changes in the growth rate of the cells or in bulk
protein synthesis. Early in the course of these experiments (Tashjian, Bancroft and Levine, 1970), it
was observed that within 4 hr after adding medium containing an active extract, the cells appeared, under
the low power phase contrast microscope, to be more
highly stretched (or adherent to the plastic surface),
and few rounded-up cells could be seen. This finding
will be documented later in quite a different context,
namely stimulation of prolactin production by TRH.
    Because of the unexpected nature of the effect of
crude hypothalamic extracts on GH cells, the specificity of the donor tissue was examined. It was found
that similar extracts of bovine cerebral cortex, kidney and liver were also active (Tashjian, Bancroft and
Levine, 1970). The liver is an abundant source of
this material and the prolactin-stimulating factor has
been partially characterized from that source (Bancroft
and Tashjian, 1970). It differs from the known hypothalamic releasing factors in that it appears to be of
molecular weight greater than 10,000 by its behavior
during dialysis and gel filtration. It is also heat
labile, losing most of its biological activity during
incubation at 60° for 10 min. Although the fractionated liver factor is still grossly heterogeneous on
polyacrylamide disc gel electrophoresis, it stimulates
prolactin production 3- to 4-fold at dose levels as
low as 10 µg protein/ml culture medium and antagonizes
the effects of HC on both prolactin and GH production
in GH cells. It does not, however, induce prolactin
production in the GH_12C_1 strain of cells which do not
produce the hormone under normal culture conditions.
In recent experiments it has been shown that similar
extracts of bovine serum do not affect prolactin production when added to medium of GH_3 cells, a finding
which demonstrates that the effect is not due merely

to some nonspecific, universally dispersed bovine material.

The physiological significance, if any, of the prolactin-stimulating factor extracted from several bovine tissues remains uncertain and must await further studies in the living animal. Nevertheless, these results indicate that caution must be used in ascribing effects of crude brain extracts on pituitary function to the action of specific hypothalamic factors.

Effects of Thyrotropin Releasing Hormone on Growth Hormone Cells

Until two months ago (Matsuo *et al.*, 1971), the only hypothalamic factor for which the structure was known was TRH. It is a tripeptide, L-(pyro)Glu-L-His-L-Pro-NH (Burgus *et al.*, 1969; Folkers *et al.*, 1969). TRH causes the rapid release of pituitary thyrotropin both *in vivo* and *in vitro* and consistent effects of TRH on the synthesis or release of other pituitary hormones had not been described until Tashjian, Barowsky and Jensen (1971) reported a direct effect of TRH on prolactin production by GH cells.

When synthetic TRH, at concentrations of 0.10 to 10 ng/ml medium, was added to cultures of two strains of GH cells, there was a stimulation of the production of prolactin and an inhibition of GH production by the cells (Tashjian, Bancroft and Jensen, 1971). There was no effect on cell growth. Stimulation of prolactin production reached a maximum level, 2- to 5-times control, at 24-48 hr, and the effect persisted for at least 20 days in the continued presence of TRH. In contrast to the rather long lag period before the effects of HC on GH cells are observed, the stimulation of prolactin production by TRH is seen as early as 3 hr after the tripeptide is added to the culture medium (Fig. 8). Furthermore, also in contrast to the HC effect which begins to wane as soon as the steroid is removed from the medium (Bancroft, Levine and Tashjian, 1969), the effect of TRH can be initiated by a short exposure to the peptide; the effect can then persist for at least 7 days despite washing the cells and addition of fresh medium lacking TRH (Fig. 9).

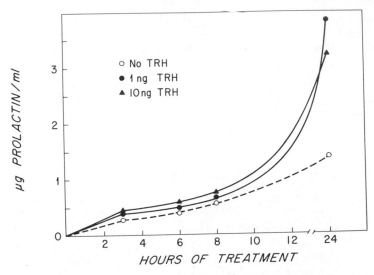

Figure 8. Early effects of TRH on prolactin production by GH₃ cells. Duplicate
dishes were used for each point. At zero
time, fresh medium either containing TRH
(1 or 10 ng/ml) or lacking TRH was added to
4 sets of replicate culture dishes. Medium
was collected at intervals from treated and
control dishes and frozen for hormone assay.
Although the effects of TRH at 3, 6 and 8
hr were small, there was no overlap of the
prolactin values between control and
treated cultures, and the consistency of
the effect in this and 3 other independent
experiments lead us to conclude that the
differences are significant.

The effects of TRH on prolactin and GH production
by GH₃ cells are not merely due to an action on the
hormone release mechanism. The duration and magnitude
of the effects are too great to be explained by release alone, and the data in Figure 10 show that the
prolactin that appears in the medium exceeds that
present in the cells and that treatment with TRH also
increases intracellular and medium prolactin by a
similar ratio over that found in unstimulated, control

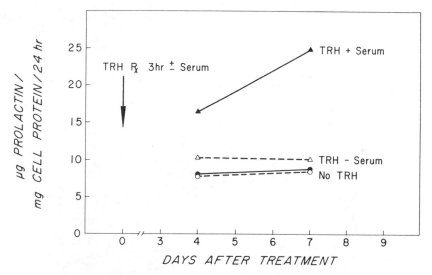

Figure 9. Effects on prolactin production by GH_3 cells of a short exposure to TRH in medium with and without serum. Duplicate cultures were used for each point. At zero time, fresh medium (with or without serum supplements) either containing TRH (1.0 ng/ml) or lacking TRH was added to 4 sets of replicate culture dishes. The cultures were incubated for 3 hr, the medium was removed, and all cultures were washed twice with about 4 ml of fresh medium containing serum but lacking TRH. The washes were dis-discarded, and fresh medium supplemented with serum but lacking TRH was added, and the cultures incubated for 4 days. The medium was then collected and fresh medium containing serum but lacking TRH was added for an additional 3 days. The data shown are for both the 4- and 3-day periods after a 3-hr exposure to TRH. The solid and dashed lines connect points in which the cells were exposed to medium with or without serum, respectively, for the first 3-hr period.

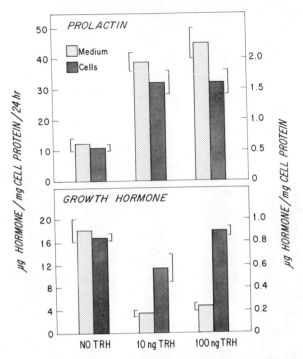

Figure 10. Intracellular versus extracellular levels of prolactin and GH in control and TRH-treated cells. Duplicate cultures were used. The bars give the mean values and the brackets give the range. Fresh medium either containing TRH (10 or 100 ng/ml) or lacking TRH was added to replicate dishes of GH_3 cells. The cultures were incubated for 72 hr and the medium was collected for hormone assays. The cells were washed 3 times with saline, scraped from the dishes into 2 ml of distilled water and treated for 5 min at 1-2° in a Raytheon model DF 101 sonic oscillator. After removal of aliquots for determination of protein, the cell sonicates and the 72-hr medium were assayed for prolactin and GH.

cultures. In addition, the decrease in the appearance
of GH in medium is not accompanied by an accumulation
of GH in the cells, a finding that reveals that TRH is
not merely suppressing GH release.

As shown previously (Tashjian, Barowsky and Jensen,
1971), TRH does not affect the growth of GH cells.
Nor can the changes in prolactin and GH production
after treatment for 24 hr be explained by changes in
bulk synthesis in the cells (Table 1). Also presented

TABLE 1

Effects of thryotropin releasing hormone and bovine
liver factor (HV-15) on protein synthesis and
prolactin and growth hormone production by GH_3 cells*

Treatment	Relative amino acid incorpor. (treated: control)	Relative prolactin production (treated: control)	Relative GH production (treated: control)
TRH (10 ng, 24 hr)	1.0	3.6	0.88
HV-15 (100 μg prot./ml, 72 hr)	1.1	9.2	0.46

*Cells at moderate density were incubated in con-
trol medium or in medium containing TRH or liver factor for 24 and 72 hr, respectively. During the last
70 min of treatment the incorporation of ^{14}C-labeled
amino acids into trichloroacetic acid-insoluble material was measured in duplicate dishes. Medium was
collected for measurements of hormone concentrations.
Relative incorporation and production figures were
calculated by dividing the appropriate values obtained
in treated cultures by those obtained in controls.

in Table 1 are the data which show that the effects of
the bovine liver factor on prolactin and GH production
are not accompanied by a significant change in amino
acid incorporation into general cell protein. As with

the effects of HC on GH and prolactin production
(Tashjian, Bancroft and Levine, 1970), continued protein synthesis in GH$_3$ cells is required for the effects
of TRH to be observed. Cycloheximide (3.6 x 10^{-5} M)
inhibited by greater than 92% the rate of appearance
of prolactin in the medium of both control and TRH-
treated cells, and the effect of the inhibitor of protein synthesis was reversible following wash-out of
the cycloheximide.

It was mentioned previously that hypothalamic and
other tissue extracts produced changes in the gross
appearance, under phase contrast optics, of GH$_3$ cells
which seemed to correlate temporally with the observed
stimulation of prolactin production. Similar changes
were seen when the cells were incubated in medium containing TRH, which also stimulates prolactin production. These changes are clearly visible in cells examined with the scanning electron microscope (Figs. 11
and 12). The functional significance of these rather
large changes in the appearance of GH$_3$ cells treated
with TRH are as yet unknown; however, since they appear to correlate temporally with increased prolactin
production induced by two different signals, we conclude that further study of the details of the relationship between cell ultrastructure and function
should be profitable in this system.

Although we have emphasized that the GH cell system
is a useful *in vitro* model, we should point out that
information derived from study of this system may have
wider biological significance. For example, these
studies of TRH show that the hypothalamic hypophysiotrophic factors may not be entirely hormone specific,
and that they affect processes more general than
merely release, namely biosynthesis or turnover or
both. Finally, the effect of TRH on prolactin production by GH cells in culture may help to explain the
clinical observation of gynecomastia and/or galactor-
rhea in some patients with hypothyroidism (Ingbar and
Woeber, 1968), provided it can be demonstrated that
human prolactin production can be enhanced by TRH.

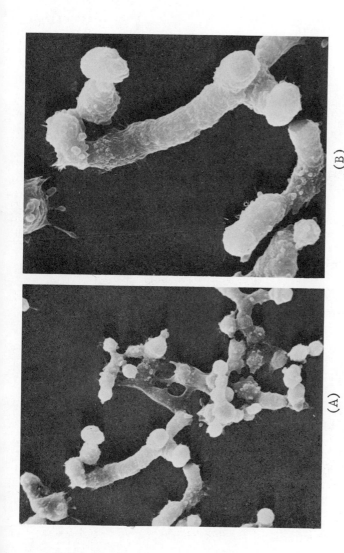

(A)

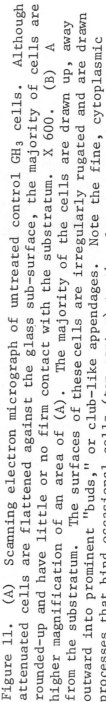

(B)

Figure 11. (A) Scanning electron micrograph of untreated control GH₃ cells. Although attenuated cells are flattened against the glass sub-surface, the majority of cells are rounded-up and have little or no firm contact with the substratum. X 600. (B) A higher magnification of an area of (A). The majority of the cells are drawn up, away from the substratum. The surfaces of these cells are irregularly rugated and are drawn outward into prominent "buds," or club-like appendages. Note the fine, cytoplasmic processes that bind occasional cells (top center) to the glass substratum. X 1500.

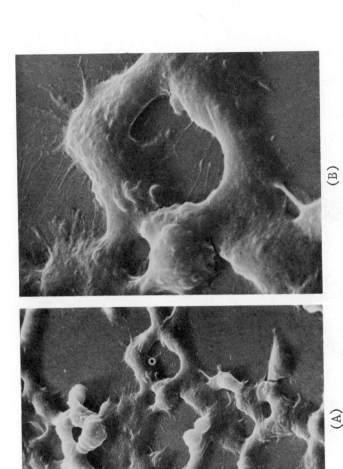

(A)

(B)

Figure 12. (A). Scanning electron micrograph of GH₃ cells treated with TRH, 10 ng/ml, for 20 hr. Most of the cells are flattened firmly against the substratum. These cells are not attenuated, as are the occasional firmly bound cells in control cultures. X 900. (B) A higher magnification of an area of (A). The cells are not attenuated; they are substantial in appearance and are bound firmly to the substratum by numerous fine, branching cytoplasmic processes that extend outward from the basal regions of each cell. The surfaces of these cells are quite smooth in contrast to those of control preparations (Fig. 11A & B). Bud-like surface projections are seen only rarely. X 2400.

*Evidence that the Control of Prolactin
and Growth Hormone Production are not
Invariably Related in a Reciprocal
Manner*

It has been shown that both HC and TRH affect prolactin and GH production; when the production of one
hormone was increased the other was decreased. We
should like to conclude by describing three sorts of
experiments which show that the production of these
two proteins is not necessarily reciprocally linked.

For some years it has been recognized that certain
ergot alkaloids, ergocornine for example, inhibit prolactin secretion *in vivo* (Shelesnyak, 1954; Wuttke,
Cassell and Meites, 1971). It has remained uncertain,
however, whether this effect is at the level of the
hypothalamus or whether prolactin secretion is influ-
enced by a direct action on the pituitary gland. We
undertook the present studies in GH cells with the
synthetic ergot alkaloid 2-Br-α-ergocryptine, which is
a more potent antiprogestational agent in the rat than
ergocornine, in order to test the hypothesis that the
drug does have a direct effect on the pituitary and to
see whether that effect is specific for prolactin.
The results show that ergocryptine does inhibit prolactin production in GH_3 cells in a dose-related manner and that there is no significant effect, at the
dose levels tested, on the production of GH (Fig. 13).
In experiments not shown, it has also been found that
GH_3 cells fail to secrete any prolactin when the incubation medium contains ergocryptine (1 μg/ml) plus HC
(5×10^{-6} M), and that ergocryptine can inhibit the
prolactin-stimulating effect of bovine liver factor.
Lu, Koch and Meites (1971) have shown direct inhibition by ergocornine of prolactin release from freshly
explanted rat pituitary glands.

The second stimulus which affects GH cells in such
a way that GH and prolactin production are not reciprocally modified is db-cAMP. In this case, the cyclic
nucleotide stimulates the production of both hormones
(Fig. 14). It is known that GH_3 cells contain the enzyme, adenyl cyclase, and that prolactin production is
affected by treatment of the cultures with several

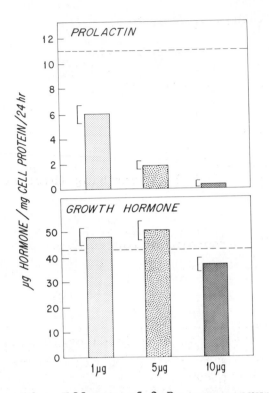

Figure 13. Effects of 2-Br-α-ergocryptine
on prolactin and GH production by GH₃
cells. Duplicate cultures were used. The
bars give the mean values and the brackets
give the range. Fresh medium either containing ergocryptine (1, 5 or 10 μg/ml) or
lacking the drug was added to replicate
dishes of cells. Cultures were incubated
for 3 days, the medium was collected, and
fresh medium either with or without ergocryptine was added for an additional 3 days.
The data shown are for the collection period 3-6 days. Essentially the same results
were observed in the 0-3 day collection
period. The dashed horizontal lines give
the levels of prolactin and GH production
in control cultures not treated with ergocryptine. There was no effect of ergocryptine on cell growth at 1 and 5 μg/ml; at
10 μg/ml total cell protein was only 92% of
control after 6 days of treatment.

382

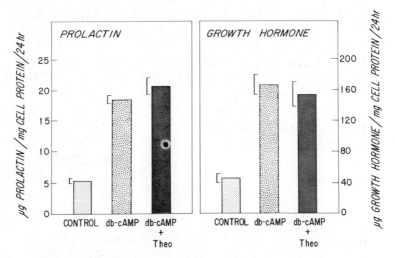

Figure 14. Stimulation by db-cAMP of both
prolactin and GH production by GH_3 cells.
Duplicate cultures were used. The bars
give the mean values and the brackets give
the range. Fresh medium containing db-cAMP
(10^{-3} M), db-cAMP (10^{-3} M) plus theophyllin
(10^{-4} M), or lacking drugs was added to
replicate dishes of cells. Cultures were
incubated for 3 days, the medium was collected, and fresh medium with the appropriate drugs was added for an additional 4
days. The data shown are for the collection period 3-7 days. Theophyllin alone
(10^{-4} M) had little or no effect on prolactin or GH production or on cell growth. In
cultures treated for 7 days with db-cAMP or
db-cAMP plus theophyllin, total cell protein was only 70% of control.

prostaglandins known to affect adenyl cyclase in other
tissues. Additional studies in GH cells will be required, however, before the endogenous adenyl cyclase-
cAMP system can be considered to play an important
role in hormone production.
     Lastly, calcium ions have been implicated in the
control of hormone release from many endocrine cells

(Rubin, 1970) and possibly may be involved in the
mechanism of action of the hypophysiotrophic factors
of the hypothalamus (Geschwind, 1970). In contrast to
the stimulation of both prolactin and GH production by
cAMP, we have found that EGTA, a chelating agent with
a high degree of specificity for calcium, suppresses
both GH and prolactin production in GH cells. These
effects which are completely reversed by calcium, are
observed at EGTA concentrations of 2-6 x 10^{-4} M, levels
which do not affect cell growth for a period of 7 days.

SUMMARY

An animal cell system, composed of multipotent rat
pituitary tumor cells, has been established in culture. The cells have several features which make them
a promising model for studying the control of expression of cell-specific functions in eucaryotic cells.

1. Several clonal strains synthesize and secrete
   into the culture medium two different protein
   hormones, prolactin and growth hormone. One
   strain produces only growth hormone.
2. The production of prolactin and growth hormone
   can be stimulated or suppressed by a variety of
   signals added to the culture medium. These
   agents include steroid hormones, proteins, small
   peptides, nucleotides, prostaglandins, ergot
   alkaloids and calcium-chelating compounds.
3. Specific changes in the function of these cells
   are accompanied by morphological alterations
   visible by both phase contrast and scanning
   electron microscopy.
4. The cells synthesize glutamine synthetase and
   the activity of the enzyme is stimulated by hydrocortisone.
5. One of the two hormones produced by the cells
   would appear to be appropriate and the other inappropriate in the sense that, in the normal
   animal, growth hormone and prolactin seem to be
   produced by different types of cells. This system may therefore be useful in probing the mech-

anism underlying the "ectopic humoral syndromes"
(Liddle *et al.*, 1969).

ACKNOWLEDGMENTS

The authors wish to thank Misses N. J. Barowsky,
D. K. Jensen and Y. E. Santo, and Mr. E. F. Voelkel
for expert assistance, and Dr. R. S. Karpinski for
advice on preparing material for scanning electron
microscopy. We thank Dr. M. S. Anderson of Abbott
Laboratories, North Chicago, Illinois for the synthetic TRH, and Dr. K. Saameli and Prof. E. Flükiger
of Sandoz Ltd., Basle, Switzerland for the methanesulfonate of 2-Br-α-ergocryptine. Dr. S. Ito of Harvard Medical School kindly made available the use of
the scanning electron microscope. The original investigations described in this report were supported
in part by Research grant AM 11011 from the National
Institute of Arthritis and Metabolic Diseases. R. F.
H. was a postdoctoral research fellow of the Pharma-
ceutical Manufacturer's Association Foundation.

REFERENCES

Bancroft, F. C., and Tashjian, A. H., Jr. (1970). *In Vitro 6*, 180.
Bancroft, F. C., and Tashjian, A. H., Jr. (1971). *Exp. Cell Res. 64*, 125.
Bancroft, F. C., Levine, L., and Tashjian, A. H., Jr. (1969). *J. Cell Biol. 43*, 432.
Burgus, R., Dunn, T. F., Desiderio, D., and Guillemin, R. (1969). *C. R. Acad. Sci. (Paris) 269*, 1870.
Eagle, H., Piez, K. A., Fleischmann, R., and Oyama, V. I. (1959). *J. Biol. Chem. 234*, 592.
Folkers, K., Enzmann, F., Bøler, J., Bowers, C. Y., and Schally, A. V. (1969). *Biochem. Biophys. Res. Commun. 37*, 123.
Fujita, T., Tokunga, J., and Inone, H. (1971). In *Atlas of Scanning Electron Microscopy in Medicine*, (New York: Elsevier), pp. 1-5.
Geschwind, I. I. (1970). In *Hypophysiotrophic Hormones*

of the Hypothalamus: *Assay and Chemistry*, J. Meites, ed. (Baltimore: Williams and Wilkins Co.), pp. 289–319.

Ham, R. G. (1963). *Exp. Cell Res. 29*, 515.

Ingbar, S. H., and Woeber, K. A. (1968). In *Textbook of Endocrinology*, R. H. Williams, ed., *4th Ed.* (Philadelphia: W. B. Saunders Co.), pp. 05–286.

Kahwanago, I., Heinrichs, W. L., and Herrmann, W. L. (1970). *Endocrinology 86*, 1319.

Liddle, G. W., Nicholson, W. E., Island, D. S., Orth, D. N., Abe, K., and Lowder, S. C. (1969). *Recent Progr. Hormone Res. 25*, 283.

Lowry, O. H., Rosebrough, N. F., Farr, A. L., and Randall, R. J. (1951). *J. Biol. Chem. 193*, 265.

Ly, K. -H., Koch, Y., and Meites, J. (1971). *Endocrinology 89*, 229.

Malven, P. V., and Hage, W. R. (1971). *Endocrinology 88*, 445.

Matsuo, H., Baba, Y., Nair, R. M. G., Arimura, A., and Schally, A. V. (1971). *Biochem. Biophys. Res. Commun. 43*, 1334.

McCann, S. M., and Porter, J. C. (1969). *Physiol. Rev. 49*, 240.

Meites, J., ed. (1970). *Hypophysiotrophic Hormones of the Hypothalamus*: *Assay and Chemistry* (Baltimore: Williams and Wilkins Co.).

Morris, J. E., and Moscona, A. A. (1970). *Science 167*, 1736.

Notides, A. C. (1970). *Endocrinology 87*, 987.

O'Malley, B. W. (1971). *New Engl. J. Med. 284*, 370.

Penman, S., Rosbash, M., and Penman, M. (1970). *Proc. Nat. Acad. Sci. USA 67*, 1878.

Puck, T. T., Marcus, P. I., and Ciecura, S. J. (1956). *J. Exp. Med. 103*, 273.

Reif-Lehrer, L. (1968). *Biochim. Biophys. Acta 170*, 263.

Rubin, R. P.,(1970). *Pharm. Rev. 22*, 389.

Samuels, H. H., and Tomkins, G. M. (1970). *J. Mol. Biol. 52*, 57.

Shelesnyak, M. D. (1954). *Amer. J. Physiol. 179*, 301.

Sonnenschein, C., Richardson, U. I., and Tashjian, A. H., Jr. (1970). *Exp. Cell Res. 61*,,121.

Sonnenschein, C., Richardson, U. I., and Tashjian, A.

H., Jr. (in press). *Exp. Cell Res.*

Takemoto, H., Yokoro, K., Furth, J., and Cohen, A. I. (1962). *Cancer Res. 22*, 917.

Tashjian, A. H., Jr., Bancroft, F. C., and Levine, L. (1970). *J. Cell Biol. 47*, 61.

Tashjian, A. H., Jr., Barowsky, N. J., and Jensen, D. K. (1971). *Biochem. Biophys. Res. Commun. 43*, 516.

Tashjian, A. H., Jr., Levine, L., and Wilhelmi, A. E. (1968). *Ann. N. Y. Acad. Sci. 148*, 352.

Tashjian, A. H., Jr., Yasumura, Y., Levine, L., Sato, G. H., and Parker, M. L. (1968). *Endocrinology 82*, 342.

Tomkins, G. M., Gelehrter, T. O., Granner, D., Martin, D., Jr., Samuels, H. H., and Thompson, E. B. (1969). *Science 166*, 1474.

Waelsch, H. (1955). In *Methods in Enzymology*, S. P. Colowick and N. O. Kaplan, eds., *Vol. 2* (New York: Academic Press), p. 267.

Wasserman, E., and Levine, L. (1961). *J. Immunol. 87*, 290.

Williams-Ashman, H. G., and Reddi, A. H. (1971). *Ann. Rev. Physiol. 33*, 31.

Wuttke, W., Cassell, E., and Meites, J. (1971). *Endocrinology 88*, 737.

The Inheritance of Differentiative Capacity

Introduction

The classical embryologist Driesch recognized many years ago that the fertilized egg is the only totipotent cell that arises during the life cycle of an animal and that embryogenesis represents the progressive restriction of differentiative capacity. That these restrictions do not arise from permanent, irreversible modification of the genome was first demonstrated in general by experiments involving the tranplantation of nuclei from somatic cells to enucleated eggs. It was confirmed for specific markers by the study of heterocaryons made between different somatic cells and the segregant clones derived from them. These experimental systems now show promise of great fruitfulness in delineating the genetic elements of the regulatory apparatus that are responsible for these restrictions.

As mentioned in the Preface, it is also true that cells within the developing embryo yield clones which

appear to inherit the capacity to develop along restricted pathways. The evidence has come from three
experimental approaches. The first employs synthetic
mice produced by mixtures of blastomeres containing
appropriate genetic markers. Two others not represented in this volume involve somatic segregations in
insects (Hadorn, E. (1966). *25th Symp. Soc. Dev. Biol.*,
p. 85) and x-chromosome mosaicism in human females
(Gandini *et al*. (1968). *Proc. Nat. Acad. Sci. USA 61*,
945). That differentiative capacity is clonally inherited is also apparent from the existence in animal
embryos of stem cells whose progeny can be shown to
give rise to glandular, hemopoietic, immunogenic, myogenic, chondrogenic or other tissues. In recent years
the ability to cultivate clones *in vivo* or *in vitro*
which inherit the capacity to perform somatic functions under permissive environmental conditions has
provided a tool of remarkable power for the eventual
unraveling of the mechanism(s) of inheritance. Two
notable examples of the latter approach are included
here.

Some Characteristics of
Gene Expression as Revealed by
a Living Assay System

**H. R. Woodland, C. C. Ford, J. B. Gurdon,
and C. D. Lane**

*Department of Zoology
Oxford University
Oxford, England*

INTRODUCTION

One of the more important phenomena involved in a
developmental program is the interaction between those
cellular components which carry developmental information and those parts of the cell that process it; and
the most fully studied aspect of this process is the
interaction between nucleus and cytoplasm. In the
animal which we have studied, the frog *Xenopus laevis*,
all the available evidence indicates that the nuclear
genetic information is the same in different types of
cells (Gurdon, 1962; Gurdon and Laskey, 1970; Laskey
and Gurdon, 1970). Thus the production of differences
between cells must initially involve heterogeneity in
the processing apparatus. In order to understand how
the capacity to follow the developmental program is
inherited it is therefore necessary first to understand the characteristics of the processing system.

Genetic activity in early development may be classified under three headings, DNA synthesis, RNA synthesis and protein synthesis. The work reported here is orientated towards finding out as much as possible about the nature and control of these processes. Much biochemical work must, perforce, utilize cell-free systems, but there must always be doubt as to the relation of results obtained in this way to events occurring in living cells. To avoid such doubts the method we have used involve the micro-injection of cell components, or other substances, into living oöcytes or eggs. These two types of cells are biochemically interesting because they are metabolically very different from each other, particularly as regards their nucleic acid metabolism; they are embryologically interesting, because they represent the parent cell of the embryo in its formative and mature phases (for our purposes the sperm can be regarded as of minor importance, and in numerous species it is dispensible); and lastly they are very convenient because of their large size. In this paper we summarize what we know so far about the three classes of genetic activity which occur in these cells, and compare our results with what is known from studies of cell-free systems. We have not attempted a comprehensive review of work done *in vitro*, but confine ourselves to those aspects of *in vitro* work which are directly relevant to that done *in vivo*, and wherever possible that which uses the same or closely related animal species.

DNA SYNTHESIS

*Characteristics of Normal Oöcytes
and Eggs*

The oöcyte of *Xenopus* is a growing, nondividing cell, intensely active in RNA synthesis and protein accumulation. It does not synthesize DNA, except for a short while at the beginning of oogenesis when rDNA[*]

---

[*]Abbreviations: rRNA, ribosomal RNA; mRNA, messenger RNA; Hb, haemoglobin; SDS, sodium dodecyl sulfate;

is amplified (Gall, 1969). The full grown oöcyte is stimulated by hormone to undergo the first meiotic reduction division. Various other changes also occur at this time (Smith and Ecker, 1970), after which it is called an egg. The stimulus of fertilization (or artifical activation) leads to the completion of meiosis and the formation of a female "pronucleus." The egg and sperm pronuclei replicate their DNA between 20 and 40 min after fertilization (Graham, 1966). The cell divides after about 90 min and then enters a phase of rapid and frequent cell division, in which the rate of DNA synthesis is faster, and the duration of the S-phase shorter, than found even in rapidly growing bacteria (Graham and Morgan, 1966; Gurdon, 1968b). The egg makes little, if any, nuclear RNA (Gurdon and Woodland, 1969) and is therefore a cell involved primarily in DNA synthesis, in contrast to the oöcyte which makes only RNA. These two cells, one of which may be converted *in vitro* into the other by hormone treatment, are therefore eminently suitable for the study of factors which cause DNA synthesis to begin.

DNA Synthesis Studied in Living Cells by Micro-Injection

Enzymes of DNA synthesis. The basic method we have used in this study is to inject a DNA template and a radioactive deoxynucleotide or deoxynucleoside into living cells. The template has been in the form either of intact nuclei or purified DNA. This contrasts with the conventional method of investigation, which is to attempt to extract the enzymes and then to assay their activity *in vitro*. The kinds of complexities which this latter approach can lead to is well exemplified by recent studies of the DNA polymerases of *Escherichia coli* (e.g., De Lucia and Cairns, 1969; Knippers and Strätling, 1970). The aim of these in-

---

dCTP, dTTP, dGTP, dATP = deoxynucleotide triphosphates of cytidine, thymidine, guanosine and adenosine respectively; rDNA, ribosomal DNA; SSC, standard saline citrate (0.15 M NaCl, 0.015 M sodium citrate).

jection experiments is to study the control of DNA
synthesis under conditions as near as possible to
those of normal cells. This can be achieved by con-
ducting the experiments in the living cell, or eventually by developing *in vitro* systems with similar
properties.

In order to start with experiments as simply related to the normal situation as possible, intact
nuclei were injected into cells. It was found that
they behave in the same way as the resident cell
nucleus, for they synthesize DNA in eggs, but not in
oocytes. These experiments are discussed more fully
by Gurdon and Woodland (1968; 1970). They lead to the
general conclusion that the initiation of DNA synthesis is under cytoplasmic control, the cytoplasm becoming competent to induce DNA synthesis when the
germinal vesicle breaks down during maturation of the
oöcyte. In an attempt to simplify the rather complex
interaction which occurs in experiments of this sort
Gurdon, Birnstiel and Speight (1969) injected pure DNA
into eggs. They showed that injected native and denatured DNA stimulated the incorporation of ^{3}H-thymi-
dine into DNA. This DNA has the same buoyant density
as that injected, both when examined in a double-
stranded (Gurdon, Birnstiel and Speight, 1969) and a
single-stranded form (Ford, C. C. and Woodland, H.R.,
unpublished data). When denatured DNA is injected.
about half of the radioactive DNA extracted behaves
as native DNA on neutral CsCl gradients. Thus unless
single stranded molecules, parts of which are newly
synthesized, become annealed with themselves, the denatured DNA-stimulated incorporation appears to represent replication.

In an analogous study, DNA was injected into
oöcytes and in initial experiments, no replication was
observed (Gurdon and Speight, 1969). In more recent
experiments in which the extraction procedure included
extensive protein digestion (Ford, C. C. and Woodland,
H. R., unpublished data), it was found that in oocytes,
denatured DNA markedly stimulates DNA synthesis (Table
1 and Fig. 1). With the same extraction methods
stimulation was not observed when native DNA was injected (Table 1). The DNA synthesized in these ex-

TABLE 1

DNA synthesis *in vivo* in response to injected templates

Type of cell	Radioactive precursor	DNA counts/min nucleotide counts/min		
		Native DNA template (%)	Denatured DNA template (%)	Denatured native
Oöcytes*	³H-dATP	0.03	24.8	830**
Oöcytes†	³H-thymidine	0	10.3	∞**
Unfertilized eggs	³H-thymidine	1.10	9.4	8.6

Batches of 30 oöcytes or eggs were injected with DNA (60 µg/ml) and radioactive precursor as described by Gurdon, Birnstiel and Speight (1969) and incubated for 90 min (eggs) or 6 hr (oöcytes). DNA was extracted by procedures described elsewhere (Ford, C. C. and Woodland, H. R., unpublished data). These methods give recoveries of over 50% judged by absorbance. Phosphorylation levels were measured by the methods described by Woodland (1969). Acid insoluble counts within the OD marker on CsCl gradients were adjusted for recovery to give the total DNA counts/min. Values for control samples have been subtracted.

*Full grown oöcytes from a female that had not laid for several months.

**The high value of control samples does not allow detection of small amounts of native primed synthesis (i.e., that which would have produced a denatured:native ratio greater than 25:1).

†Full grown oöcytes from a female that had laid eggs three days previously.

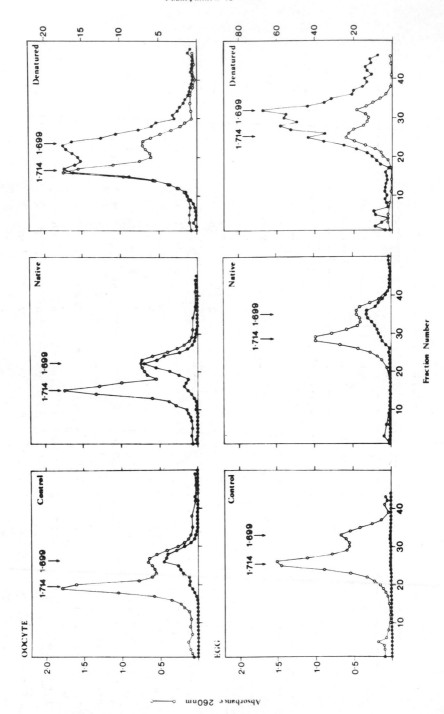

Count$\sqrt{}$min. $\times 10^{-2}$

Absorbance 260 nm

398

periments has been characterized on CsCl isopycnic
gradients, and it is found that the product behaves
as if similar in base composition to the injected denatured DNA and it includes both radioactive native
and denatured types of DNA (Fig. 1). The reason why
this denatured component was not observed previously
(Gurdon, Birnstiel and Speight, 1969) was probably
that a different extraction and purification procedure
was used.

Since we know the precursor pool sizes in these
cells (see next section) it is possible to calculate
the mass of DNA synthesized and hence the proportion
of the injected DNA which is replicated (Table 2). It
can be seen that in respect to denatured DNA-dependent
synthesis, eggs and oöcytes are comparable. The capacity of eggs to replicate both denatured and native
DNA falls off quite markedly as the amount of injected
DNA is increased.

The results outlined so far show that the natural
replication of DNA, characteristic of eggs but not
oöcytes, can be copied by injecting nuclei, and also
by injecting purified native DNA. Clearly it is now
of interest to know if a cell-free system can be prepared with the same properties as the living cell.
Initially the aim has been to move in small steps from
in vivo conditions, and therefore the DNA synthetic

Figure 1. (Opposite page.) CsCl gradient centrifugation of DNA synthesized after injection of DNA and
labeled precursor into oöcytes and eggs. DNA was extracted as described elsewhere (Ford, C. C. and Woodland, H. R., in preparation). Native and denatured
calf thymus DNA were added as optical density markers.
Samples were dissolved in 1/10 SSC and CsCl added to
give a final refractive index of 1.400 and a final
volume of 5.0 ml. These were centrifuged for 94 hr at
43,000 rpm in an MSE 10 x 10 ml angle rotor. Fractions were collected by downward delivery and, after
refractive index and absorbance measurements were
taken, the fractions were precipitated with carrier
DNA in 5% TCA. Precipitates were collected on Millipore filters, washed, dried and counted in liquid
scintillant.

TABLE 2

Percent of template DNA replicated
in vivo and *in vitro*

| Assay | Type of cell | ng Injected per cell or added per cell extracted | % Replication | |
			Native DNA template	Denatured DNA template
*In vivo**	Oöcyte	30	0	3.3
	egg	30	0.6	5.0
	egg	2	3.4	7.7
	egg	0.2	10.0	12.3
*In vitro***	Oöcyte	1000	0.025	0.611
	egg	1000	0.161	1.52

*DNA injection, extraction, and estimation of to-
tal counts per minute were as Table 1. Pool sizes
were taken from values in Table 4. Values for control samples have been subtracted.
**Extracts were prepared and assays performed as
described elsewhere (Ford, C. C. and Pestell, R. Q.
W., in preparation). Control values have been subtracted. Incorporation into DNA was measured as
acid insoluble counts (shown to be 97% sensitive to
DNase). Apart from this, percent replication values were calculated in the same way as the *in vivo*
values. ^{3}H dTTP was assumed to have remained 100%
phosphorylated.

characteristics of crude extracts of eggs (Ford, C. C.
amd Pestell, R. Q. W., in preparation) and oöcytes
(Ford, C. C., unpublished data) have been investigated.
as one might expect, the egg extract shows incorporation of deoxynucleotide triphosphates into DNA, dependent both on the presence of DNA and the four
deoxynucleotide triphosphates, two of the criteria
normally applied to identify DNA polymerases *in vitro*.
Incorporation is stimulated both by native and de-

natured DNA, but as in many such extracts (Keir, 1965) the latter is much more effective (Table 3). This re-

TABLE 3

DNA polymerase activity in oocytes and eggs, assayed *in vitro*

Source of extract	pMoles dTMP/µg protein per 20 min/10 µg DNA		Denatured native
	Native DNA	Denatured DNA	
Oöcyte	25	812	32.5
Egg	60 ± 23	388 ± 203	6.5

The procedures used were the same as for Table 2, except that the values are derived from 20 min incubations. Denatured:native ratios calculated from Table 2 differ from those shown in this table because denatured DNA-primed synthesis is not linear over 1 hr, whereas native-primed synthesis is linear. The large standard error for egg extracts may reflect variability between eggs of different females.

sult correlates well with that obtained *in vivo* at fairly high DNA concentrations (Tables 1 and 2). In preliminary studies the same procedures have been used to study oöcyte extracts and similar results were obtained, except that the preference for denatured DNA was even higher (33:1, Table 3). In oöcytes the stimulation of incorporation by native DNA is so low compared to that by denatured DNA that it probably does not represent synthesis on a native template. It is more likely that the native DNA contained or acquired sufficient nicks to have allowed a denatured DNA-dependent activity to have produced this amount of synthesis. The denatured DNA-dependent activity is present in both oöcytes and eggs in roughly similar amounts (Tables 2 and 3). This high denatured DNA-dependent activity of both oöcyte and egg extracts, contrasted with high native DNA-dependent activity in

eggs but not oöcytes, parallels the results obtained
with living cells (Table 1). The results obtained *in
vitro*, although of a preliminary nature, are therefore
in agreement with those obtained *in vivo*.

The value of the assay systems which we have outlined depends on their accurately producing the pattern of DNA synthesis observed in normal cells. For
the following reasons we believe this is true for
cells injected with DNA.

1. A similar amount of DNA synthesis takes place if
   the same quantity of DNA is injected in the form
   of intact nuclei or as purified DNA (Gurdon,
   Birnstiel and Speight, 1969). If only one nucleus is injected the DNA synthesis observed
   enables nuclear division and development to proceed in a normal fashion. The implication is
   that pure DNA and whole nuclei behave similarly
   as regards the amount of synthesis they stimulate.
2. When native DNA (but not denatured) is injected
   into eggs there is a lag of about 20 min before
   a stimulation of DNA synthesis is observed
   (Gurdon, Birnstiel and Speight, 1969). This lag
   correlates well with the observation that the
   normal egg and sperm nuclei begin to synthesize
   DNA 20 min after fertilization (Graham, 1966).
3. Native DNA, in contrast to denatured DNA, does
   not stimulate DNA synthesis in oöcytes (Ford, C.
   C. and Woodland, H. R., in preparation). This
   observation fits well with the absence of DNA
   synthesis in normal oöcyte nuclei and in nuclei
   injected into oöcytes (Gurdon, 1967).

It therefore seems a reasonable hypothesis that the
microinjection assay for DNA synthesis gives results
which are relevant to the control processes which
operate in normal cells. Since the *in vitro* assays
give rather similar results with native DNA, it seems
possible that they also reflect events which happen in
normal nuclei. The significance of the results using
denatured DNA is not immediately obvious, but it seems
that the synthesis dependent on this type of molecule

is regulated in a different way from that which governs
the replication of normal nuclei. Again this conclusion is supported by both *in vivo* and *in vitro* assays.
It therefore seems likely that both the living and the
test tube assays will prove useful in studying the
control of DNA synthesis.

DNA Precursors

The ability of a cell to support DNA synthesis relies on its containing the four common deoxynucleotide
triphosphates. These molecules are therefore capable
in principle of controlling DNA synthesis. We have
estimated the amounts of these molecules in eggs and
oöcytes both by a conventional type of method and by
an *in vivo* assay.

The total deoxynucleotide triphosphate content of
eggs and oöcytes has been determined by Woodland and
Pestell (in preparation). The procedure used is summarized in the legend to Figure 2. It can be seen
from this figure that eggs yield four major peaks of
deoxyribose-containing material. One co-elutes with
^{3}H-dTTP, the others are in the positions expected of
the other three common deoxynucleotide triphosphates
and contain the appropriate bases. The data presented
in Table 4 indicate that there are similar amounts of
each in the unfertilized egg. These amounts are sufficient to enable the synthesis of about 2,500 nuclei.
If a similar estimation is carried out on oöcytes,
somewhat less pyrimidine and no purine deoxynucleotide
triophosphates are detected (Table 4). Although one
can never prove a substance to be absent, at face
value this result suggests that precursors might play
some role in the regulation of DNA synthesis.

We have already described experiments involving the
injection of denatured DNA, which indicated that this
conclusion is not justified, for oöcytes are able to
support DNA synthesis stimulated by denatured DNA
(Tables 1 and 2). It is believed that the newly synthesized product has the same base composition as the
denatured injected DNA (see above), and therefore it
must represent the incorporation of the three deoxynucleotide triphosphates other than the radioactive

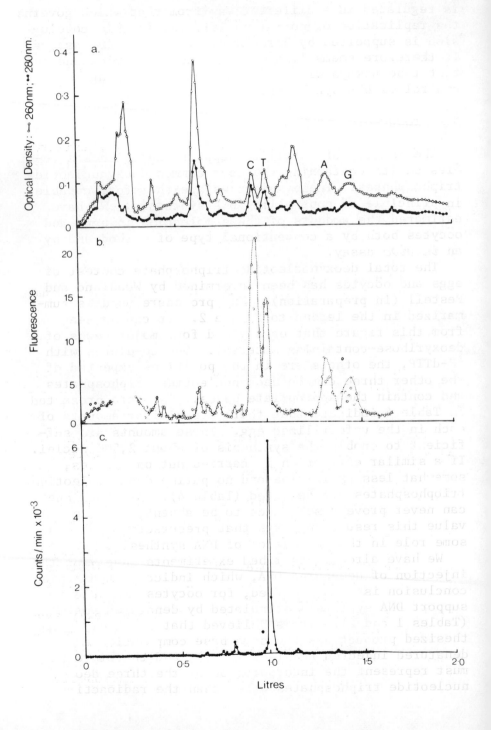

one. Since the same result is obtained with ^{3}H-dATP
and ^{3}H-thymidine as precursors, all four deoxynucleoside triphosphates seem to be present, and they are
not rate limiting, for the injection of further deoxynucleoside triphosphates does not raise the amount of
DNA synthesis observed (Ford, C. C. and Woodland, H.
R., in preparation). The amount of incorporation into
DNA in these experiments indicates that oocytes contain enough DNA precursors to synthesize at least
125-150 diploid nuclei. This amount of purine triphosphate would have been less than that detectable
by the chemical methods of estimation employed.

It might be that enzymes other than those which
operate on denatured DNA need much higher levels of
precursors than the enzyme which replicated denatured
DNA. But it is found that injecting further precursors fails to stimulate both native DNA-dependent
(Ford, C. C. and Woodland, H. R., in preparation) and
nuclear (Woodland, H. R. and Gurdon, J. B., unpublished data) DNA synthesis.

The experiments using living cells therefore argue
strongly against a regulatory role of DNA precursors

Figure 2. (Opposite page.) Dowex-1 chromatography
of an extract of 24,000 unfertilized eggs. The extract was first treated with periodate and methylamine,
a process which destroys all ribonucleotides with un-
substituted 2'- and 3'-OH groups (Pestell, R. Q. W.
and Woodland, H. R., in preparation). The degradation
products were then washed through the column with
water. This part of the elution profile is not shown.
Nucleotides were separated by eluting with a 2 liter
linear gradient from 0-1.0 M NH_4HCO_3. (a) UV absorption at 260 nm (O) and 280 nm (●); the positions of
dCTP, dTTP, dATP and dGTP are indicated by C, T, A and
G respectively. (b) Fluorescence in arbitrary units
resulting from the deoxyribose assay of Kissane and
Robins (1958), preceded by bromination to labilize the
pyrimidine glycosidic bond. (c) Radioactive counts
per min of 0.5 ml aliquots from each fraction, derived
from ^{3}H-dTTP added to the original mixture in order to
estimate deoxynucleotide recoveries (from Pestell, R.
Q. W. and Woodland, H. R., in preparation).

TABLE 4

Deoxynucleotide content of eggs and oöcytes

Type of cell	Method of estimation	pMoles/cell			
		dATP	dCTP	dGTP	dTTP
Eggs*	Absorbance at 260 nm	13	19	12	12
	Fluorescence	13	16	11	9
Oöcyte** Expt. A	Fluorescence	<2	8	<1	7
Oöcyte† Expt. B	Fluorescence	<1	2	<1	7

In the experiment with eggs there was sufficient UV absorbance to calculate the content of the deoxynucleotide peaks both by UV absorption as well as by fluorescence. This was not possible in the experiments with oöcytes. (From Woodland, H. R. and Pestell, R. Q. W., in preparation.)

*These results are from the chromatography of an extract of 24,000 unfertilized eggs as shown in Figure 1.
**These results are from the chromatography of an extract of 18,900 large oöcytes taken from two female frogs 3 days after they laid eggs.
†These results are from the chromatography of 12,300 large oöcytes taken from one frog a week after it laid eggs.

in the appearance of the cytoplasmic state which induces DNA synthesis when oöcytes mature to eggs. They also show how microinjection can provide a valuable assay for the availability of substances under the conditions which exist inside normal cells.

RNA SYNTHESIS

RNA Synthesis in Normal Eggs and Oöcytes

Oögenesis is a phase of development involving intense RNA synthesis, in particular rRNA, but also of the other main classes of RNA (Brown, 1967; Gurdon, 1968b; Davidson, 1968). Towards the end of oögenesis the rate of rRNA synthesis is apparently reduced (Crippa, 1970), although quite active RNA synthesis by the nucleoli may be detected by autoradiographic (Gurdon, 1968a; Smith and Ecker, 1970) and biochemical methods. In contrast, the egg makes only minute amounts of RNA, none of which may be detected in the nucleus (Gurdon and Woodland, 1969). Oöcytes and eggs are therefore favorable types of cells for the study of RNA synthesis, in just the same way as they are for the study of DNA synthesis.

RNA Synthesis Studied by Microinjection into Living Cells

Many of the experiments designed to study DNA synthesis in oöcytes and eggs have their exact parallel in the study of RNA synthesis. Thus, we began our study by injecting nuclei of various sorts into these cells. This type of work has shown that in eggs and oöcytes, injected nuclei conform to the activities of the endogenous nuclei, no matter what they did before injection (Gurdon and Woodland, 1968; Gurdon and Woodland, 1970). For example, blastula nuclei, which lack nucleoli and do not synthesize RNA at a detectable rate, are induced to synthesize RNA in oöcytes, and to form typical nucleoli at the same time. On the other hand, neurula nuclei, which are intensely active in RNA synthesis, make no RNA in eggs and their nucleoli vanish (Gurdon, 1968a). These results, which are discussed more fully elsewhere (Gurdon and Woodland, 1968; Gurdon and Woodland, 1970), lead to the general conclusion that the transcription of nuclear DNA is under cytoplasmic control in eggs and oöcytes

of *Xenopus*, and that the injected DNA in an intact nucleus is transcribed normally by the host cell.

The interaction between nucleus and cytoplasm is
likely to be very complex. Studies involving transcription *in vitro* have indicated that it is the protein component of chromosomes which is responsible for
the regulation of genetic activity. We have therefore
attempted to establish the feasibility of analyzing
transcriptional control in slightly simpler experiments in which purified DNA and nuclear proteins are
injected separately.

When purified DNA and ^{3}H-uridine are injected into
oöcytes, a stimulation of incorporation is observed,
an effect not seen after injection into eggs (Gurdon
and Woodland, 1970). This extra incorporation is
RNase sensitive and heterogeneous in size (Knowland,
Ph.D. thesis). It is not yet certain that it is synthesized on the injected template, but the difference
between oöcytes and eggs seems to reflect normal cell
function. Preliminary indications are therefore that
the injection of DNA into eggs and oöcytes may provide a living assay system for RNA polymerase activity
which would prove useful in studying the regulation of
transcription. These experiments also seem to indicate that only the DNA component is necessary for the
nucleus to make the appropriate response to the cytoplasm in its transcriptional as well as its replicative function.

Of the nuclear proteins, we have investigated the
effect only of histones. It appears that in both
oocytes and eggs, molecules of this type rapidly accumulate in nuclei, but do not immediately reduce the
synthesis of RNA (Gurdon, 1970). This finding is consistent with the inability of added histones to inhibit completely the template function of isolated nucleoli *in vitro* (Liau, Hnilica and Hurlbert, 1965).
The indication from *in vitro* experiments is that the
nonhistone proteins, possibly with associated RNA, are
the important agents of genetic regulation (Bekhor,
Kung and Bonner, 1969; Huang and Huang, 1969; Paul,
1970). Although the RNA synthesized *in vitro* on a
chromatin template resembles in some respects that
made *in vivo* (Paul and Gilmour, 1966; Bekhor, Kung and

Bonner, 1969; Huang and Huang, 1969) its identity is not known, and exactly how the events studied *in vitro* bear on those *in vivo* is therefore not clear.

The study of living cells by microinjection provides an opportunity to circumvent this problem, as is well illustrated by the experiments of Crippa (1970) on an inhibitor of rRNA synthesis. These experiments depend on the observation that full-grown oöcytes are less active in rRNA synthesis than immature oöcytes, suggesting that they might contain an inhibitor. Crippa was able to isolate from full-grown oöcytes a protein which bound specifically to rDNA, inhibited rRNA synthesis when injected into growing oöcytes, and which was absent from this latter type of cell. How this protein acts is now known, but it clearly has the properties of a natural rRNA synthesis inhibitor. The great attraction of this agent is that its assay occurs in the environment of a normal cell, and involves changes of genetic activity which are easily identifiable. The study of rRNA synthesis in *Xenopus laevis* presents other advantages, for the ribosomal genes may be purified quite easily, their structure is understood better than any other eukaryote gene, a system for their transcription *in vitro* has already been described (Reeder and Brown, 1970), and the enzyme responsible for their transcription *in vivo* has possibly been isolated (Tocchini-Valentini and Crippa, 1970b).

The study of eukaryote RNA polymerases is at present in a rudimentary state, indeed the investigations of eukaryote chromatin transcription *in vitro* usually utilize prokaryote enzymes. In recent years various factors which form part of bacterial RNA polymerase, and exert a positive control over transcription, have been identified (Burgess *et al.*, 1969). As yet it is not known if such factors exist in eukaryotes, but there is an indication that oöcytes are likely to prove useful in the identification of agents of this type. This comes from an experiment of Tocchini-Valentini and Crippa (1970a) in which the *E. coli* sigma factor was injected into oöcytes. A two-fold stimulation of RNA synthesis was observed. While it is difficult to interpret, this result suggests that oöcytes

may be helpful in identifying agents which affect the
activity of RNA polymerase.

PROTEIN SYNTHESIS

*Protein Synthesis in Normal Eggs
and Oöcytes*

The characteristics of protein synthesis in embryonic cells of amphibia have been reviewed by Smith and
Ecker (1970), but unfortunately our knowledge relates
mainly to *Rana pipiens*. In this species the nonhor-
mone stimulated oöcytes seem to be relatively inactive
in protein synthesis. The rate of protein synthesis
rises several-fold at maturation of the oöcyte, and
the same elevated rate is maintained through fertilization and early cleavage (Smith and Ecker, 1969).
The proteins synthesized by oöcytes and eggs seem to
differ qualitatively, as judged by gel electrophoresis
in the presence of SDS. Clearly these changes make
the protein synthetic systems of oöcytes and eggs es-
especially interesting for study, but as yet it is not
known if such changes occur in *Xenopus*. In contrast
to those of *Rana*, the oöcytes of *Xenopus* seem to be
very active in protein synthesis (Moar *et al.*, in
press). In various respects the reproductive biology
of the two anurans is rather different. The oöcytes
of *Rana* grow during the summer and autumn and lie dormant during the winter; the ovary then consists mainly
of fully grown oöcytes, which are ovulated and laid in
the spring. In contrast the ovary of *Xenopus* seems
to contain all sizes of oöcytes at all seasons; laying
can be induced at all seasons and can occur at short
intervals. In *Rana* the mature ovulated eggs are
stored in an ovisac, where they may remain dormant for
many days, but in *Xenopus* there is no ovisac and immediately after ovulation the eggs pass down the oviduct
and into the external medium. Fertilization therefore
always occurs within several hours of ovulation.
These differences in the growth of the oöcytes and in
events after ovulation make it dangerous to assume *a*

priori that the characteristics of protein synthesis
in *Rana* and *Xenopus* are similar.

*Protein Synthesis Studied in Living
Cells by Microinjection*

   Translation of natural mRNA molecules. Most of our
work on natural mRNA has used the putative Hb mRNA,
which is the purest RNA of this sort available in
large amounts (Chantrenne, Burny and Marbaix, 1967).
In the first experiments the main classes of RNA found
in rabbit reticulocytes were injected into oöcytes,
and as expected only the 9S fraction produced any detectable change in protein synthesis (Lane, Marbaix
and Gurdon, in press). This it did by stimulating the
synthesis of a ^{3}H-histidine-labeled protein which was
characterized as Hb by the following criteria:

1. co-elution with marker Hb on G-100 Sephadex columns (Fig. 3);
2. co-electrophoresis with Hb on acrylamide gels;
3. after removal of haem the dissociated radioactive subunits co-elute with rabbit α- and β-glo-
   bin chains from CM-cellulose columns;
4. peptides prepared from purified α- and β-chains
   labeled with ^{3}H-histidine in the frog oöcyte,
   co-elute from an ion-exchange column with those
   prepared from globin chains labeled
   with ^{14}C-histidine (Fig. 4).

The evidence that the 9S RNA of rabbit reticulocytes
is the only RNA able to direct Hb synthesis in living
cells is therefore fairly conclusive.
   Reticulocyte 9S RNA has also been translated successfully *in vitro*. Perhaps the best system so far
developed to do this is the reticulocyte lysate described by Lockard and Lingrel (1969). This system is
already making haemoglobin at a rapid rate, so it has
limitations in the study of translational control.
Experiments in which mRNA is translated by the protein
synthetic apparatus of another cell have been reported
by Heywood (1969; 1970). It was found that myosin
mRNA could only be translated by a reconstituted cell-
free system of reticulocytes if certain factors were

washed off the blood cell ribosomes and replaced by
similar factors from the ribosomes of muscle cells.
Two results obtained *in vitro* which seem to differ
from these are that Hb mRNA can be translated by an
ascites cell-free system (Mathews, in press), and that
a putative immunoglobulin mRNA from myeloma cells can

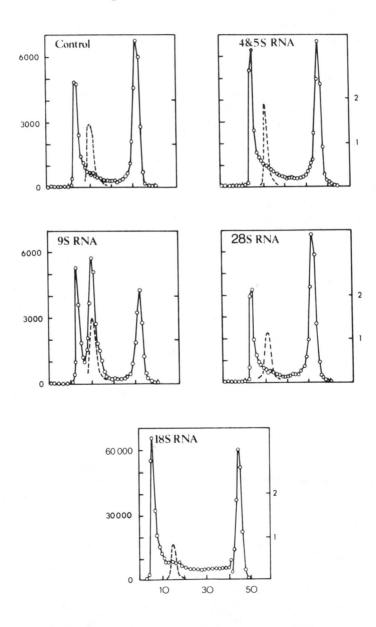

be translated by the reticulocyte lysate (Stavnezer
and Huang, 1971), in both cases with no added ribosome
wash. These results may be abnormal because both the
ascites cell-free system and the myeloma RNA are derived from cancerous cells, but they agree with the
ability of oocytes and eggs to translate both an Hb
mRNA and the immunoglobulin mRNA (Gurdon *et al.*, in
press). It may be that this catholic taste for mRNAs
displayed by oöcytes and eggs reflects a difference
between normal differentiated cells and undifferentiated or cancerous cells. One explanation of the difference between Heywood's results and ours may be that
oöcytes and eggs contain Hb mRNA-specific initiation
factors. If this were so, they can only be of primary
importance in red blood cell differentiation if they
are localized in some particular region of the egg.

In order to compare the various systems for mRNA
translation in a meaningful way, it is of obvious importance to know how efficient they are. We have
attempted to estimate the rate of Hb synthesis in eggs
and oöcytes, but unfortunately this type of measurement presents many problems (Gurdon *et al.*, in press).
In order to calculate the specific activity of the
radioactive amino acid injected into the cell we were
forced to assume that the extractable amino acid pool,
measured by amino acid analysis, was the same as the
actual pool used for protein synthesis. If this were
not so our estimates would be a maximum of six times
too high. We were also forced to make assumptions
which might have led to our having *under*estimated the

Figure 3. (Opposite page.) Batches of 20 oöcytes
were injected with haemin and the RNA indicated at
1000 µg/ml (50 mµg/cell) and incubated in ^{3}H-histidine
(1 mCi/ml) for 7 hr. The protein synthesized was analyzed on G-100 Sephadex columns in the presence of
marker rabbit Hb. The ordinates are counts/min (left)
and absorbance at 415 nm (right). The right hand radioactivity peak is reduced x 10^{-3} in the 18S RNA result. The abscissae represent fraction number. Only
the 9S RNA produces a significant stimulation of Hb
synthesis. (--), absorption at 415 nm; ($\bullet$), counts/
min (redrawn from Lane, Marbaix and Gurdon, in press).

rate of synthesis, per injected mRNA molecule, by
about two-fold (Gurdon *et al*., in press). The efficiency of translation of Hb mRNA, calculated on the
basis of these assumptions, is presented in Table 5,
together with data from other systems for comparison.
It can be seen that each injected mRNA is translated
several times per hour, and since synthesis continues
for over a day, each must have been translated several
hundred times in total. The table also shows that the
translation of Hb mRNA in living cells is more rapid

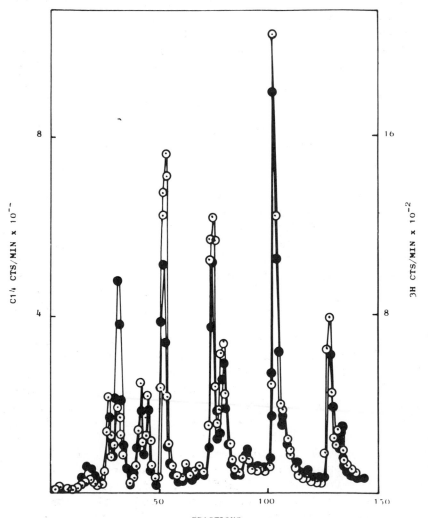

FRACTIONS

TABLE 5

Efficiency of Hb mRNA translation in various systems

Translational systems	Temper-ature of incubation (°C)	# of globin chains synthesized/ mRNA molecule/hr	Half-life of translational system
Cell-free system (lysate using added mRNA)*	25	0.8	1 hr
Rabbit reticulocytes in culture (using endogenous mRNA)**	20	80	2-3 hr (at 37°)†
Injected oocytes	19	24††	26 hr

*Calculated from data from Lockard and Lingrel (1969).

**Hunt, Hunter and Munro, 1969.

†Armentrout, Schinkel and Simmons, 1965.

††This high value is obtained by the injection of only 0.01 ng of Hb mRNA into the cell. Lower values may be obtained when more is injected.

than in the best cell-free system available, and it compares favorably with that observed in intact retic-

Figure 4. (Opposite page.) Peptide analysis of a tryptic digest of purified β-chains synthesized in oöcytes of frogs and labeled with ^{3}H-histidine (O), and of purified β-chains synthesized in intact rabbit reticulocytes labeled with ^{14}C-histidine (●). The synthesis of β-chains in oöcytes was stimulated by rabbit Hb mRNA injection. The separation is achieved on a 75 x 0.6 cm Technicon type P Chromobead column. Similar results are obtained when the purified α-chains are analyzed in this way (Lane, C. D. and Marbaix, G., in preparation).

ulocytes. This high efficiency of translation,
coupled with the long life of Hb mRNA in this cell,
gives the oöcytes the peculiar advantage of great sensitivity in the assay of mRNA; quantitites of mRNA in
the ng-pg range may readily be identified (Lane, Marbaix and Gurdon, in press). The high efficiency also
has an advantage in the study of the translation process, because data derived from systems working at
sub-optimal rates may be misleading.

One respect in which the behavior of reticulocyte
9S RNA differs in the test tube from its behavior in
living cells is that in the former it generally lowers
the overall rate of protein synthesis (J. Lingrel,
personal communication), whereas in eggs and oöcytes
the rate is stimulated (Moar *et al.*, in press). This
stimulation may be as much as 100%, the endogenous
proteins being made at the same rate as in controls,
and the extra being Hb. It therefore seems likely
that the inhibition *in vitro* represents an interaction
of an abnormal type, produced by some nonphysiological
characteristic of the cell-free system. It is not yet
known if the inhibiting agent is the mRNA itself or
some untranslated component of the RNA preparation
added.

The stimulation of the overall rate of protein
synthesis by injected mRNA *in vivo* provides us with
information concerning the living cell. It suggests
that the amount of protein made in the unmanipulated
cell may be limited by the availability of mRNA. Injecting increasing amounts of Hb mRNA has revealed
that the translational capacity of the cell is not
unlimited, and that it becomes saturated by about 10
ng Hb mRNA per cell (Fig. 5). The cellular component
which is limiting is not yet known. These experiments
also reveal the surprising fact that the injected message and the endogenous message do not compete for the
components limiting translation (except at extremely
high mRNA concentrations). This is evident from Figure 5, for the ratio of synthesis on endogenous mRNAs
to that on Hb mRNA reaches a plateau, and it is known
that the endogenous incorporation is not decreased at
these mRNA inputs. The reason for this lack of competition is a matter for conjecture at present, but

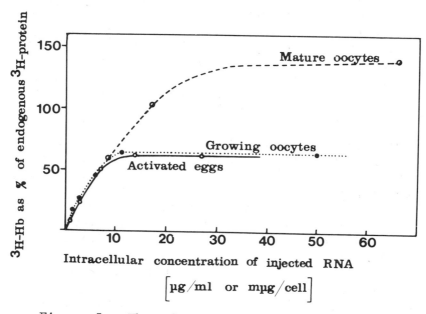

Figure 5. The effect of injecting increasing amounts of Hb mRNA into oöcytes and
eggs. The cells were incubated in ^{3}H-his-
tidine for 10 hr and the proteins synthesized were analyzed as described in the
legend to Figure 3. It is not yet certain
that the growing and mature oöcytes always
differ by the amount shown here. "Endogenous" protein is that made on endogenous
mRNAs (Moar *et al.*, in press).

the experiment is especially important for the purpose
of this article in that it shows that the use of living cells can reveal important phenomena which have
not been observed in cell-free systems. It is in
revealing phenomena of this type and enabling their
further study that injected eggs and oöcytes are likely to be of advantage in the study of the control of
translation.

   *Effects of artificial RNA molecules in living
cells.* The second problem that we have investigated
relates to the characteristics of polyribonucleotides

which confer upon them the ability to be recognized as a message. In prokaryotes, genetic evidence and work using cell-free systems supports the view that the translated part of mRNA molecules begins ApUpGp, or GpUpGp (Clark and Marcker, 1968; Bretscher, 1969). A similar conclusion has been reached regarding eukaryotes through the use of mammalian cell-free systems, although the evidence of GpUpGp as an initiator is much weaker than that for ApUpGp (Smith and Marcker, 1970; Brown and Smith, 1970). We have attempted to find whether these conclusions are also valid for living cells (Woodland, H. R. and Ayres, S. E., in preparation). Polyribonucleotides were injected into eggs or oöcytes, together with the appropriate radioactive amino acids. As might be expected, polymers which did not begin with an initiator codon did not stimulate the incorporation of the appropriate amino acid (Table 6): e.g., (Up)n did not stimulate phenylalanine incorporation, nor did (Ap)n stimulate lysine incorporation, and homopolymers of the amino acids coded by these polymers could not be detected. (Up)n might fail to produce an effect because it is rapidly degraded, but this seems not to be the reason because its destruction has been shown to be very slow in oöcytes and eggs. Much more surprising is the result obtained with ApUpGp(Up)$_{\bar{n}}$, which is the best known artificial mRNA in the test tube (Brown and Smith, 1970). It is found that this molecule is an extremely effective inhibitor of protein synthesis, even when small amounts are injected (Table 6). The small amount of residual phenylalanine incorporation does not include detectable amounts of polyphenylalanine synthesis. Unless polyphenylalanine is completely degraded immediately as it is synthesized, ApUpGp(Up)$_{\bar{n}}$ does not therefore seem to be acting as a message. In addition, its inhibitory effect on endogenous protein synthesis contrasts markedly with the behavior of natural mRNA, which fails to compete with endogenous mRNA. Table 6 also shows that GpUpGp(Up)$_{\bar{n}}$ has neither a marked stimulatory nor an inhibitory effect. There is therefore no indication that it is acting as an mRNA, although its effect is clearly different from that of ApUpGp(Up)$_{\bar{n}}$.

TABLE 6

The effects of various polynucleotides on protein synthesis in unfertilized eggs

Sample injected	Inhibition produced with ^{3}H-Phe as precursor (%)	Inhibition produced with ^{3}H-lys as precursor (%)
ApUpGp$(Up)_{\bar{n}}$ 12.5 ng/cell	95	99
ApUpGp$(Up)_{\bar{n}}$ 1.25 ng/cell	98	99
$(Up)n$ 12.5 ng/cell	32	-5
$(Up)n$ 1.25 ng/cell	-14	6
GpUpGp$(Up)_{\bar{n}}$ 10 ng/cell	6	-6
GpUpGp$(Up)_{\bar{n}}$ 1 ng/cell	32	5
ApUpGp 10 ng/cell	24	7

The percentage inhibition is computed from the ratio of acid insoluble to total radioactivity in samples of 10-15 eggs injected with polymers, as compared with those injected with radioactive amino acids alone. In small samples of cells such ratios are very variable, even in controls. An inhibition or stimulation of as much as 30% is therefore not significant. (Woodland, H. R. and Ayers, S. E., in preparation.)

The the results obtained from living cells are quite different from those predicted from experiments conducted *in vitro*, and they provide another illustration of the advantages of using a system which is al-

tered from the normal cell as little as possible.
The reason that living cells and test-tube systems
do not behave in the same way is obscure at present,
but small differences in the way in which bacterial ribosomes react with natural mRNA and with
ApUpGp have been reported (Revel, Herzberg and Greenspahn, 1969; Brawerman, 1969). It is known that these
bacterial virus messages have ApUpGp initiators within
the RNA molecule, so it seems that other nucleotide
sequences are important in the initiation of protein
synthesis. Our results are consistent with this
interpretation being applied to eukaryotes as well as
prokaryotes. Since living cells seem to show more
stringent initiation requirements than existing cell-
free systems they would seem peculiarly suitable for
the further investigation of this problem.

CONCLUSIONS

In this paper we have described some experiments
designed to study DNA, RNA and protein synthesis in
living oöcytes and eggs of amphibia. Even though most
of the lines of investigation followed are in their
preliminary stages, they indicate that this type of
approach will prove of some value. It should assist
in the development of cell-free systems which parallel
as accurately as possible phenomena occurring in
living organisms. Cell-free systems should be especially useful in studying the detailed mechanism of
processes occurring in living cells, whereas the use
of oöcytes and eggs may have a direct application in
the study of regulatory phenomena.

ACKNOWLEDGMENTS

The authors are indebted to Drs. R. Q. W. Pestell,
G. Marbaix and M. B. Mathews for permission to quote
their work before it has been published, and to S. E.
Ayers, V. A. Moar and P. Lyons for excellent technical
assistance. They are grateful to the Medical and the
Science Research Councils of Great Britain for finan-
cial support.

REFERENCES

Armentrout, S. A., Schinkel, R. D., and Simmons, L. R. (1965). *Arch. Biochem. Biophys. 112*, 304.

Bekhor, I., Kūng, G. M., and Bonner, J. (1969). *J. Mol. Biol. 39*, 351.

Brawerman, G. (1969). *Cold Spring Harbor Symp. Quant. Biol. 34*, 307.

Bretscher, M. S. (1969). *Progress in Biophysics 19*, 177.

Brown, D. D. (1967). *Current Topics in Devel. Biol. 2*, 47.

Brown, J. C., and Smith, A. E. (1970). *Nature 226*, 610.

Burgess, R. R., Travers, A. A., Dunn, J. J., and Bautz, E. K. F. (1969). *Nature 221*, 43.

Chantrenne, H., Burny, A., and Marbaix, G. (1967). *Progress in Nucleic Acid Research and Molecular Biology 1*, 173.

Clark, B. F. C., and Marcker, K. A. (1968). *Scientific American 218*, 36.

Crippa, M. (1970). *Nature 227*, 1138.

Davidson, E. H. (1968). *Gene Activity in Early Development* (New York: Academic Press).

De Lucia, P., and Cairns, C. (1969). *Nature 224*, 1164.

Gall, J. G. (1969). *Genetics 61, Suppl. 1*, 121.

Graham, C. F. (1966). *J. Cell Sci. 1*, 363.

Graham, C. F., and Morgan, R. W. (1966). *Devel. Biol. 14*, 439.

Gurdon, J. B. (1962). *J. Embryol. Exp. Morphol. 10*, 622.

Gurdon, J. B. (1967). *Proc. Nat. Acad. Sci. USA 58*, 545.

Gurdon, J. B. (1968a). *J. Embryol. Exp. Morphol. 20*, 401.

Gurdon, J. B. (1968b). *Essays in Biochemistry 4*, 25.

Gurdon, J. B. (1970). *Proc. Roy. Soc. Ser. B (London) 176*, 303.

Gurdon, J. B., and Laskey, R. A. (1970). *J. Embryol. Exp. Morphol. 24*, 27.

Gurdon, J. B., and Speight, V. A. (1969). *Exp. Cell Res. 55*, 253.

Gurdon, J. B., and Woodland, H. R. (1968). *Biol. Rev. Cambridge Phil. Soc. 43*, 233.

Gurdon, J. B., and Woodland, H. R. (1969). *Proc. Roy. Soc. Ser. B (London) 173*, 99.

Gurdon, J. B., and Woodland, H. R. (1970). *Current Topics in Devel. Biol. 5*, 39.

Gurdon, J. B., Birnstiel, M. L., and Speight, V. A. (1969). *Biochim. Biophys. Acta 174*, 614.

Gurdon, J. B., Lane, C. D., Woodland, H. R., and Marbaix, G. (in press). *Nature*.

Heywood, S. M. (1969). *Cold Spring Harbor Symp. Quant. Biol. 34*, 799.

Heywood, S. M. (1970). *Proc. Nat. Acad. Sci. USA 62*, 1782.

Huang, R. C., and Huang, P. C. (1969). *J. Mol. Biol. 39*, 365.

Hunt, T., Hunter, T., and Munro, A. (1969). *J. Mol. Biol. 43*, 123.

Keir, H. M. (1965). *Progress in Nucleic Acid Research, 4*, 82.

Kissane, J. M., and Robins, E. (1958). *J. Biol. Chem. 233*, 184.

Knippers, R., and Strätling, W. (1970). *Nature 226*, 713.

Knowland, J. S., D. Phil. Thesis, Bodleian Library, Oxford.

Lane, C. D., Marbaix, G. and Gurdon, J. B. (in press). *J. Mol. Biol.*

Laskey, R. A., and Gurdon, J. B. (1970). *Nature 228*, 1332.

Liau, M. C., Hnilica, L. S., and Hurlbert, R. B. (1965). *Proc. Nat. Acad. Sci. USA 53*, 626.

Lockard, R. E., and Lingrel, J. B. (1969). *Biochem. Biophys. Res. Commun. 37*, 204.

Mathews, M.B. (in press). *Nature*.

Moar, V. A., Gurdon, J. B., Lane, C. D., and Marbaix, G. (in press). *J. Mol. Biol.*

Paul, J. (1970). *Current Topics in Devel. Biol. 5*, 317.

Paul, J., and Gilmour, R. S. (1966). *J. Mol. Biol. 16*, 242.

Reeder, R. H., and Brown, D. D. (1970). *Proc. Lepetit*

Colloquium on RNA Polymerase, L. Silvestri, ed. (Florence), p. 249.

Revel, M., Herzberg, M., and Greenspahn, H. (1969). *Cold Spring Harbor Symp. Quant. Biol. 34*, 261.

Smith, A. E., and Marcker, K. A. (1970). *Nature 226*, 607.

Smith, L. D., and Ecker, R. E. (1969). *Develop. Biol. 19*, 281

Smith, L. D., and Ecker, R. E. (1970). *Current Topics in Devel. Biol. 5*, 1.

Stavnezer, J., and Huang, R. C. (1971). *Nature 230*, 172.

Tocchini-Valentini, G. P., and Crippa, M. (1970a). *Nature 226*, 1243.

Tocchini-Valentini, G. P., and Crippa, M. (1970b). *Nature 228*, 993.

Woodland, H. R. (1969). *Biochim. Biophys. Acta 186*, 1.

Expression and Re-Expression of Tissue Specific Functions in Hepatoma Cell Hybrids

Mary C. Weiss, Roger Bertolotti, and Jerry A. Peterson

Centre de Génétique Moléculaire du C.N.R.S.
Gif-sur-Yvette, France

INTRODUCTION

Our purpose here is to describe a genetic approach to the study of cell differentiation. The tool which is being used in these studies is hybridization of somatic cells (Barski, Sorieul and Cornefert, 1960), the only method of genetic analysis which is so far applicable to mammalian somatic cells (Ephrussi, 1965), and which permits the combination, in a single hybrid nucleus, of parental genomes which differ in the expression of genes controlling tissue-specific functions. We shall begin by saying a few words about cell hybridization in order to familiarize you with the advantages as well as the limitations of the material with which we are working. Cell hybridization is usually performed with cells cultivated *in vitro*, which are maintained by serial transfer, and which can be rendered homogeneous by cloning. Most cells grown

in vitro are in certain respects abnormal. Although
normal diploid cells can be propagated *in vitro*, they
grow slowly, clone rather poorly, and live only for a
finite number of generations, usually of the order of
20 to 50 (depending primarily upon the species of origin). By contrast, cells of heteroploid permanent
lines grow more rapidly, clone well and can be propagated for an indefinite number of generations. Such
cells are frequently derived from tumors, or have
undergone "spontaneous transformation" *in vitro*: they
are not diploid, and frequently contain a subtetra-
ploid chromosome complement (for a complete review,
see Harris, 1964).

Hybrid cells can be derived from nearly any cross
between two different cell types, although for most
hybrids thus far studied, at least one of the parents
has been a cell of a permanent line. The isolation
of the hybrids is greatly facilitated by the use of a
selective system, inactivated Sendai virus (which induces cell fusion), or both. Hybrid cells can be derived from crosses of cells of the same or of different species; their identification is confirmed by
analysis of the karyotype and by the presence of isozymes or antigens of both parental types. Initially,
most kinds of hybrid cells contain essentially the
complete chromosome complements of both parents; long
term propagation is accompanied by a reduction in
chromosome number, which may amount to loss of 5 to
50% of the chromosomes, and which may involve the
chromosomes of only one parental type or of both, depending upon the species of origin of the parental
cells. With some exceptions, which will be discussed
below, the joint expression of both parental genomes
can usually be detected. (This is a brief summary of
experiments performed between 1961 and 1969, by many
different investigators. A complete review of the
literature can be found in the book by Ephrussi, 1972.)

Returning to cell hybridization for the study of
differentiation, it should be emphasized that in all
of the experiments involving propagating hybrid cells
which have been performed, the parental cell type
characterized by the production of a tissue-specific
protein has been derived from a differentiated tumor,

and possessed the properties of permanent lines described above, including an abnormal karyotype. The use of neoplastic differentiated cells is dictated by purely technical considerations, such as those mentioned above, and including the stability of expression of tissue-specific proteins by such cells, a stability which is generally not observed in cultures of normal diploid cells. Therefore, extrapolations of results obtained with neoplastic cells to processes of normal development must be made with great caution until similar experiments are carried out with normal diploid cells. If similar results are not found, these comparisons may show how control mechanisms of normal and neoplastic cells differ.

MELANOMA x FRIBROBLAST HYBRIDS

The first study in which cell hybridization was used for the analysis of the control of a tissue-specific function was carried out by Davidson, Ephrussi and Yamamoto (1966). In these experiments, crosses were performed between pigment producing hamster melanoma cells and mouse fibroblasts, and the resulting hybrids were examined for the presence of both parental chromosome sets, the production of "household" enzymes (the term introduced by Ephrussi to refer to those enzymes which are indispensible for the growth and survival of all cells), and the production of pigment and the "luxury" (nonessential) enzyme required for its synthesis. The result of these studies was that the hybrids contained essentially all of the chromosomes of both parents, that joint expression of parental "household" enzymes was detected (Davidson, Ephrussi and Yamamoto, 1968), but the "luxury" function of the melanoma parent was abolished: all hybrids were unpigmented, and no activity of the enzyme dopa oxidase could be detected.

In the very first report of these results (Davidson, Ephrussi and Yamamoto, 1966), the authors pointed out some of the relevant questions to be answered in future experiments, namely: (a) will abolition or "extinction" (Ephrussi, 1972) of differentiated functions prove to be a general phenomenon in such hybrids; (b) is the continuous presence of certain genes of the nondifferentiated parent required for the maintenance of extinction; and (c) will the behavior of differentiated functions in somatic hybrids vary, depending upon whether the hybrids are intra- or inter-specific. [The latter possibility was soon thereafter examined by Silagi (1967), who found a similar extinction of pigment production in mouse melanoma x mouse fibroblast hybrids.]

The experiments to be described below were designed to answer some of these questions. Moreover, by using cells derived from a well-differentiated rat hepatoma, characterized by the production of numerous proteins specific to or characteristic of hepatic differentiation, we have been able to examine simultaneously several parameters of liver differentiation in somatic hybrids. Our aims in this work have been to examine the inheritance of the potentiality and commitment to express a given tissue-specific function, and the coordinate or independent expression of multiple tissue-specific functions by a given cell population.

HEPATOMA HYBRIDS

The rat hepatoma cells we are using were derived by Pitot *et al*. (1964) from the Reuber (1961) H35 hepatoma. All of the hepatoma populations we will be dealing with are clones, and are characterized by a stable karyotype comprised of 52 chromosomes (rat diploid number, 42), and the stable production of a number of liver-specific products. Those we have analyzed (Table 1) include: high baseline of tyrosine aminotransferase (EC 2.6.1.5; TAT*) and its inducibility, aldolase B (EC 4.1.2.7) and serum albumin.

The fates of these tissue-specific functions have been examined in hybrids from two different crosses. The characteristics of the parental cells and the crosses performed are shown in Table 2. The hepatoma

---

*Abbreviations: TAT, tyrosine aminotransferase; Dex, dexamethasone.

TABLE 1

Tissue-specific proteins analyzed in hepatoma hybrids

Product	Tissue specificity	Present in hepatoma clones (Fu5; Fu5-5)
TAT*: high baseline	Liver	+
TAT: glucocortico-steroid mediated induction**	Liver	+
Aldolase B†	Liver > intestine, kidney	+
Serum albumin††	Liver	+

*
Analyzed by a modification (Hayashi, Granner and Tomkins, 1967) of the spectrophotometric method of Diamondstone (1966).

**
Dexamethasone at 1 µM has been used as the inducer.

†
Analyzed by the electrophoretic method of Penhoet, Rajkumar and Rutter (1966) and the quantitative spectrophotometric method of Blostein and Rutter (1963).

††
Analyzed by double immunodiffusion in agar (see Crowle, 1961) and microcomplement fixation (Wasserman and Levine, 1961).

parental cells have already been described. 3T3 mouse fibroblasts (Todaro and Green, 1963) have a nearly tetraploid chromosome complement and were originally derived from mouse embryos. These cells synthesize collagen at a level comparable to diploid fibroblasts (Green, Goldberg and Todaro, 1966). BRL-1 cells are diploid and were derived (by primary cloning) from the liver of a young Buffalo rat (Coon, 1968; 1969). These epithelial cells, established in culture by Dr. Hayden Coon, and kindly provided by him, may be "de-differentiated" hepatocytes or they may have been derived from some other epithelial component of the

TABLE 2

Characteristics of parental cells
and crosses performed

Cell line	Species	Number of chromosomes (diploid equivalents)	Cell type	Expresses liver functions
Fu5 and Fu5-5	Rat	52 (1.24)	Neoplastic hepatocyte	+
3T3*	Mouse	76 (1.9)	Fibroblast	−
BRL-1**	Rat	42 (1.0)	Epithelial liver cell	−

Crosses: 3T3 x Fu5 ⟶ hybrid series 3F[†]

BRL-1 x Fu5-5 ⟶ hybrid clone BF5[††]

[*]Clone 3T3-4E, deficient in thymidine kinase and resistant to 30 µg/ml of 5-bromodeoxyuridine (Matsuya and Green, 1969).

[**]BRL-1 is a subclone of BRL-1D1 (Coon, 1968; 1969).

[†]Isolation of these hybrids is described by Schneider and Weiss (1971).

[††]Isolation described by Weiss and Chaplain (1971).

liver; since these cells do not produce detectable amounts of any of the liver-specific proteins analyzed, it is not possible to distinguish between these two possibilities.

In the two crosses with which we are concerned, the hepatoma was the common parent (Table 2). The 3T3 and BRL-1 parental cells differ in three important respects:

1. 3T3 is a mouse cell line, and when crossed with rat hepatoma cells, gives rise to an interspecific hybrid; BRL-1 cells, like the hepatoma, are from rat;

2. 3T3 is a heteroploid cell line, and contributes more than a diploid set of chromosomes; BRL-1 is diploid, and contributes to the hybrid fewer chromosomes than the hepatoma, and

3. 3T3 is a fibroblast line and is far removed ontogenetically from the liver; BRL-1 was derived either from liver mesenchyme [which during the course of embryogenesis induces the differentiation of hepatocytes (Le Douarin, 1961)] or from a derivative of hepatoblasts.

In summary, the cross of 3T3 by hepatoma is analogous to the melanoma cross described above (melanoma x fibroblast) while that of BRL-1 by hepatoma involves two more closely related cell types.

Interspecific Hepatoma x Fibroblast Hybrids

Hybrids from the cross 3T3 x Fu5 (hybrid series 3F) were isolated as colonies (presumed clones) following fusion by UV-inactivated Sendai virus and growth in selective medium, which permits survival only of hybrid and Fu5 cells. Karyological analysis showed that the individual hybrid clones contained very similar numbers and kinds of chromosomes: all were characterized by the presence of slightly fewer chromosomes than expected, most likely owing to loss of some of the rat chromosomes. Table 3 shows that approximately half of the rat chromosomes can be distinguished from those of the mouse, owing to differences in their lengths and shapes, and that the hybrids contain somewhat fewer of these rat markers than expected.

The baseline activity and inducibility of TAT in the parental and hybrid cells have been measured (Schneider and Weiss, 1971). As seen in Table 4, the hepatoma parental cells (Fu5) have high activity, and after 24 hr in medium supplemented with 1 μM dexamethasone (Dex), this activity increases by a factor of six. 3T3 fibroblasts show neither high activity nor inducibility, and the same is true of the hybrid cells. Thus, in these hybrid cells, there is simul-

TABLE 3

Numbers of chromosomes in parental and hybrid cells

Cell line	Total	Number of rat markers
Fu5	52	26
3T3	76	0
Hybrids:		
Expected	128	26
Observed*	107–118	17.2–22.4

Data summarized from Schneider and Weiss (1971).
*Range of mean values from eight hybrid clones.

TABLE 4

TAT in parental and hybrid cells

Cell line	TAT-specific activity	
	Baseline	Induced for 24 hr (1 μM Dex)
Fu5	24.7	154.0
3T3	0.8	0.77
Hybrids (avg.)	1.6 (0.9–1.8)	1.6 (0.7–1.9)
3F11	1.6	1.5
3F14	1.8	1.5

Data summarized from Schneider and Weiss (1971).

taneous extinction of high TAT baseline and inducibility, and at no time after the addition of Dex nor
during the growth cycle can expression of these hepatoma functions be detected. However, in the hybrid
cells there is measurable TAT activity, and this activity is about twice as high as in 3T3 cells. Heat
inactivation experiments showed that in the hybrid
cells, half of the TAT activity is inactivated at a
rate similar to that of enzyme from 3T3 cells and half

like that of Fu5 enzyme. Thus, it was concluded that
although the high baseline TAT activity of the hepatoma parent is not expressed in the hybrid cells, this
is not due either to loss of the rat structural gene
specifying TAT, nor to its total inactivation. Moreover, the presence in the hybrid cells of some rat TAT,
and the total absence of its inducibility, are compatible with the hypothesis (of Tomkins *et al.*, 1969)
that there is a specific gene which mediates TAT induction, and that the activity of this gene cannot be
detected in the hybrid cells (Schneider and Weiss,
1971).

Similarly, studies of aldolase isozymes (for a review see Penhoet, Kochman and Rutter, 1969) have been
carried out on the parental and hybrid cells (Bertolotti and Weiss, 1972). While both 3T3 and Fu5 parental cells contain aldolase A, the "household" form of
the enzyme, only the hepatoma cells contain aldolase B,
the form present only in the liver, kidney and intes-
tine. These different forms of aldolase can readily
be separated by electrophoresis (Penhoet, Rajkumar
and Rutter, 1966). The 3F hybrid cells contain only
aldolase A, and heat inactivation studies show that
aldolase A of both rat and mouse type are present. No
aldolase B can be detected in the hybrids during any
phase of the growth cycle, and mixing experiments suggest that if the hybrid cells synthesized 10% as much
aldolase B as the hepatoma parent, this would be
detected.

Analyses of hepatoma x fibroblast hybrids have
shown extinction of two independent liver functions:
TAT and its inducibility, and aldolase B. These results are similar to those obtained with the melanoma
x fibroblast hybrids. However, different results were
obtained when albumin synthesis was examined in the
3T3 x Fu5 hybrids (Peterson and Weiss, in preparation).

The Fu5 hepatoma cells synthesize albumin and secrete it into the medium: the presence of albumin can
be detected by immunodiffusion and the amount measured
by microcomplement fixation. Table 5 shows that the
hepatoma cells secrete 4 µg of albumin per 10^6 cells
per 72 hr and that the 3T3 cells produce none. (Neither mouse albumin nor anything which cross reacts

TABLE 5

Rat serum albumin production by parental
and hybrid cells

Cell line	Immunodiffusion	Microcomplement fixation[*]
		(μg RSA/10^6 cells/72 hr)
Fu 5	+	4 (3.8–4.7)
3T3	–	< 0.03
Hybrid clones:		
3F11	+	1.5 (1.4–1.6)
3F14	+	0.6 (0.5–0.7)

Data of Peterson and Weiss (unpublished). Rabbit
anti-rat albumin antiserum was employed. For immunodiffusion analyses, growth medium in which cells had
been cultivated for 72 hr was collected, dialyzed
(against 0.001 M KPO$_4$, pH 7.4) and lyophilized. For
microcomplement fixation, growth media were subjected
to heating at 60° for 20 min (which destroys the anti-
complementary material present in fetal calf serum),
and the rat a bumin (RSA) measured as described by
Richardson, Tashjian and Levine (1969).

[*]The values shown include the range and average determinations obtained in three independent experiments.
The value shown for 3T3 is the minimal amount of mouse
albumin which could be detected by immunodiffusion
with anti-mouse albumin antiserum and therefore signifies the sensitivity of this method. Medium from 3T3
cells contains no material which reacts with anti-
mouse albumin or anti-rat albumin antisera in immunodiffusion or complement fixation.

with anti-rat albumin can be detected in the medium or
in cell extracts.) All of the hybrid clones secrete
rat albumin, and while the amount produced by each
clone is constant, there is significant variation from
one clone to another: the hybrids secrete between 5

and 30% of the amount produced by the hepatoma parental cells. These results show a clear dissociation
between the expression of the tissue-specific enzymes
(TAT and aldolase B) and the specialized protein (albumin) destined for secretion. The former are completely extinguished in the hybrids, while the latter
continues to be produced, albeit at a level lower than
in the hepatoma parent. These results suggest that
there are different mechanisms by which these functions are regulated. However, before reaching these
conclusions, let us examine the same three products in
the BRL-1 x hepatoma hybrid, which involved two rat
parental cells, one of them diploid, and both of them
epithelial cells.

Intraspecific Hepatoma x Epithelial
Cell Hybrids

From the BRL-1 x Fu5-5 cross, only one hybrid clone
(BF5) was isolated. The expected number of chromosomes was 93-95, and the observed number, 91-93, a
range which remained stable over 75 cell generations
(4 months) of continuous culture (Weiss and Chaplain,
1971). Thus, it can be said that in this hybrid
clone, chromosome loss was negligible.
Table 6 gives a summary of the results of tests for

TABLE 6

Characteristics of parental and hybrid cells

Cell line	Number of chromosomes	TAT		Aldo-lase B	Serum albumin
		Base-line	In-duced		
BRL-1	42	−	−	−	−
Fu5-5	52	+	+	+	+
BF5	92 (91-93)	−	−	−	+

Summarized from data of Weiss and Chaplain (1971),
Bertolotti and Weiss (1972), and Peterson and Weiss
(unpublished data).

the three liver functions (TAT and its inducibility,
aldolase B and albumin production) in the BF5 hybrid
cells. It can be seen that the tissue-specific enzymes which were absent in the 3F hybrids are similarly absent in the BF5 hybrid, and albumin, which was
produced by the 3F hybrids, is likewise produced by
the BF5 hybrid. These striking similarities of the
two kinds of hybrids reinforce the hypothesis that
there are different mechanisms by which these various
hepatic functions are regulated in the hybrid cells
and suggest the following conclusions regarding the
extinction of differentiated functions:

1. It is not merely the consequence of species differences.
2. It does not require the introduction of a hyper-
   diploid complement.
3. And it may occur even when the nonexpressing
   parent is derived from a closely related cell
   type.

It should be added that factors such as hybridization
itself and the consequent increase in ploidy cannot be
invoked as the cause of extinction since similar
assays have been performed on several clones of $2s$
hepatoma cells, isolated following fusion with Sendai
virus, and in every case, the liver-specific properties continued to be expressed (published and unpublished experiments of the authors). With regard to
the mechanism of extinction, we still know nothing,
nor do we know at what level the block occurs. However, we can now begin to rule out some hypotheses,
as will become clear from the following sections.

PHENOTYPIC EXPRESSION IN
DIFFERENTIATED FUNCTIONS
IN OTHER HYBRIDS

There is little doubt that extinction of differentiated functions is the most commonly observed result
in crosses such as those described above. There exist
other similar cases of extinction which have not yet

been mentioned, including the studies of Benda and Davidson (1971) on the glial cell protein S-100, and the inducible enzyme glycero-3-phosphate dehydrogenase (Davidson and Benda, 1970) in rat glial cell x mouse fibroblast hybrids; those of Sonnenschein, Tashjian and Richardson (1968) on growth hormone secretion in rat pituitary cell x mouse fibroblast hybrids; those of Coffino *et al.* (1971) on antibody secretion in mouse myeloma x mouse fibroblast hybrids. On the other hand, there are examples of continued expression of hybrid cells of tissue-specific functions (although usually at a level lower than that which characterized the parental cells): Green *et al.* (1966) found that fibroblast lines, which differed in their production of hyaluronic acid and collagen, when crossed gave rise to hybrids showing intermediate levels of production; Mohit and Fan (1971) have described the continued production of free Kappa chains by hybrids obtained from a cross of antibody producing mouse myeloma and nonproducing mouse lymphoma cells; and Minna *et al.* (1971) found that mouse neuroblastoma x mouse fibroblast hybrids continued to demonstrate electrophysiological properties of the neuroblastoma. (It will be noticed that none of the products belonging to the latter class are enzymes, and most of them are secreted by the cells.)

In some experiments mentioned here, the analyses of hybrids for tissue-specific functions have been complemented by demonstration of the co-expression of parental "household" enzymes.

RE-EXPRESSION

In the cases where extinction of specialized functions has been observed, before beginning to think about mechanisms, it is essential to know whether the maintenance of extinction requires the continuous presence of certain chromosomes of the nondifferentiated parent. There are now three cases in which an answer to this question has been supplied.

Esterase-2

In principle, an answer to this fundamental question can be obtained by examining the properties of hybrids such as those described above but which differ in one important respect: they must undergo a preferential loss of the chromosomes derived from the non-differentiated parent. The first experiments along these lines, performed by Klebe, Chen and Ruddle (1970), made use of a permanent cell line of mouse cells, derived from a renal adenocarcinoma, and characterized by the production of an eserine-insensitive esterase, ES-2, which is found in the kidney, digestive tract, liver and plasma. Cells of this line were crossed with normal human diploid fibroblasts; these hybrids, like all human-mouse hybrids, showed preferential loss of the human chromosomes. Eight hybrid clones were isolated, and seven of the eight were found to produce ES-2 at a level comparable to that of the mouse parental cells. The one clone which did not show ES-2 activity was subcloned; among eight subclones isolated, one showed re-expression of ES-2. A detailed karyological analysis of ES-2$^+$ and ES-2$^-$ hybrids led the authors to conclude that extinction of ES-2 is correlated with retention of the human C-10 chromsoome. Further, they interpreted the data to indicate that the mouse genome possesses an inherently stable epigenetic mechanism, which permits the re-expression of ES-2 upon loss of the human genetic regulator element.

Again, in this case, like in that of the first melanoma x fibroblast cross, the question arose of whether this was a unique case, and whether the difference between species may have played a role in this phenomenon.

Tyrosine Aminotransferase Inducibility

In the above section, where the BRL-1 x Fu5-5 hybrids were described, the stability of the karyotype of the BF5 hybrid was emphasized. This karyotypic stability was observed under conditions of regular serial transfer: i.e., cultures were trypsinized and reinoculated at low density as soon as confluence was reached.

The BRL-1 and hepatoma parental cells differ in an important respect: the former are subject to contact inhibition of growth and cease dividing as soon as confluence is reached, whereas the hepatoma cells, derived from a tumor, are characterized by growth in multiple layers. The hybrid cells (BF5) like BRL-1 are contact inhibited. We reasoned that by selecting for variant hybrid cells which grow in multiple layers like the hepatoma cells, we might be selecting for hybrids which had lost BRL-1 chromosomes.

Confluent cultures of BF5 were therefore maintained for one to three months with frequent renewals of medium, and indeed, after serveral weeks, focal areas of such varient hybrids, growing in multiple layers, were observed. Numerous foci, presumably clonal in origin, were isolated and subcloned, and from them, 17 subclones have been derived. All of the 17 variant subclones had greatly reduced chromosome numbers, as can be seen in Table 7, where several karyotypes are shown. Instead of the 91–93 chromosomes of BF5, these variants contained 52 to 80 chromosomes, and the ones shown in the table contained 52-63 chromosomes. Since the hybrids are rat x rat, it is not possible to determine whether Fu5-5 or BRL-1 chromosomes have been lost.

All of these chromosomally reduced variant hybrid subclones have been tested for TAT activity and inducibility, and for aldolase B. None of them contained high TAT activity levels, but one of them, BF5-1, showed increased TAT activity upon exposure to the steroid hormone dexamethasone (Dex), as shown in Table 8. However, since the "induced" activity in these hybrid cells is very low by comparison with that found in the hepatoma cells, and since as will be seen later, the kinetics of induction are different in the two kinds of cells, a detailed comparison of the parameters of TAT induction in hepatoma and BF5-1 cells was undertaken.

To begin with, BF5-1 cells were subcloned in order to determine whether the population was a mosaic one, composed of cells with very low and with high activity. Among 12 subclones isolated, all possessed the same

TABLE 7

Karyotypes of parental and hybrid cells

Cell line	Total	LSM	LM	SM	Telo
Fu5-5	52 (51-53)	2.0	1.0	22.6	26.1
BRL-1	42	2	0	20	20
BF5(22)*	91.9(91-93)	4.0	1.0	41.8	45.1
BF5(75)	91.4(89-93)	4.0	1.0	38.6	47.8
BF5-C(51)	92.0(91-93)	4.0	1.0	42.8	44.2
BF5-1(65)	63.0(61-68)	2.8	0.8	27.8	32.0
BF5-1-1(80)	63.1(60-66)	3.1	1.0	27.7	31.3
BF5-1-1b(110)	55.1(52-57)	2.0	1.0	24.4	27.3
BF5-γ-5	52.1(49-54)**	2.0	1.0	26.3	22.7

Mean values are given, as well as the range (in parentheses) of the total number of chromosomes. Chromosomes are grouped in the following classes: LSM, large submetacentric; LM, large metacentric; SM, small metacentric; Telo, telocentric and including the large satellited chromosomes.

*The numbers in parentheses after the strain designation indicate the calculated total number of cell generations at the time of karyotypic analyses.

**In this population, 15% of the cells have double this number of chromosomes ($2s$).

properties: low TAT baseline and some inducibility. The subclones showing the greatest inducibility, BF5-1-1 and BF5-1-1-b, were used for further study (Table 8).

In Figure 1, the kinetics of TAT induction of Fu5-5 and BF5-1-1 cells are shown: in both cases the induction curves from early logarithmic and late logarith-

TABLE 8

Activity and inducibility of TAT in parental
and hybrid cells

Cell line	Early logarithmic cultures			Stationary phase cultures		
	Base-line	5 hr Dex	24 hr Dex	Base-line	5 hr Dex	24 hr Dex
BRL-1	0.2	0.3	0.2	0.7	-*	0.6
BF5-C	0.4	0.4	0.4	0.4	0.6	0.5
BF5-1-1	0.5	4.3	3.9	0.4	5.7	1.3
BF5-1-1-b	1.1	1.6	9.2	0.2	3.0	8.4
BF5-γ-5	-	-	-	0.2	0.3	0.2
Fu5-5	16.9	144.5	583.0	41.6	171.0	224.0

Values given are enzyme specific activity units
(m moles p-hydroxyphenylpyruvate/min/mg protein at
37°); data summarized from Weiss and Chaplain (1971)
Cultures were induced with 1 μM Dex, and harvested
for assay after 5 and 24 hr.

*Not assayed.

mic cultures are shown. As can be seen, the kinetics
of TAT induction, and the maximum activity obtained,
differ in the two stages of growth, and this is true
for both hepatoma and hybrid cells. Moreover, the
time course of induction in late log phase hepatoma
cells is similar to that of early log phase BF5-1-1
hybrid cells. A subclone of BF5-1-1 (BF5-1-1-b, see
Table 8) shows induction curves even more like those
of the hepatoma, for high activity is maintained for
at least 24 hr in the presence of Dex, even in late
log or stationary phase. Since cells of this subclone
have lost six more chromosomes, the peculiar "bell-
shaped" induction curve, characteristic of late log
phase BF5-1-1 cells, appears to be a property which
disappears upon further loss of chromosomes.

Figure 2 shows the early phase of TAT induction in

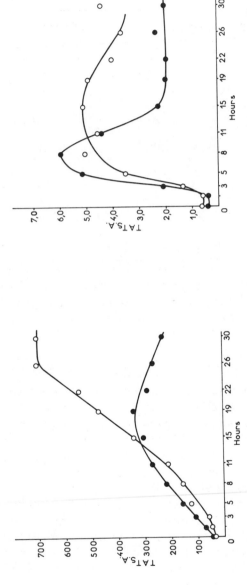

Figure 1. Time course of induction of TAT in hepatoma (Fu5-5, left) and hybrid (BF5-1-1, right) cells. For both cell types, the kinetics of enzyme induction are shown for early log phase (○) and late log phase (●) cultures. It will be noticed that there are striking differences in the rate of increase in enzyme specific activity (TAT S.A.) and the maximum induced activity of cultures in the different growth phases, and that the pattern of enzyme induction is similar in late log phase hepatoma and early log phase hybrid cultures. Cells were induced at time 0, with 1 μM dexamethasone and harvested for assay at the times indicated.

442

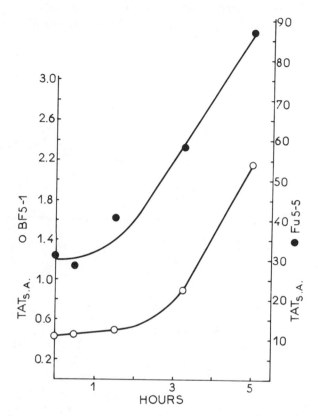

Figure 2.   Early phase of TAT induction in
FU5-5 (●) and BF5-1 (O) cells. Dexamethasone (1 μM) was added to a series of cultures of the two cell types at time 0, and
cells were harvested for assay at the times
indicated. Notice that the scales (ordinates) for the enzyme specific activity
(TAT S.A.) are different for the two cell
types.

hepatoma and hybrid cells: in both cases, there occurs
a lag before enzyme activity begins to rise, as would
be expected if *de novo* synthesis of enzyme is involved
in induction, and as has been shown by others (Kenney,
1962; Granner *et al.*, 1968) for TAT induction in rat
liver and in hepatoma cells.
    In both hepatoma and hybrid cells, TAT induction is

inhibited by both cycloheximide (Fig. 3) and actinomycin D (Table 9), which suggests that synthesis of both

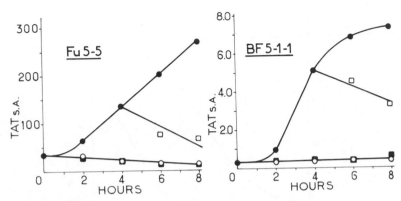

● dex
○ cycloheximide
■ dex + cycloheximide
□ dex 4 hours, then cycloheximide added

Figure 3.   Inhibition by cycloheximide of
TAT induction in hepatoma (FU5-5) and hybrid (BF5-1-1) cells. A series of cultures was treated at time 0 with the combinations of reagents indicated in the
figure, and samples were harvested for a
assay every 2 hr for the next 8 hr.   The
concentration of cycloheximide used (0.1
mM) inhibits protein synthesis by 96 and
99% in Fu5-5 and BF5-1-1 cells respectively.   Dexamethasone was used at 1 M.   Enzyme specific activity (TAT S.A.) is shown
on the ordinates.

protein and RNA is required for induction. Moreover,
it has been found that the concentrations of Dex that
cause TAT induction are similar for hepatoma and hybrid cells (Fig. 4), although the hepatoma cells show
a slight response at a concentration (10^{-9} M) which
fails to elicit induction in the hybrid cells.
    Finally, the physical properties of TAT from paren-
tal and hybrid cells, both induced and noninduced,
have been compared. It has been shown by others that

TABLE 9

Inhibition of TAT induction by actinomycin D

1 µM Dex	µg/ml Actinomycin D	Fu5-5		BF5-1-1	
		Specific activity	% Inhibition of induction	Specific activity	% Inhibition of induction
–	0	27.5		0.4	
+	0	121.0		5.1	
+	0.1	61.0	64	2.0	66
+	0.5	43.1	83	0.9	89
+	1.0	34.8	92	0.7	14
–	1.0	36.1		0.4	

Cells collected and assayed after 5 hr exposure
to the agents indicated. Data of Weiss and
Chaplain (1971).

both baseline and induced hepatoma cell TATs have the
same physical (Thompson, Tomkins and Curran, 1966) and
serological (Granner *et al.*, 1968) properties. It has
been found that the heat stabilities of enzymes from
induced hybrid cells and from hepatoma cells are the
same (Fig. 5). However, the noninduced hybrid cell
enzyme is different (more similar to that from BRL-1),
and this suggests that the "baseline" enzyme in the
hybrid cells is either a modified form of TAT, or is
another enzyme which also possesses TAT activity (see
Miller and Litwack, 1971).
   All of these similarities between the hepatoma and
the segregated hybrid cells leave little doubt that
the TAT induction observed in the hybrid cells is real,
and represents re-expression of hepatoma genes involved in TAT induction as a consequence of elimina-
tion of certain BRL-1 chromosomes.

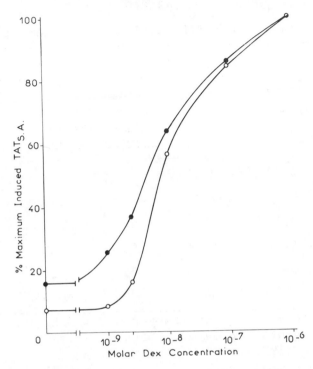

Figure 4. Induction of tyrosine aminotransferase in Fu5-5 (●) and BF5-1-1
cells (O) by different concentrations of
dexamethasone. The indicated concentrations of Dex were added to the cultures
5 hr before harvesting the cells for
enzyme assays. The maximum enzyme specific
activity was found in both hepatoma and hybrid cells treated with 10^{-6} M Dex; activities found at all other concentrations
are plotted as percent of this maximum.

Aldolase B

Aldolase isoenzymes have also been examined in all
of these segregated hybrid subclones. Those described
above, BF5-1 and its descendants, do not re-express
aldolase B *in vitro*, but a different and independent
subclone, BF5-γ-5 (Table 7) does produce aldolase B.
This is shown in Figure 6, which includes the aldolase

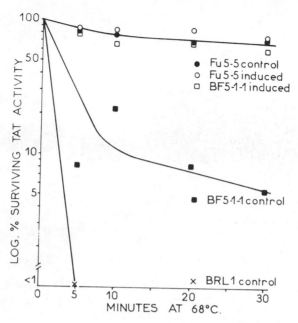

Figure 5. Heat inactivation of tyrosine aminotransferase from noninduced (control) and induced parental and hybrid cells. Cell extracts were subjected to heating at 68°, and at the times indicated, the extracts were agitated, aliquots were removed and rapidly cooled in ice. After a brief centrifugation, all samples were assayed. No activity was detected in the BRL-1 extract that was heated for 5 min; the amount of activity remaining in the control BF5-1-1 extracts was barely measurable.

patterns of Fu5-5 and BF5-γ-5, both of which display pure aldolase A (near the origin) as well as two of the AB heterotetramers, and of BF5 (the chromosomally complete hybrid) and BRL-1, for which only a single band of activity, corresponding to pure aldolase A, can be seen. Moreover, specific activity and substrate activity ratio measurements show that cells of this clone produce even more aldolase B than the

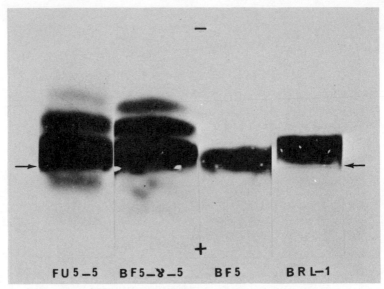

Figure 6.   Aldolase zymograms of parental
(Fu5-5, BRL-1) and hybrid (BF5-γ-5 and BF5)
cells (see text for details).

parental hepatoma cells (Bertolotti and Weiss, 1972).
And finally, no induction of TAT can be detected in
these cells (Table 8).

CONCLUSIONS

We are now faced with the question of what can be
concluded from these diverse data, obtained from the
study of somatic hybrids arising from crosses of
parental cells which differ in the expression of tissue-specific functions (reviewed by Davidson, 1971 and
Ephrussi, 1972). (In order to permit a general discussion, the parental cells will be referred to as
"expressing" and "nonexpressing", rather than differentiated and undifferentiated, since all of the cell
types with which we have dealt must have been, when
still in the organism, committed to a specific path of
differentiation.)
A first conclusion is that the extinction of tissue-
specific functions in somatic hybrids is a very gener-

al, but not a unversal phenomenon; none of the exceptions to this general rule concerns an enzyme, and most of them have been described in studies of products destined for secretion. In the absence of more extensive data, we can postulate, as a working hypothesis, that at some level, the regulation of these products for secretion differs from that of tissue-specific enzymes.

Moreover, the number of cases in which extinction has been observed are now sufficiently numerous and include enough controls, to make it highly unlikely that this phenomenon is due to obvious trivial causes, such as species differences, cell fusion or changes in ploidy. However, more subtle, but perhaps also trivial causes cannot be ruled out, since in no case is there evidence which shows that the mechanism by which a given tissue-specific function is extinguished is the result of the action of genes which play a role (direct or indirect) in the normal (i.e., developmental) regulation of synthesis of that protein. Nevertheless, irrespective of whether the disappearance of tissue-specific enzymes in somatic hybrids reflects abnormal or normal regulation (the latter implying that the mechanism which causes extinction is the same as that, or one of those, which insures the absence of that enzyme in the nonexpressing parent), we now have at hand an experimental system for the study of regulation in mammalian somatic cells.

A second conclusion is that the simultaneous extinction or expression of all tissue-specific proteins is not necessarily observed in a given hybrid: in the hepatoma hybrids, there is a clear difference in the behavior of the enzymes, TAT and aldolase B, and that of albumin. These results suggest the existence of multiple controls, within cells of a given tissue type, which regulate, qualitatively and quantitatively, the spectrum of products characteristic of that tissue.

The third conclusion is that tissue-specific enzymes, which are extinguished in hybrid cells, can reappear when chromosomes of the nonexpressing parent are lost. The fact that genes controlling these enzymes are maintained through numerous hybrid cell generations, in the absence of their expression, shows

that extinction is not due to loss of the relevant
structural genes, and suggests that re-expression is
the consequence of a stable epigenetic change which
occurred in the normal precursor of the differentiated
tumor cell (Ephrussi, 1972).

The observation that TAT inducibility and aldolase
B are independently re-expressed reinforces the above
suggestion of "multiple controls." If a single element extinguished these two properties, they should be
re-expressed simultaneously, unless the structural
genes for one of them had been lost. [The latter possibility can be excluded in the case of the segregated
hybrids which re-express TAT inducibility but not aldolase B, because these cells do produce aldolase B
when grown as tumors in rats (Bertolotti and Weiss
unpublished experiments).]

The phenomenon of re-expression by segregated hybrids may provide us with a whole new set of experimental models of gene expression, since in some cases
re-expression seems not to be correlated with a re-
version to the exact phenotype of the expressing par-
parent, probably owing to the persistence of some
chromosomes of the nonexpressing parent. Thus, such
hybrids may be expected to contain, and to express,
constellations of genes which are not normally simultaneously active. This suggestion finds support in
the properties of the hybrids which re-express enzyme
(TAT) inducibility in the absence of an appreciable
baseline activity, an unusual situation for mammalian
cells.

To conclude, although the mechanisms by which cells
are committed to and maintain the synthesis of tissue-
specific proteins remains obscure, the observation
that re-expression of these products does occur, and
most likely as the result of loss of a specific chromosome (as indicated by the studies of ES-2), suggests
that we may soon arrive at an understanding of the
mechanism by which cells can restrict their synthetic
activities. Whether the latter will tell us anything
about commitment to and maintenance of a program of
synthetic activity remains unknown, but it is surely
one important aspect of the overall problem posed by
cellular differentiation.

SUMMARY

The expression of three functions characteristic of liver cells has been examined in hybrids resulting from crosses of well differentiatied rat hepatoma cells with mouse fibroblasts and with diploid epithelial cells derived from rat liver. These liver-specific proteins, produced only by the hepatoma parental cells, behave similarly in the two kinds of hybrids: the enzymes, tyrosine aminotransferase (and its inducibility with steroid hormones) and aldolase B, are greatly diminished or absent, while some serum albumin continues to be produced. In one group of hybrids (rat hepatoma x rat epithelial liver cells), loss of large numbers of chromosomes has been correlated with the independent re-expression of tyrosine aminotransferase inducibility and aldolase B.

ACKNOWLEDGMENTS

It is a pleasure to acknowledge the participation of Dr. Jerry Schneider, Mme Michèle Chaplain and Melle Anne Debon in various phases of this work. The authors are particularly grateful to Professor Boris Ephrussi for his continuous encouragements, and for many thought provoking discussions of interpretation. This work has been supported by a grant to Dr. Ephrussi from the Délégation Générale à la Recherche Scientifique et Technique.

REFERENCES

Barski, G., Sorieul, S., and Corenfert, F. (1960). *C. R. Acad. Sci., Paris 251*, 1825.

Benda, P., and Davidson, R. L. (1971). *J. Cell. Physiol.* (in press).

Bertolotti, R., and Weiss, M. C. (1972a). In *Cell Differentiation, Proceedings of the First International Conference on Cell Differentiation*, R. Harris and D. Viza, eds. (Copenhagen: Munksgaard).

Bertolotti, R., and Weiss, M. C. (1972b). *J. Cell. Physiol.* (in press).

Blostein, R., and Rutter, W. J. (1963). *J. Biol. Chem. 238*, 3280.

Coffino, P., Knowles, B., Nathenson, S. G., and Scharff, M. D. (1971). *Nature New Biol. 231*, 87.

Coon, H. G. (1968). *J. Cell Biol. 39*, 29a.

Coon, H. G. (1969). *Carnegie Inst. Wash. Yearbook 67*, 419.

Crowle, A. J. (1961). *Immunodiffusion* (New York and London: Academic Press), p. 209.

Davidson, R. L. (1971). *In Vitro 6*, 411.

Davidson, R. L., Ephrussi, B., and Yamamoto, K. (1966). *Proc. Nat. Acad. Sci. USA 56*, 1437.

Davidson, R., Ephrussi, B., and Yamamoto, K. (1968). *J. Cell. Physiol. 72*, 115.

Davidson, R. L., and Benda, P. (1970). *Proc. Nat. Acad. Sci. USA 67*, 1870.

Diamondstone, T. I. (1966). *Anal. Biochem. 16*, 395.

Ephrussi, B. (1965). In *Developmental and Metabolic Control Mechanisms and Neoplasia* (Baltimore: The Williams and Wilkins Co.), p. 486.

Ephrussi, B. (1972). *Hybridization of Somatic Cells* (Princeton: Princeton University Press).

Granner, D. K., Hayashi, S., Thompson, E. B., and Tomkins, G. M. (1968). *J. Mol. Biol. 35*, 291.

Green, H., Ephrussi, B., Yoshida, M., and Hamerman, D. (1966). *Proc. Nat. Acad. Sci. USA 55*, 41.

Green, H., Goldberg, B., and Todaro, G. J. (1966). *Nature 212*, 631.

Harris, M. (1964). *Cell Culture and Somatic Variation* (New York: Holt, Rinehart and Winston).

Hayashi, S., Granner, D. K., and Tomkins, G. M. (1967). *J. Biol. Chem. 242*, 3998.

Kenney, F. T. (1962). *J. Biol. Chem. 237*, 3495.

Klebe, R. J., Chen, T., and Ruddle, F. R. (1970). *Proc. Nat. Acad. Sci. USA 66*, 1220.

Le Douarin, N. (1964). *Bull. Biol. Fr. et Belg. 98*, 544.

Matsuya, Y., and Green, H. (1969). *Science 163*, 697.

Miller, J. E., and Litwack, G. (1971). *J. Biol. Chem. 246*, 3234.

Minna, J., Nelson, P., Peacock, J., Glazer, D., and

Nirenberg, M. (1971). *Proc. Nat. Acad. Sci. USA 68*, 234.

Mohit, B., and Fan, K. (1971). *Science 171*, 75.

Penhoet, E. E., Kochman, M., and Rutter, W. J. (1969). *Biochemistry 8*, 4391.

Penhoet, E. E., Rajkumar, J., and Rutter, W. J. (1966). *Proc. Nat. Acad. Sci. USA 56*, 1275.

Pitot, H. C., Peraino, C., Morse, P. A., and Potter, V. R. (1964). *Nat. Cancer Inst. Monogr. 12*, 229.

Reuber, M. D. (1961). *J. Nat. Cancer Inst. 26*, 891.

Richardson, U. I., Tashjian, A. H., Jr., and Levine, L. (1969). *J. Cell Biol. 40*, 236.

Schneider, J. A., and Weiss, M. C. (1971). *Proc. Nat. Acad. Sci. USA 68*, 127.

Silagi, S. (1967). *Cancer Res. 27*, 1953.

Sonnenschein, C. A., Tashjian, A. H., Jr., and Richardson, U. I. (1968). *Genetics 60*, 227.

Thompson, E. B., Tomkins, G. M., and Curran, J. F. (1966). *Proc. Nat. Acad. Sci. USA 56*, 296.

Todaro, G. J., and Green, H. (1963). *J. Cell Biol. 17*, 299.

Tomkins, G. M., Gelerhter, T. D., Granner, D., Martin, D., Jr., Samuels, H., and Thompson, E. B. (1969). *Science 166*, 1474.

Wasserman, E., and Levine, L. (1961). *J. Immunology 87*, 290.

Weiss, M. C., and Chaplain, M. (1971). *Proc. Nat. Acad. Sci. USA* (in press).

Clonal Differentiation
in Early Mammalian Development

Beatrice Mintz

Institute for Cancer Research
Philadelphia

INTRODUCTION

In the present "state of the art," the organism itself is the only place where the *determinative* events
leading to cell specialization in the development of
higher vertebrates can be experimentally investigated.
Cells presently capable of differentiation in culture
[e.g., myoblast cell development into muscle fibers
(Konigsberg, 1961)] were in fact already determined or
programmed for that kind of development in the embryo
before they were explanted. Teratomas, containing as
they do pluripotential stem cells (Stevens, 1967;
Pierce, 1967), offer promise of providing a means of
examining determination outside the organism under
controlled conditions; it is not yet possible, however,
to channel their line of differentiation at will.
Thus, in multicellular species such as mammals, we are
still confronted by the necessity of finding ways to

study the problem within the complex framework of the organism.

It has recently become possible to trace the origins of mammalian cell diversification *in vivo* by several techniques that bring to light the clonal lineages of which tissues are comprised. The term *clone* is used here to refer to the mitotic progeny of one cell in which a tissue-specific constellation of gene loci first became active, derepressed, or capable of being transcribed, and which has transmitted this specific functional capacity to its cellular progeny (Mintz, 1970). (Transcription need not occur at all times in the cell cycle and may require the intervention of inducers or other agents.) By tracing clonal histories back to their inception, we presumably identify the cells in which a given functional genetic specialization was initiated.

The first mammalian clonal history to be ferreted out was that of the melanoblasts in mice (Mintz, 1967). This was done by producing viable mouse embryos with cellular genotypic mosaicism—in this case, at melanoblast loci. Blastomeres from two genotypically different embryos are assembled so as to form a single composite, by methods that have been recently reviewed in detail (Mintz, 1971a). Any cellular genotypic differences, including immunogenetic disparities (Mintz and Palm, 1969), can be incorporated into single individuals in this way. The two cellular phenotypes ultimately found within any tissue of these *allophenic* mice are indicative of separate cell lineages in the ontogeny of the tissue. Inasmuch as both cell strains are present in the embryo before cellular diversification has occurred, both participate in, and ultimately serve to reveal, cellular genealogies. When the two cell lines differ allelically at, say, a pigmentary locus, the clonal components of the melanoblast system are visualized as discrete patches of one or the other color. Other markers, appropriate to other kinds of specialized cells, have been used to trace clonal origins in a variety of tissue *in vivo* in allophenic mice.

In another approach also based on cellular genotypic mosaicism, mouse embryo cells of one genotype

are injected into the blastocyst cavity of an embryo
of another genotype, where they become incorporated
and yield mice essentially resembling allophenic ones
(Gardner and Lyon, 1971).

In still another technique, functional (rather than
structural) genetic mosaicism has been used in mammalian females heterozygous for X-linked genes (Gandini
and Gartler, 1969). Single-allele activity per cell
at these loci (Lyon, 1962; Russell, 1962) leads to
cellular phenotypic differences useful as markers
within some tissues.

Studies on allophenic mice have led us to certain
generalizations and hypotheses concerning mammalian
differentiation. The points to be discussed here are
the following:

1. Each working program for a specialized mammalian
   cell type appears to be initiated in a relatively small number of clonal initiator cells; the
   number is always at least two and is tissue-specific.
2. There is often (or possibly always) a time gap
   between *determination* of a specific cellular
   developmental program and *expression* of that
   program.
3. In ordinary mice, where the genotype is presumably generally the same in all cells of an individual, there are phenotypic differences among
   cells within a specific differentiated tissue
   that appear to be *clonal* in nature; i.e., among
   the various clones that comprise a population of
   specifically differentiated cells there can be
   small but significant molecular microheterogen-
   eities, even when the locus responsible for the
   product is homozygous.

CLONAL ORIGINS

Two phenotypically and genotypically different
populations of cells have been found within each of
the many tissues thus far examined in allophenic mice
that bear appropriate allelic markers in their two

cell strains. Therefore, each specialized kind of adult cell must be multiclonal, having originated from at least two genetically determined cells (Mintz, 1970). The actual clonal numerology is, however, quite tissue-specific, irrespective of the markers used to identify the clones, and is thus evidently genetically fixed. Some examples will be briefly reviewed.

In allophenic mice with cells from two different pigment cell color strains, an orderly coat color pattern is found. Although there are many individual variations in the pattern, all seem to be changes rung upon a single, or archetypal, developmental theme. The archetype consists of a series of 17 transverse, mediolateral bands down each side of the head, body, and tail. Each band has been interpreted as a clone descended from a single genetically determined precursor cell (Mintz, 1967). The total clonal composition is visible at once only in those allophenic individuals in which, by chance, each clone happens to differ in genotype, and therefore in color, from its neighbors; if two or more neighboring clones are of the same color strain, they constitute a relatively wide transverse stripe. The developmental individuality of each of the melanoblast cell lineages is nevertheless verifiable by the fact that each of the 34 bands varies in color type independently of the adjacent clones, in at least some individuals. Left- and right-side bands often vary in color independently of one another and are sometimes out of register, indicating that they arise from separate clonal initiator cells. The maximum number of bands visible in an individual is 34 and 34 precursor cells therefore are responsible for forming all of the pigment cells in the coat. This basic clonal plan of melanoblast differentiation is subject to many modifications, depending upon the genes involved. Many changes can be accounted for by clonal selection, due to a selective advantage of one clonal phenotype over the other; the latter is then displaced or replaced to varying degrees (Mintz, 1969; 1970). Even single-allele differences on coisogenic backgrounds in the two component cell strains are, in some genotypes, a sufficient basis for selection to act at the cellular level.

The clonal history of the hairs in the coat, as judged from cellular allelic hair marker experiments in allophenic mice, turns out to be quite different from that of the melanoblasts in those hairs. This is not surprising, in view of their diverse origins: melanoblasts arise from neural crest cells (Rawles, 1947), while the development of hairs seems to depend chiefly on their mesodermal component of somite origin (Mintz, 1969; 1970). Although the hair itself is an ectodermal outgrowth, the hair follicle contains an important dermal component that appears to exercise an inductive influence on the development of the adjacent ectoderm. In allophenic mice with morphological or other hair genotypic differences in their two cell strains, a series of transverse stripes, independent on left and right sides of the body, is again formed, but the frequency and contribution of these hair bands is unrelated to that of the melanoblast bands. Hair clone frequency corresponds, instead, to the frequency of somites, in the region of the body where the serially repeated bilateral somite blocks were visible in the embryo, signifying the probable origin of each hair follicle clone from one clonal initiator cell in the dermatome portion of a somite. The total hair follicle clonal number is estimated at approximately 170.

The somite, in addition to its dermatome derivatives, also gives rise to muscle, from its myotome component, and bone, from its sclerotome component. The somites first become visible on late day 7 of embryonic life (counting the vaginal plug date as day 0). Allophenic studies indicate that the somite itself is multiclonal and, at the time of its first appearance, may already be a collection of determined cells destined to give rise to these various derivatives. When allophenic embryos were made from cells of two strains with different allelic electrophoretic variants (Carter and Parr, 1967) of the dimeric enzyme glucosephosphate isomerase (GPI), individual somites were in fact found to contain both electrophoretic types of GPI (Gearhart and Mintz, 1971). In order to examine further the history of the myogenic component

of the somite, the genotypic composition of the ex-
trinsic eye muscle was analyzed. These muscle are
unusual in that each arises from the myotome of only
one somite, or from only a part of the myotome of one
somite (literature reviewed by Goodrich, 1930).
Therefore, each of the eye muscle of allophenic mice
originating from two electrophoretically different GPI
strains should contain only one GPI type if the myogenic component in its somite of origin developed as a
single clone; if the muscle developed from two or more
precursor cells in that somite, it should comprise
both genotypes in at least some cases. Myogenesis *in
vivo* occurs by the fusion of uninucleated myoblasts
and, in multinucleated muscle fibers of allophenic
mice with nuclei of two pure strains, each nucleus
continues to code for its own allelic form of an enzyme, and the hybrid or heteropolymeric form of the
enzyme then results from combinations of the different
monomeric subunits in the cytoplasm (Mintz and Baker,
1967). No hybrid enzyme has been found in allophenic
tissues other than skeletal muscle. In the eye
muscles of allophenic mice with GPI cellular differences, not only were both pure-strain variants of GPI
present, but hybrid enzyme was also formed (Fig. 1).
Therefore, the uninucleated myoblast cell population
of one somite, even before myoblast cell fusion, was
apparently already a mixture of predetermined cells of
two genotypes. The myotome of one somite thus arises
from at least two precursor cells.

   To investigate the origins of the sclerotome component of somites, the vertebral columns of allophenic
mice with cells from two strains with morphological
vertebral differences were anlayzed, in order to identify the smallest unit capable of independent morphological variation in strain phenotype; this was presumed to represent a clone (Moore and Mintz, 1972).
Among the numerous permutations and combinations found
in the relative proportions and distributions of the
two strains, it was evident that a right or left half-
vertebra sometimes differed in strain-type independently of the contralateral half or the rest of the
axial skeleton. In addition, some right or left-half-
vertebrae were of intermediate strain-type (Fig. 2).

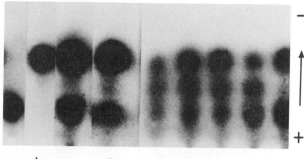

Figure 1. Composite of starch gel electrophoretic analyses of glucosephosphate isomerase phenotypes in extracts of individual eye muscles that are each of single-so-
mite origin. Slots **a** and **b** are eye muscle
samples from pure-strain controls (Gpi-1A
and Gpi-1B types). *In vitro* mixtures of
the pure types (slots **c** and **d**) do not yield
hybrid enzyme. Eye muscles from allophenic
animals (slots **e-i**) comprise both A- and B-
types (BALB/c ⟷ C3Hf strain combination)
and most also have hybrid enzyme. (Data
from Gearhart and Mintz, 1971).

The conclusions reached from these data were that left
and right sides develop from separate cell lineages,
and that two genotypically different cell lineages on
one side would account for an intermediate phenotype
on that side. Thus, each vertebra comprises at least
four developmental cell lineages: two on the left and
two on the right side. The most likely conclusion is
that these correspond to the four intersegmentally derived components of a vertebra, i.e., the two caudal
sclerotomite elements from a pair of somites and the
two cranial elements from the next succeeding pair.
The proposed archetype of vertebral ontogeny is either
that these four sclerotomite elements are themselves
the clonal units that can be separately genetically
controlled or, alternatively, that the entire sclerotome from each somite is the clonal unit. The former

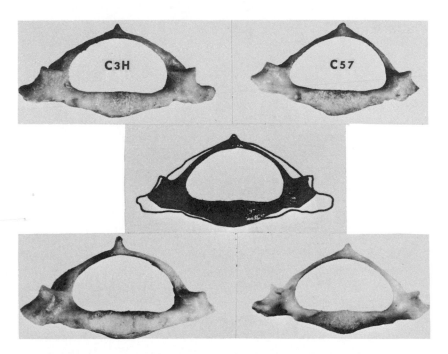

Figure 2. Phenotypes of first thoracic
vertebrae from C3H and C57BL/6 pure-strain
controls (upper row) and from two C3H ⟷
C57BL/6 allophenic mice (lower row). Out-
line drawings of the two control vertebrae
have been superimposed to show some of the
differences in size and shape of the lateral masses; additional differences in the
third dimension are not visible here. The
first allophenic case has left-right
strain-type asymmetry (C3H on viewer's
left, C57 on right). The other allophenic
vertebra has intermediate phenotypes with
characteristics of both strains on each
side. (Data from Moore and Mintz, 1972.)

possibility we considered the more likely (Moore and
Mintz, 1972), in which case the entire vertebral
column (body and tail) may consist of some 240 clones.
    A comparison of the genotypic composition of each
of the various somite derivatives (hair follicles,

muscle, and vertebrae) in a given axial region within single allophenic individuals showed that the three tissues in that region sometimes differed from each other in their cellular genotypes. Therefore, these three major somite derivatives are likely to have arisen from separate, rather than identical, precursor cells within the somite (Moore and Mintz, 1972).

Summarizing the preceding lines of evidence, the dermatome of a somite comes from one precursor cell that will form one hair follicle clone, and probably from other precursor cells that will form other dermal elements in the skin; the myotome comes from an irreducible minimum of two myogenic precursor cells; and the sclerotome may arise from a mere two chondrogenic precursor cells. However, at the time of its first appearance, on day 7 of embryonic life, a somite contains a much larger number of cells than these relatively modest clonal members in its major components would indicate. Therefore, the somite may, when first visible, already be a collection of genetically determined cells some mitotic generations beyond the (presomite) stage when each clone was programmed in one initiator cell.

The example of the retina further illustrates the tissue-specific nature of clonal numerologies and histories and shows that even closely associated kinds of specialized cells can have independent clonal controls. We have produced allophenic mice in which the pigment cell layer in the retina consists partly of pigmented (genotype C/C) and partly of albino (c/c) cells. In flat preparations of such retinas, radially arranged streams of pigmented or albino cells are seen (Mintz and Sanyal, 1970; Mintz, 1971b). This pattern is evidently based upon radial proliferation of clones from initiator cells at the center. The ontogeny of the visual layer of the retina was also analyzed, with other markers: some cells in the photoreceptor layer were from a standard strain with normal cell viability and vision ($+/+$) while others were from a retinal degeneration (rd/rd) strain in which the visual cells form but are inviable and degenerate postnatally. In mosaic retinas with both genotypes, each cell strain behaves autonomously and the location of null areas left a er *rd/rd* cells disappear can be used to map the clonal origins of the photoreceptor layer. In reconstructions made from serial sections, many variations were found, but again a developmental archetype could be found: it consisted of 10 radiating clonal sectors per retina. Thus, the neural retina of each eye appears to proliferate radially from a small circlet of 10 initiator cells. When cell genotypes differ for both pigmentation and retinal degeneration simultaneously, the unrelatedness of the patterns in the respective retinal layers demonstrates that the clones arise independently in the two layers.

Some kinds of specialized cells have, however, been found to be traceable to a common origin. One of the most interesting examples is in the hematopoietic system. Analyses in allophenic mice lead to the conclusion that both red and white blood cells originate from a common pool of precursor cells. Red and white circulating blood cells from BALB/c ⟷ C57BL/6 animals were separated and analyzed genotypically by means of their allelic electrophoretic variants of glucosephosphate isomerase. The genotypic compositions of the two kinds of blood cell populations were almost perfectly correlated within individuals, despite the fact that some other tissues in the same individuals had quite different genotypic compositions (Mintz, 1971b, summarized from data collected by R. Niece and B. Mintz). Other experiments, dealing with differentiation within hematopoietic colonies of the spleen after injection of marrow cells into irradiated recipient mice (Till and McCulloch, 1961), have in fact given evidence for pluripotential stem cells in blood development. Further allophenic studies have also provided data indicative of shared origins of blood and antibody-producing cell precursors (Mintz and Palm, 1969; Wegmann and Gilman, 1970).

DETERMINATION AND EXPRESSION
OF CELL TYPE

From the sorts of clonal studies just discussed, it is likely that the cells of which each specialized

kind of adult tissue is composed are merely mitotic
descendants of a much smaller number of precursor
cells in which tissue-specific genetic function first
occurred. The time when functional genetic specialization was initiated in these cells can be deduced
from certain data in the allophenic experiments, and
can be shown, in at least some cases, to precede
detectable differentiation of those cells.

   Cell differentiation may perhaps begin temporally
with the determination of the embryo itself. From the
fact that allophenic mice can contain two cellular
genotypes, the embryo, in the definitive sense, must
be detemined at some *post*zygotic stage from at least
two cells, rather than from the single-celled zygote.
The results of blastomere rearrangement experiments
have led to the conclusion that all cells are labile
and equipotential during the preblastocyst period
(Mintz, 1965). Some cells (not necessarily all) in
the inner cell mass will later form the embryo proper
while other cells form extraembryonic structures. Although some mice that originate as composites of two
genotypic kinds of blastomeres are actually genotypic
cellular mosaics, others are found that are nonmosaics;
they have cells of only one or the other of the two
possible genotypes. This loss of mosaicism in some
embryos occurs after the composites are transferred
to the uterus of a surrogate mother. In a large experiment involving two inbred cell strains with many
markers used to test genotypes in many tissues, approximately 75% of the adult individuals had both cell
strains while 25% had only one strain. This 75% frequency of mosaicism was explained on a three-cell
model of embryo origin: if each of the three cells had
an equal chance of being of one or the other strain,
the statistical expectation, on expansion of the bi-
nomial, would yield 75% mosaic individuals (Mintz,
1970). Thus, "embryo-determining" genes may be among
the first to act differentially in some cells only and
may possibly become active in as few as three cells.
From these, "embryo clones" would form. Progressuve
genetic specialization then occurs, perhaps in many
systems; an example has been given above, for a common hematopoietic origin from which various kinds of
blood cells have diverged.

Many kinds of cellular specialization of gene function seem to be initiated in the early postimplantation period between days 5-7 of embryonic life, before any substantial morphogenetic changes have taken place (Mintz, 1970; 1971c). In the coat melanoblast system, for example, we can arrive at an estimate of when the loci for pigment cell differentiation first became active in the 34 melanoblast clonal initiator cells of the neural crest. Inasmuch as clones on the left and right sides of the dorsal midline can be seen to behave autonomously in allophenic mice, they must have originated before the neural crest cells on the two sides could pass across the midline—i.e., before the longitudinal neural folds on the two sides came into contact and closed the gap running anteroposteriorly down the center of the embryo. The neural folds start to fuse on day 8 of development and day 7 would therefore be the latest time at which melanoblast determination might have occurred. To estimate the earliest possible time, we can point to the necessity for the embryo to have at least 34 cells at the time in question in order for it to set aside 34 primordial melanoblasts; presumably it would also need some additional cells as a source of other kinds of tissues. When the blastocyst stage is first reached, on day 3 of development, the embryo has only about 32 cells, thereby effectively ruling out days 1-3, and probably also day 4, as the time of melanoblast determination, because of inadequate numbers of cells. Thus, melanoblast determination, as judged from the clonal patterns and clonal numbers, appears to occur very early (in the day 5-7 period) in prenatal life, but the pigment cells are not distinguishable in the skin and do not form melanin until a few days after birth—a delay of over two weeks.

For some other kinds of cells, there may be little or no time lapse between determination of cell type and expression of cell type. Blood cells are probably among the first cells to be phenotypically diagnosable; they are evident in the yolk sac by day 7, and judging from equipotentiality of blastomeres as late as day 3 (Mintz, 1965), the hematopoietic cells were probably determined no earlier than day 4 of development.

Indirect evidence obtained from other *in vivo* clonal analyses is consistent with the interpretation
that many other kinds of specialized cells were determined shortly after implantation, in the day 5-7 period. Hair patterns in allophenic mice show left-right
asymmetry, as do melanoblast patterns, and this fact,
which suggests origin from cells prevented from crossing the midline, leads to the conclusion that the hair
follicle initiator cells were determined in the
somites before closure of the neural folds, and therefore before day 8. The clonal number (approximately
170) again indicates that day 5 is the earliest time
at which these initiator cells could have been set
aside; there would be insufficient numbers of cells in
earlier embryo stages. Somites of myogenic and of
chondrogenic derivatives of somites (Gearhart and
Mintz, 1971; Moore and Mintz, 1972) have also shown
that the critical genetic decisions for these cell
specializations probably occurred in the presomite
stages after implantation.

Recent interesting observations on teratoma induction by Stevens (1970) provide independent and novel
evidence in support of this veiw that determination of
cell types is largely completed by day 7 of embryonic
life in the mouse. Stevens found that embryos 6 days
of age or younger, when grafted to an ectopic site,
in the adult testis, developed into mixtures of various differentiated tissues as well as pluripotential
proliferative cells capable of giving rise to many
cell types. When embryos 8 days of age or older were
similarly grafted, although they produced growths
that contained many kinds of differentiated tissues,
they no longer yielded proliferating undifferentiated
embryonic cells. He concluded that after 6 days of
embryonic life "most if not all the cells have already
become determined."

It is also noteworthy that mouse embryos of the
early postimplantation period cannot be cultured *in
vitro* with any appreciable success, although both
earlier- and later-stage embryos undergo considerable
spans of normal development in culture. This refractoriness of the stages after implantation and before
organogenesis may well be related to initiation of

many functional genetic cellular specializations during that time, as the embryo's developmental and nutritional requirements might then be very exacting and relatively more difficult to duplicate *in vitro*.

What triggers specific determinations of cell type remains for the present obscure. Equally baffling is what mechanisms are responsible for "counting out" specific numbers of clonal initiator cells (e.g., 34 presumptive melanoblast cells, 20 presumptive photoreceptor cells, etc.), or for activating specific loci in cells located in certain places in the embryo (e.g., the melanoblast precursor cells appear to be in a longitudinal array of 17 cells on each side of the dorsal midline, the photoreceptor precursors in a circlet of 10 cells on each side of the head). If we project the clonal patterns of these and other tissues backward into the mammalian embryo, we obtain a kind of "fate map" whose fixed architecture is as striking as its etiology is mysterious. Presumably, local physiological conditions differ in the various parts of a mass of embryonic cells, and gradients of food, oxygen, and metabolites exist. While such regional disparities may well play important and consistent roles in triggering specific genetic functions, we are still lacking in precise knowledge of how this orderly early "fate map" or architecture of clonal determination is controlled.

The developmental beginnings of clonal lineages, in the sense of inception of specialized gene function in cells, may bear significant relationships to certain chromosomal changes involving compacted (Schultz, 1965) or heterochromatic regions. In female mammals with two X-chromosomes, including the mouse, it is generally believed that only one gene is active per cell at each locus, thereby effectively making the gene dosage similar to that in the male, with only one X (Lyon, 1961; Russell, 1961). The inactive X present in the female's cells and not in the male's cells is represented by the condensation referred to as sex chromatin and is seen in adult interphase cells of females. According to DeMars (1967), this sex chromatin body is first visible in the blastocyst stage of the mouse embryo, when there are about 50 cells.

If we turn for a moment to the clonal models of coat melanoblast and of hair follicle development found in allophenic mice, we find that the same sorts of patterns seen in allophenics with allelic genotypic cellular differences are also found in ordinary *single*-genotype mice that are female and heterozygous for X-linked genes affecting melanoblasts and/or hair follicles. The *tortoiseshell* (*To/+*) and *tabby* (*Ta/+*) X-linked genotypes are examples: these females display mottling patterns indistinguishable from the corresponding patterns observed in allophenic mice with homozygous genotypic cellular differences for any genes, including autosomal ones, affecting color and/ or hairs. The different cell phenotypes in the allophenic case are due to genotypic differences, those in the X-linked heterozygous females to differential function of one or the other allele per cell despite the cellular identity of genotype. Thus, the clonal models of melanoblast and also of hair follicle development, seen first in allophenic mice where the cell markers are mutually exclusive and unambiguous, is also applicable to single-genotypic animals (Mintz, 1970; 1971c). If we surmised, from details of the allophenic patterns, that melanoblast clones and hair clones were initiated by specific-locus gene action in the day 5-7 period of embryonic life, then we would have to extend this same conclusion to the X-linked heterozygotes, inasmuch as they display the same clonal numbers and patterns as do the allophenics. In short, according to this speculation one allele per cell at these X-linked loci turns on in the day 5-7 period of development, initiating not only the clones of the tissues in question, but also the differentiation into two separate clonal phenotypes (Mintz, 1970). The experiments of Gardner and Lyon (1971) with injection into mouse embryos of cells from embryo donors with X-linked color markers confirm that this differential allelic activity at X-linked loci has not yet occurred at the blastocyst stage.

Returning to the observations of DeMars (1967), we can now conclude that single X-compaction, as represented by condensed sex chromatin first seen ontogenetically in the blastocyst (day 3), seems to precede clonal differentiation controlled by X-linked genes (day 5-7). While the evidence is still limited, we may conjecture from these time relations that the visible coiling or heterochromatinization of one X is a mechanism that *prevents* inactive loci from becoming active at some later time, rather than a mechanism that causes inactivation of previously active loci. Ohno (1969), on other grounds, also suggested that single-allele activity per cell of X-linked genes might occur through single-allele activation (following inactivity at the locus) rather than inactivation. Inasmuch as compaction also occurs in some autosomal chromosome regions as well, it seems reasonable to add that its function there may also be one of "depotentiating" gene action at the loci in those chromosomal regions, that is, of preventing future activation or derepression of those loci.

CLONAL HETEROGENEITY

The fact that allophenic clonal patterns (e.g., in melanoblast and hair follicle systems) are found in single-genotype females heterozygous for X-linked markers, is explicable, as already stated, by singly active alleles, for which other kidns of convincing evidence exist. What is surprising, however, is that similar mottling patterns, indistinguishable from some in the allophenics, are also seen in many autosomal single-genotypic mice, including not only heterozygous but also homozygous genotypes (Mintz, 1970; 1971c). These clonal phenotypic variants *within* a tissue have been termed *phenoclones* (Mintz, 1971b). Mottling genotypes have long been known in mice and other animals but their close resemblance to either archetypal or modified clonal patterns in allophenic mice now justifies regarding the cellular phenotypic differences, as clonal differences. For such phenoclones in autosomal heterozygous genoytpes (e.g., two melanoblast clonal phenotypes in Mi^{wh}/+ or two hair clonal phenotypes in A^{vy}/+ mice), it is conceivable that single-allele activity per cell might be occurring at

these autosomal loci even though there is no other
evidence for this. But for autosomal homozygotes,
such an explanation is not relevant. Examples of
autosomal homozygous mice with ostensible phenoclones
are s/s (*piebald*) and a^m/a^m (*mottled agouti*). It is
unlikely that such cases are due to somatic cell mutations, as the genotypes are true-breeding. It is
similarly unlikely that the issue is one of residual
heterozygosity in these highly inbred strains.

The most attractive hypothesis is that mechanisms
exist—albeit still unknown—that can generate phenotypic differences among genotypically identical and
homozygous clones of a tissue. The differences,
established in the clonal initiator cells and locked
into their mitotic progeny, may be small but significant as raw material for clonal selection and for
developmental and physiological adaptation. Clonal
phenotypic differences need not be limited to melanoblasts and hairs. Obvious advantages would be con-
ferred on any system of a multicellular organism by
increased cellular phenotypic variability. These advantages would be analogous to the role of variation
among individuals in evolution. Although a great deal
of cellular heterogeneity may exist within specific
tissues of individuals, it would obviously require
special methods to detect. Much of the tissue phenotypes that we conventionally describe may in fact obscure the existence of molecular microheterogeneity
among cells, and may overlook clonal selection before
the "total" tissue phenotype is achieved. Clonal
heterogeneity has of course often been discussed in
the immune system as a probable basis for antibody
diversity, and has perhaps been considered a unique
attribute of that system. On the "phenoclone" view,
it now seems more likely that clonal heterogeneity may
be an attribute of all tissues of higher organisms and
may even have been increasingly selected for in the
course of evolution. The extent of heterogeneity may,
of course, vary greatly from one kind of tissue to
another, and the mechanisms that bring it about may
differ among some specialized tissues.

What mechanisms could possibly account for molecular microheterogeneity if the same alleles are active

in all the clones? It has been proposed (Mintz, 1970;
1971c) that gene duplications or genetic redundancy
(Britten and Kohne, 1968) may be involved. The possibility that duplications and translocations of genes
may have occurred in evolution has been suggested for
hemoglobins (Ingram, 1963). If any "repetitious" DNA
units occurred as tandem duplications in which at
least one tandem member of a linear assemblage had a
mutational change, and was therefore not an exact
duplicate, the potential for heterogeneity would exist
at the transcriptional level. If the number of tandem
sections that were transcribed varied from clone to
clone, this "variable reading length" would sometimes
include the mutated section(s) and sometimes not
(Mintz, 1970; 1971b; 1971c). Some of the resultant
protein differences might lead to differences in molecular and cellular efficiency which would confer
upon some of the phenotypically variant cells the possibility of selective advantage.

The problem then becomes one of controlling factors
responsible for determining the starting and stopping
points of genetic transcription, and therefore the
extent and constancy of transcription. The discovery
of factors which cause termination of RNA synthesis at
distinct sites on DNA templates in microbial systems
(Roberts, 1969) raises intriguing possibilities that
comparable control mechanisms may exist in cells of
higher organisms and may be subject to still further
refinements in them. Evidence of RNA molecules of
different sizes from one Balbiani ring in the giant
chromosomes of *Chironomus* salivary glands has in fact
recently led Daneholt *et al.* (1970) to suggest that
the size differences might be due to serial repetitions of DNA and transcription of varying numbers of
unit templates. Comparable data in mammals would be
difficult to obtain. At least one specific possibility for investigation has been proposed (Mintz, 1971c),
namely that the so-called *diffuse* allele at the mouse
hemoglobin locus may produce its phenotype—a diffuse
band or collection of bands after electrophoresis—by
transcription of variable numbers of tandem units with
each variation expressed in a different clonal initiator cell and perpetuated in its cellular progeny.

While there is presently no real knowledge of the
extent of molecular microheterogeneity among cells of
a particular specialized kind within an individual,
n r of the mechanisms involved, the clonal models from
allophenic mice appear to be applicable to single-genotype animals and to furnish a basis for interpreting
all "variegating" genotypes in the latter as being due
to fine-focus clonal phenotype differences. The clone
may thus be thought of as the critical developmental
unit at which phenotypic variation is controlled. In
light of these observations, it seems useful to examine further the possibility that there is far more
cellular diversification in higher organisms than conventional classifications have led us to expect.

ACKNOWLEDGMENTS

These investigations were supported by United
States Public Health Service grants HD-01646, CA-06927,
and RR-05539, and by an appropriation from the Common-
wealth of Pennsylvania.

REFERENCES

Britten, R. J., and Kohne, D. R. (1968). *Science 161*,
    529.
Carter, N. D., and Parr, C. W. (1967). *Nature 216*, 511.
Daneholt, B., Edström, J. -E., Egyházi, E., Lambert,
    B., and Ringborg, U. (1970). *Cold Spring Harbor
    Symp. Quant. Biol. 35*, 513.
DeMars, R. (1967). *J. Nat. Cancer Inst. Monogr. 26*,
    327.
Gandini, E., and Gartler, S. M. (1969). *Nature 224*,
    599.
Gardner, R. L., and Lyon, M. F. (1971). *Nature 231*,
    385.
Gearhart, J. D., and Mintz, B. (1971). Submitted to
    Develop. Biol.
Goodrich, E. S. (1930). *Studies on the Structure and
    Development of Vertebrates* (New York: Dover Publi-
    cations).

Ingram, V. M. (1963). *The Hemoglobins in Genetics and Evolution* (New York: Columbia University Press).

Konigsberg, I. R. (1961). *Proc. Nat. Acad. Sci. USA 47*, 1868.

Lyon, M. F. (1961). *Nature 190*, 372.

Mintz, B. (1965). In *Ciba Foundation Symposium on Preimplantation Stages of Pregnancy*, G. E. W. Wolstenholme and M. O'Connor, eds. (London: J. and A. Churchill), pp. 145-155.

Mintz, B. (1967). *Proc. Nat. Acad. Sci. USA 58*, 344.

Mintz, B. (1969). *Genetics 61* (*Suppl.*), 41.

Mintz, B. (1970). In *Symp. Int. Soc. Cell Biol.*, Vol. 9 (New York: Academic Press), pp. 15-42.

Mintz, B. (1971a). In *Methods in Mammalian Embryology* J. Daniel, Jr., ed. (San Francisco: W. H. Freeman), pp. 186-214.

Mintz, B. (1971b). *Fed. Proc. 30*, 935.

Mintz, B. (1971c). In *Symp. Soc. Exp. Biol.*, Vol. 25 (England: Cambridge University Press), pp. 345-369.

Mintz, B., and Baker, W. W. (1967). *Proc. Nat. Acad. Sci. USA 58*, 592.

Mintz, B., and Palm, J. (1969). *J. Exp. Med. 129*, 1013.

Mintz, B., and Sanyal, S. (1970). *Genetics 64* (*Suppl.*), 43.

Moore, W. J., and Mintz, B. (1972). *Develop. Biol. 27*, (in press).

Ohno, S. (1969). *Ann. Rev. Genet. 3*, 495.

Pierce, G. B. (1967). In *Current Topics in Developmental Biology*, A. A. Moscona, ed., Vol. 2 (New York: Academic Press), pp. 223-246.

Rawles, M. E. (1947). *Physiol. Zool. 20*, 248.

Roberts, J. W. (1969). *Nature 224*, 1168.

Russell, L. B. (1961). *Science 133*, 1795.

Schultz, J. (1965). *Brookhaven Symp. Biol. 18*, 116.

Stevens, L. C. (1967). In *Advances in Morphogenesis*, M. Abercrombie and J. Brachet, eds., Vol. 6 (New York: Academic Press), pp. 1-31.

Stevens, L. C. (1970). *Develop. Biol. 21*, 364.

Till, J. E., and McCulloch, E. A. (1961). *Radiation Res. 14*, 213.

Wegmann, T. G., and Gilman, J. G. (1970). *Develop. Biol. 21*, 281.

Hormone Dependent
Animal Cell Strains
in Culture

Gordon Sato

Department of Biology
University of California, San Diego
La Jolla

During the past three years my laboratory has been
developing tissue culture strains which are dependent
for their growth on those hormones which are active on
the parental tissue. The potential utility of such
cultures is obvious and their feasibility is apparent
from the many examples of hormonal regulation of
growth in animal physiology.

I will describe in some detail the procedures used
to establish hormone dependent ovarian cells in culture and will only briefly comment on cultures developed from other tissues.

Luteal phase ovaries of young adult Fisher rats
were implanted into spleens of young adult female
castrates of the same strain. This procedure has been
widely used to develop gonadal transplants and tumors,
and has the following rationale (Zondek, 1934; Biskind
and Biskind, 1944): Since the spleen is drained by
the hepatic portal system, steroid hormones produced

by the implant are carried directly to the liver where
they are inactivated by enzymes which are for the most
part localized specifically in the liver. Beyond the
liver there is a deficit of gonadal steroids. This
deficit is sensed by the hypothalamic-pituitary axis,
which secretes large quantities of gonadotropins. The
gonadal implant is stimulated to grow by this means
and may develop large tumor-like growths. This stra-
tegy was adopted to obtain large masses of hormone-re-
sponsive tissue for the initiation of cultures. Six
months after inoculation, the large masses were explanted from the spleens to culture medium supplemented with an impure preparation of luteinizing hormone
from the NIH (LH NIH-B7).

For several months there was little visible sign of
growth in the cultures. Four months after the initiation of these cultures, numerous foci of growth were
noted and from these were developed an established
cell strain. This cell strain, designated 31A, has
several desirable features. It grows with a generation time of approximately 24 hr and has a plating
efficiency of better than 90%. It requires hormones
for growth (Clark, J., Jones, K., Gospodarowicz, D.
and Sato, G., unpublished data). To demonstrate this
requirement it was necessary to remove hormones from
the serum used in the medium. This was accomplished
by passing the serum through a column of Sepharose to
which anti-ovine LH gamma globulin was attached using
the cyanogen bromide technique. When 31A cells are
inoculated into media made with hormone-deficient sera,
no growth ensues. If the medium is supplemented with
the luteinizing hormone preparation, good growth is
obtained and interestingly enough this growth is markedly enhanced by steroid hormones (Fig. 1). Clones
of varied morphology have been derived from 31A. Each
clone requires luteinizing hormone. The different
clones can have their growth enhanced by different
steroid hormones. For instance, one clone may have
its growth enhanced by progesterone and not by dexamethasone, while another clone is stimulated by dexamethasone and not by progesterone. This finding may be
relevant to normal ovarian physiology.

In attempts to select for cells more responsive to

hormones, 31A cells were plated in hormone-deficient
media in the presence of FUDR. It was thought that
autonomous cells would enter the S phases and be
killed by the FUDR blockade of DNA synthesis. On the
other hand, hormone-dependent cells were expected to
remain quiescent and be unaffected by FUDR. Contrary
results were obtained. When cells surviving the
treatment were grown up and tested they were found to
be autonomous. We think that the amount of killing is
too slight to explain the result on the basis of sele-
tion for pre-existing autonomous cells. Rather, we
think that hormone-dependent cells are converted to
autonomy by FUDR and that this phenomenon may be related to the BUDR induced loss of differentiated
function (Abbott and Holtzer, 1968).

We have previously reported the establishment of
clonal strains of hormone-dependent rat mammary cells
(Posner et al., 1970). At the time of the report we
felt that an appreciable amount of the cells were
being transformed to autonomy to explain experimental
variability. With improved techniques for defining
the hormonal environment of cells in culture, we find
that the requirement for hormones by these cells is
much more stringent than previously thought.

We have been working with steroid-dependent hamster
tumors developed by Dr. Hadley Kirkman (1959). Cul-
ture-hardy tumors were selected by the method of alternate animal culture passage (Buonassisi, Sato and
Cohen, 1962). When primary cultures are subcultured
in control media and in steroid supplemented media,
growth is substantially better in steroid supplemented
media (Fig. 2).

While our experiments are still in a preliminary
state, some conclusions can nevertheless be drawn and
some speculative comments enumerated.

1. It is possible to develop culture strains which
   are dependent for their growth on those trophic
   hormones which are active on the parental tissue.
2. In order to obtain these strains it may be necessary to define and modify the hormonal environment in culture. This is illustrated by the

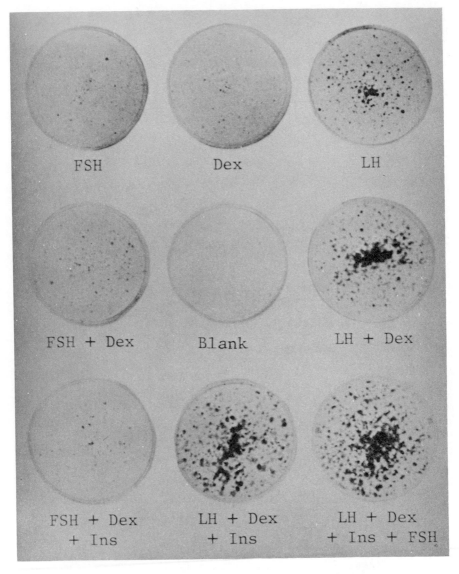

Figure 1. 5,000 31A cells were inoculated into 5.0 ml of LH-free medium. No hormones were added to the plate marked Blank. Dexamethasone was added at a final concentration of 1.0 gamma/ml and LH was added at a concentration of 5.0 gamma/ml. Six days after plating the cells were fixed with formalin and stained with crystal violet.

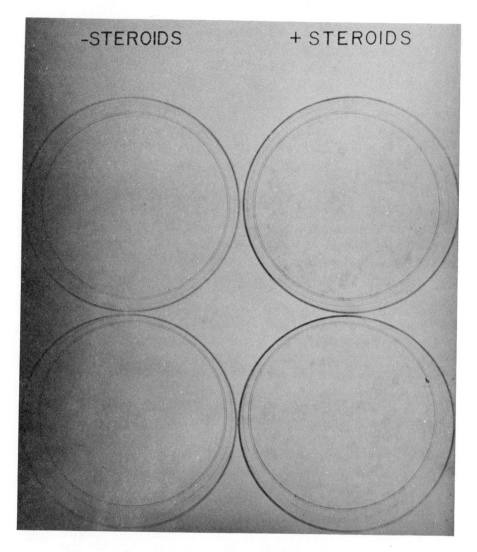

Figure 2. 0.1 ml of packed tissue from a
leiomysarcoma were plated in 10 ml of
steroid-deficient medium. The steroids
were removed from the serum component of
the medium by charcoal extraction. To one
set of plates was added 5.0 gamma/ml each
of estradiol and testosterone. Twenty
eight days after plating, the plates were
fixed with formalin and stained with
crystal violet.

      ovarian cultures, where it is necessary to pre-
pare LH-free serum.

3. The cultures offer excellent systems for the
   study of hormone mechanisms. The advantage of
   the culture system is readily seen in the case
   of the ovary where a complex of cell types, each
   responsive to gonadotropins but differentially
   responsive to steroid, is resolved by tissue
   culture cloning techniques.

4. Carcinogenesis could be studied in hormone-dependent cultures by following the conversion of
   hormone-dependent cells to autonomous cells.
   These systems should offer a valuable alternative to the assays using loss of contact inhibition as a marker.

5. The conversion of dependent cells to autonomy by
   FUDR may offer an opportunity to study the mechanisms whereby differentiated functions are
   maintained or lost. The phenomenon points up in
   a forceful way that the problem of cancer may be
   but an aspect of the general problem of embryonic differentiation. In this case the loss of a
   specialized response to a hormone results in
   loss of growth control.

6. Although immunoassays for hormones offer great
   sensitivity and technical facility, they must
   often be confirmed by bioassays. I think it is
   quite likely that the cumbersome bioassays of
   the present time will be eventually supplanted
   by tissue culture assays using growth stimulation by hormones as a marker.

7. It is our experience that hormone-dependent
   cells are harder to establish in culture than
   any others we have encountered. Part of the explanation may be found in our experience with
   the ovarian cells. That is, as the cells were
   established in culture, their hormonal responses
   were found to be more complex than initially
   expected.

ACKNOWLEDGMENTS

I thank my colleagues, Jeffrey Clark, Kenneth Jones and David Sirbasku for permission to present some of their unpublished results.

I thank Mrs. Madge Whitehead for editorial assistance.

This work was supported by grants from the United States Public Health Service No. 5 R01 GM 17019, the National Institutes of Health Genetics Program Project Grant GM 17702, and the National Science Foundation grant GB 15788.

REFERENCES

Abbot, J., and Holtzer, H. (1968). *Proc. Nat. Acad. Sci. USA 59*, 1144.

Biskind, M. S., and Biskind, G. R. (1944). *Proc. Soc. Exp. Biol. Med. 55*, 176.

Buonassisi, V., Sato, G., and Cohen, A. I. (1962). *Proc. Nat. Acad. Sci. USA 48*, 1184.

Kirkman, H. (1959). *Nat. Cancer Inst. Monogr. 1.*

Posner, M., Gartland, W., Clark, J., Sato, G., and Hirsch, C. (1970). *Dev. Biol. 4*, 114.

Zondek, B. (1934). *Scan. Arch. Physiol. 70*, 153.